Modern Perspectives in
Inorganic Crystal Chemistry

NATO ASI Series

Advanced Science Institutes Series

A Series presenting the results of activities sponsored by the NATO Science Committee, which aims at the dissemination of advanced scientific and technological knowledge, with a view to strengthening links between scientific communities.

The Series is published by an international board of publishers in conjunction with the NATO Scientific Affairs Division

A	**Life Sciences**	Plenum Publishing Corporation
B	**Physics**	London and New York
C	**Mathematical and Physical Sciences**	Kluwer Academic Publishers Dordrecht, Boston and London
D	**Behavioural and Social Sciences**	
E	**Applied Sciences**	
F	**Computer and Systems Sciences**	Springer-Verlag
G	**Ecological Sciences**	Berlin, Heidelberg, New York, London,
H	**Cell Biology**	Paris and Tokyo
I	**Global Environmental Change**	

NATO-PCO-DATA BASE

The electronic index to the NATO ASI Series provides full bibliographical references (with keywords and/or abstracts) to more than 30000 contributions from international scientists published in all sections of the NATO ASI Series.
Access to the NATO-PCO-DATA BASE is possible in two ways:

– via online FILE 128 (NATO-PCO-DATA BASE) hosted by ESRIN, Via Galileo Galilei, I-00044 Frascati, Italy.

– via CD-ROM "NATO-PCO-DATA BASE" with user-friendly retrieval software in English, French and German (© WTV GmbH and DATAWARE Technologies Inc. 1989).

The CD-ROM can be ordered through any member of the Board of Publishers or through NATO-PCO, Overijse, Belgium.

Series C: Mathematical and Physical Sciences - Vol. 382

Modern Perspectives in Inorganic Crystal Chemistry

edited by

Erwin Parthé

Laboratoire de Cristallographie aux Rayons X,
University of Geneva,
Geneva,
Switzerland

Kluwer Academic Publishers

Dordrecht / Boston / London

Published in cooperation with NATO Scientific Affairs Division

Proceedings of the NATO Advanced Study Institute on
Modern Perspectives in Inorganic Crystal Chemistry and
the 19th International School of Crystallography
Erice, Sicily, Italy
May 29–June 7, 1992

Library of Congress Cataloging-in-Publication Data

```
Modern perspectives in inorganic crystal chemistry / edited by Erwin
  Parthé.
       p.   cm. -- (NATO ASI series. Series C, Mathematical and
  physical sciences ; vol. 382)
     "Proceedings of the NATO Advanced Study Institute on "Modern
  Perspectives in Inorganic Crystal Chemistry" and the 19th
  International School of Crystallography, Erice, Sicily, Italy, May
  29-June 7, 1992"--T.p. verso.
     "Published in cooperation with NATO Scientific Affairs Division."
     Includes index.
     ISBN 0-7923-1954-0 (acid-free)
     1. Crystallography--Congresses.  2. Inorganic compounds-
  -Congresses.   I. Parthé, Erwin.  II. NATO Advanced Study Institute
  on "Modern Perspectives in Inorganic Crystal Chemistry" (1992 :
  Erice, Italy)  III. Series: NATO ASI series.  Series C, Mathematical
  and physical sciences ; no. 382.
  QD901.M62  1992
  548--dc20                                                   92-26740
```

ISBN 0-7923-1954-0

Published by Kluwer Academic Publishers,
P.O. Box 17, 3300 AA Dordrecht, The Netherlands.

Kluwer Academic Publishers incorporates the publishing programmes of
D. Reidel, Martinus Nijhoff, Dr W. Junk and MTP Press.

Sold and distributed in the U.S.A. and Canada
by Kluwer Academic Publishers,
101 Philip Drive, Norwell, MA 02061, U.S.A.

In all other countries, sold and distributed
by Kluwer Academic Publishers Group,
P.O. Box 322, 3300 AH Dordrecht, The Netherlands.

Printed on acid-free paper

All Rights Reserved
© 1992 Kluwer Academic Publishers
No part of the material protected by this copyright notice may be reproduced or utilized in any form or by any means, electronic or mechanical, including photocopying, recording or by any information storage and retrieval system, without written permission from the copyright owner.

Printed in the Netherlands

CONTENTS

Preface

Symmetry relationships between crystal structures and their practical applications.
   Gervais CHAPUIS   1

Cluster solid state chemistry : a frontier discipline between metallurgy and molecular chemistry.
   Roger CHEVREL   17

Structural and bonding principles in metal halide cluster chemistry.
   John D. CORBETT   27

Modern concepts of atom coordinations, atom volume and charge transfer in intermetallic compounds.
   Maria L. FORNASINI   57

The intergrowth concept as a useful tool to interpret and understand complicated intermetallic structures.
   Yuri N. GRIN'   77

Introduction to the electronic structure of extended systems.
   Roald HOFFMANN   97

Crystal chemical formulae for inorganic structure types.
   José LIMA-DE-FARIA   117

Crystal chemistry of complex sulfides (sulfosalts) and its chemical application.
   Emil MAKOVICKY   131

The bond valence method in crystal chemistry.
   Michael O'KEEFFE   163

Valence electron rules for compounds with tetrahedral structures and anionic tetrahedron complexes.
   Erwin PARTHÉ   177

Structural principles of silicates, phosphates, sulphates and related structures.
   Dimitri Y. PUSHCHAROVSKY   203

Crystal chemistry of high T_c superconducting oxides.
  Bernard RAVEAU, Claude MICHEL and Maryvonne HERVIEU   229

Competition between trigonal prisms and other coordination polyhedra in borides, carbides, silicides and phosphides.
  Peter ROGL   267

Index   279

PREFACE

The study of crystal structures has had an ever increasing impact on many fields of science such as physics, chemistry, biology, materials science, medicine, pharmacy, metallurgy, mineralogy and geology. Particularly, with the advent of direct methods of structure determination, the data on crystal structures are accumulating at an unbelievable pace and it becomes more and more difficult to oversee this wealth of data. A crude rationalization of the structures of organic compounds and the atom coordinations can be made with the well-known Kekulé model, however, no such generally applicable model exists for the structures of inorganic and particularly intermetallic compounds. There is a need to rationalize the inorganic crystal structures, to find better ways of describing them, of denoting the geometrical relationships between them, of elucidating the electronic factors and of explaining the bonding between the atoms with the aim of not only having a better understanding of the known structures, but also of predicting structural features of new compounds.

This book is a collection of the contributions of the invited lecturers to a summer school on "Modern Perspectives in Inorganic Crystal Chemistry" which was devoted to treating the problems mentioned above. The school, a NATO Advanced Study Institute, was held in Erice, Sicily, between May 29th and June 7th, 1992. The school was created to provide an interface between specialists who centered their research work on solving crystal chemical problems and the chemists, physicists, material scientists, metallurgists and mineralogists who wanted to obtain a state-of-the-art view of Inorganic Crystal Chemistry.

The objective of the course to stimulate a fruitful exchange in concepts, problems and goals in Inorganic Crystal Chemistry in the pleasant atmosphere of Erice, was obtained not only due to the excellent contributions of the lecturers but in very great part to the tireless efforts of Lodovico Riva di Sanseverino and his experienced collaborators, who did everything imaginable to make certain that the organization of the school would be flawless and remembered as a model to be followed.

We acknowledge the financial help and support given by the NATO Science Committee for this summer school.

E. Parthé

Geneva, June 1992

SYMMETRY RELATIONSHIPS BETWEEN CRYSTAL STRUCTURES AND THEIR PRACTICAL APPLICATIONS

GERVAIS C. CHAPUIS
Institut de Cristallographie
Université de Lausanne
B.S.P. Dorigny
CH - 1015 Lausanne
Switzerland

ABSTRACT. The space group of a structure describes the infinite set of symmetry transformations which leaves the structure unchanged. Subgroups of a particular space group are also of infinite order but consists of a subset of all symmetry transformations contained in the group. The group-subgroup relations can be used to study common features in parent crystal structures and establish a classification scheme. The latest edition of the International Tables lists the most important subgroups for each space group. For the crystal chemist, this information is a convenient tool which can facilitate the establishment of yet undiscovered structural similarities. Some examples of structure relations are given for various families of crystal structures which are derived from the highly symmetrical compounds of rutile, α-Po and perowskite. It is often useful to consider additional parameters like e.g. rotation or distortion expressing the relative displacement of structural units to refine the classification scheme based on subgoup relations.

1. Introduction

The concept of space group is firmly established in its role to characterize individual crystal structures. Its use is mainly perceived as part of a shorthand notation from which all the geometrical properties of a structure can be derived. The interesting property to relate or classify parent crystal structures by extending the concept of space group is however less familiar. The main reason for this is probably due to the difficulties to find the adequate tools in the published literature. Fortunately, in the latest edition of the International Tables for Crystallography (1983), additional data related to the field of space group relations have been included. With this information at hand, it is now easier to discover some interesting and unexpected relations between various compounds. The principles of the method along with the theoretical justification have been admirably developped and presented by Bärnighausen (1980).

The purpose of this note is to introduce briefly the subject and to illustrate practically the method with the help of a few examples. In the first part, we shall introduce the concept of subgroup of a space group along with some definitions. In the second part, the use of symmetry relationships derived from group theoretical considerations will be presented with a study of some well known families of compounds.

2. Space group relationships.

2.1 SUBGROUPS OF A SPACE GROUP

The space group of a crystal structure can be considered as the set of all the symmetry operations which leave the structure invariant. All the elements (symmetry operations) of this set satisfy the characteristics of a *group* and their number (*order*) is infinite. Of course, this definition is only valid for an ideal structure extending to infinity. For practical purpose, however, it can be applied to the finite size of real crystals. Lattice translations, proper or improper rotations with or without screw or gliding components are all examples of symmetry operations.

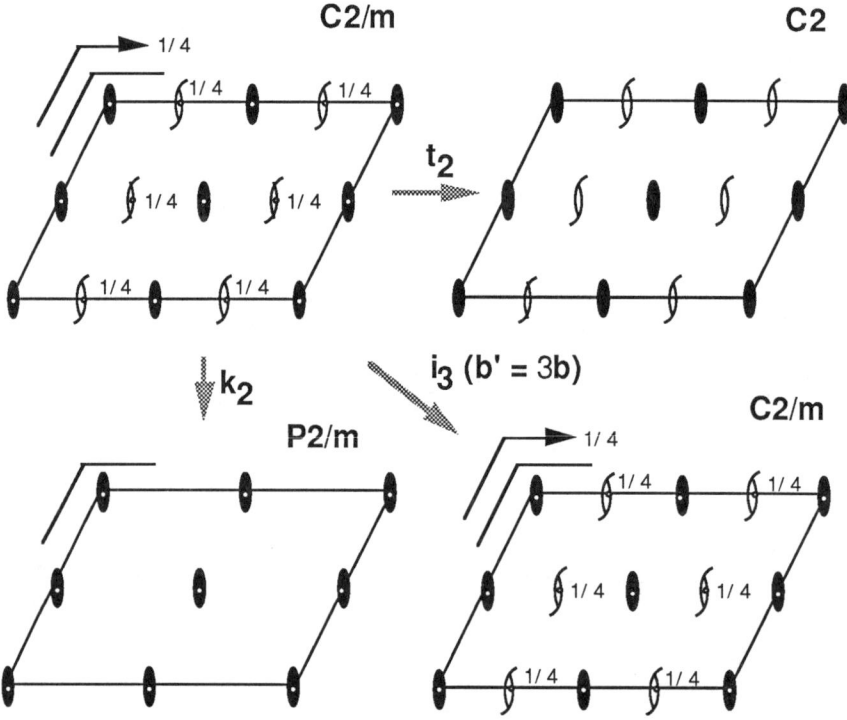

Fig 1. Three examples of maximal klassengleiche (k) and translationengleiche (t) subgroups of space group C2/m (see appendix). In two cases, the index following the type of subgroup is 2, i.e. each subgroup contains only half of the symmetry operations of space group C2/m. The third example is an isomorphic (i) subgroup, a special type of k subgroup with index 3.

In analogy to the definition of the subgroup of a group, this notion can be extended to obtain the subgroups of a space group. Of course their number is infinite but only a small finite number of them are relevant in this context for most purposes. Subgroups are conveniently characterized by their *index*, i.e. an integer indicating the number of cosets needed for the complete decomposition of the group. For a more illustrative definition, the index of a subgroup of a space space group can be defined as the inverse of the fraction of all the symmetry operations which have been conserved from the original space group. If the index of a subgroup is a prime number, the subgroup is *maximal*. In this context, only maximal subgroups will be considered.

2.2 MAXIMAL SUBGROUPS AND MINIMAL SUPERGROUPS OF A SPACE GROUP

According to group theoretical considerations, all the maximal subgroups of a space group are either "*klassengleich*" or "*translationengleich*". In the first case, the subgroup has the same crystal class as the space group. In the second case, only the translations are identical to the space group. Both are usually referred to as *k* and respectively *t* subgroups. There is a special subset of the k subgroups, i.e. the maximal subgroups which have the same space group symbol as the original space group. These are the *isomorphic* or *i* subgroups. As an example, two maximal subgroups of space group C2/m are given in Fig. 1. For clarity, only the symbols of the symmetry operations as used in the International Tables are represented. In the appendix, the subgroups of a selection of the space groups relevant for the examples treated here are listed.

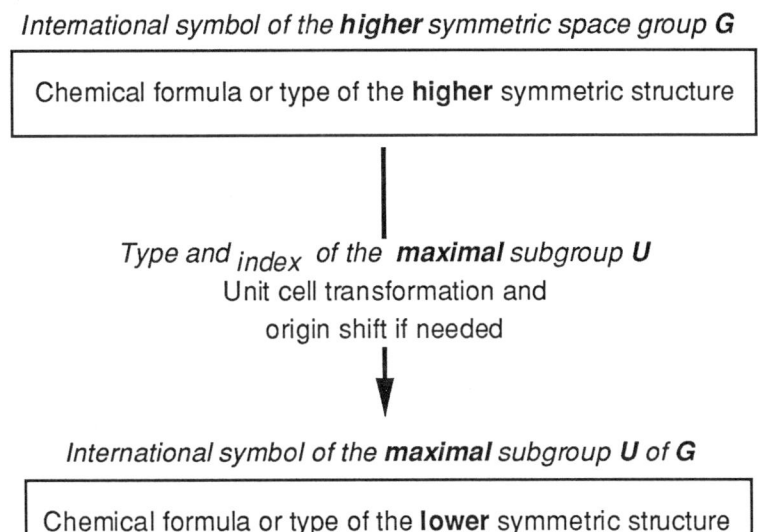

Fig 2. Schematic diagram indicating the structural relationships between two structures. The structure with the highest space group symmetry is placed on the upper part of the diagram whereas the parent structure with a lower symmetry is indicated in the lower part of the diagram.

The relationship between space group and their subgroups can also be inverted. For example, if C2 is a maximal subgroup of space group C2/m then C2/m is a *minimal supergroup* of C2. In general, they are many maximal supergroups of a given space group. Only non-isomorphic minimal supergroups are listed in the International Tables.

3. The family tree

Structural similarities between parent structures can be conveniently represented with a diagram indicating the relationship of their corresponding space group as represented in fig. 2. The space group symmetry of the structures represented in the diagram decreases from the top to the bottom. In addition, the subgroup type and index of each relation is indicated along with any additional information pertaining to a possible cell transformation and to an origin shift. In a family of compounds the structure with the highest symmetry is often referred to as the *aristotype* (Megaw, 1973). The others, less symmetrical members of the family are referred to as *hettotypes*.

3.1 THE RUTILE FAMILY

Rutile is an important aristotype from which numerous structures can be derived. In particular, the rigid columns of TiO_2 octahedra aligned along the tetragonal axis can be rotated relatively to each other. The resulting deformations are represented on fig. 3 for different rotation angles. Taking into account the additional deformation of the octahedra, an extended series of compounds can be associated to the rutile family. Table 1 lists a number of them with the rotation angle ω and a coefficient f associated with the deformation of the octahedra. The group-subgroup relations are given in fig. 4 for some of the members of the rutile family.

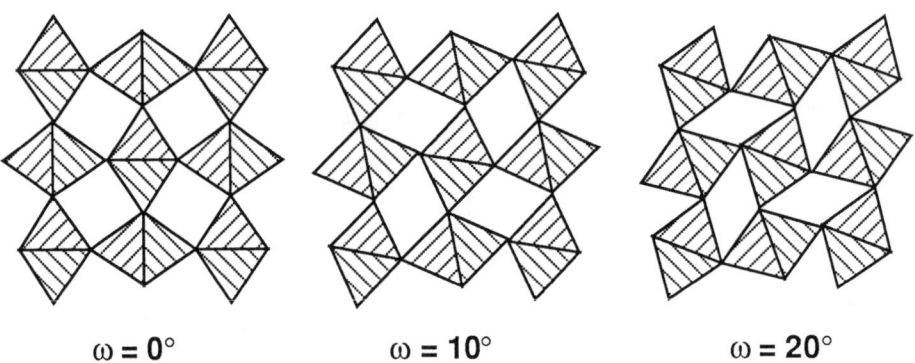

$\omega = 0°$ $\omega = 10°$ $\omega = 20°$

Fig 3. The rutile structure schematically represented by octahedra on the left. This ideal structure can be deformed by rotating the corner-sharing octahedra. For example, the hexagonal close-packed structure results from a rotation of $\omega = 9.7°$.

It should be mentioned that there is no correlation between the magnitude of the angle ω and the decrease of the symmetry of the corresponding space group. As an example, the structure obtained from a rotation of the octahedra of ≈10° is the hexagonal close-packed (hcp) structure but with orthorhombic symmetry. The resulting space group is a common subgroup of the tetragonal symmetry of rutile and the hexagonal space group.

Table 1. List of structures derived from the rutile aristotype ordered by increasing values of ω (see above figure). The coefficient f is a measure of the deformation of the octahedra.

Compound	Comments	ω(°)	f
TiO_2	Aristotype	0	1
NiF_2	at 2.5 GPa	1.8	.88
$VO_{1.7}(OH)_{.3}$		3.0	1.09
$CaCl_2$		6.3	.88
α-PdI_2	mostly quadratic planar coordination	8.7	1.38
$CaBr_2$		9.0	.9
hcp structure		9.7	1
$CrCl_2$	Jahn-Teller deformation	10.9	.99
InO(OH)	Formation of H-bonds	14.8	1.12
$FeSe_2$	Formation of covalent Se-Se bonds	16.4	1.19
FeS_2	Mineral marcasite	18.6	1.16
$FeAs_2$	Mineral loellingite	19.4	1.12
RuP_2		21.6	1.13

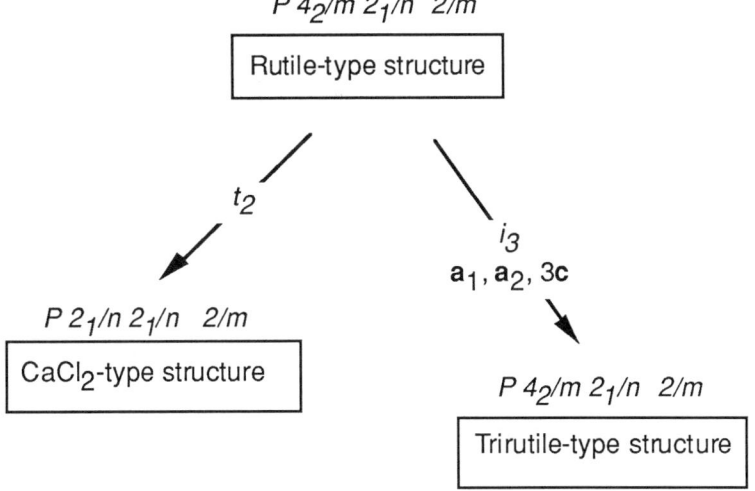

Fig 4. Structural relationships between structures parent to the rutile aristotype.

The group-subgroup relation diagrams are not only interesting from the crystal chemical point of view. They can be applied to study the structural phase transitions of compounds. In this context, the structure of $CsBr_2$ is an interesting case. Below 778 K, the structure is isomorphous to $CaCl_2$ whereas above this temperature, the structure is isomorphic to rutile (Anselment, 1985). The diagram given in fig. 4, indicates that this phase transition involves a t subgroup of index 2. As a consequence, one should expect the formation of *twin domains* due to the change of the *crystal class*. Indeed, such domains can be observed directly under the microscope. Moreover, the number of twins which can be observed must be equal to the index of the subgroup.

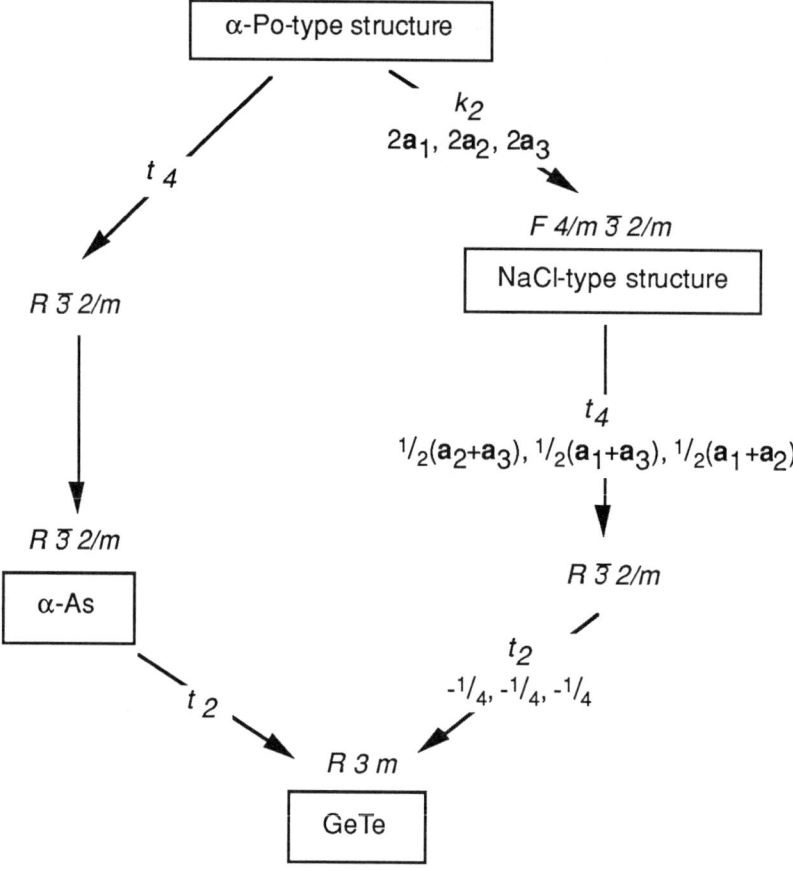

Fig 5. The α-Po family of structures.

3.2 THE α-Po FAMILY

This series of compounds reported on fig. 5 illustrates an important aspect in the relations between structures. Although the structures of NaCl and α-As are not related directly by group-subgroup relations, they share however a common (non-trivial) subgroup and super-group. This situation is often encountered with systems where many phase transitions of the same compound appear at different pressures or temperatures. The transition linked to a direct subgroup relation is often associated with a *second order* phase transition whereas a *first order* transition may share only a common supergroup (Chapuis and Zuñiga, 1988).

In the structure of α-Po, the atoms occupy the nodes of a cubic lattice. The NaCl structure can be derived from an ordered substitution of two atom types thus increasing the volume of the cube by a factor 2^3. α-As on the other hand results from a deformation of the cubic cell along the space diagonal. In fig. 6 the deformed shape of the original cube is represented for α-As. In a recent article, Beister, Strössner and Syassen (1990) have shown by x-ray diffraction and Raman spectroscopy that above a pressure of 25 GPa, the structure of α-As transforms to the α-Po type structure. In the range from 0 to 25 GPa, the volume decreases by more than 20%. Some properties of the cubic to trigonal phase transition of α-As can be readily derived from the representation given in fig. 5. The subgroup t_4 allows for the formation of fourfold twins whereas the subgroup i_2, a special type of k subgroup, induces the formation of *antiphase domains*. These properties are very useful to detect and deduce or at least contribute to identify the possible symmetry of unknown phases by optical methods.

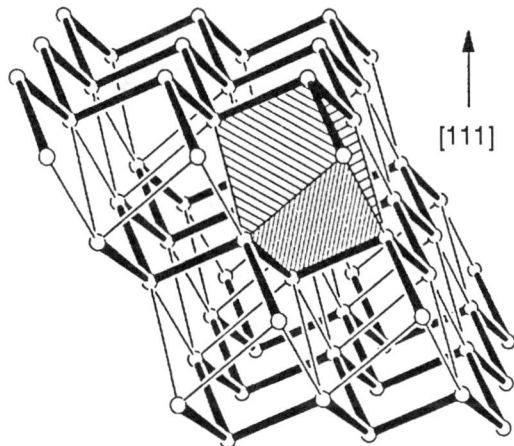

Fig 6. Structure of α-As. The hatched polyhedron results from a rhombohedral deformation of a cube

The space group of GeTe is a subgroup common to the two branches of the family tree. The structure can be obtained from the NaCl type by introducing a small asymmetry in the distance between alternating layers of Na and Cl atoms. Alternatively, the structure can be derived from α-As by an ordered substitution of As by Ge and Te in the layers parallel to the trigonal axis.

3.3 THE PEROWSKITE FAMILY

This series of structures is well known for its ferroelectrical properties i.e. the appearance of a spontaneous polarization. The general formula is ABX_3 and the structure of the aristotype consists essentially of BX$_3$ octahedra located on the nodes of a cubic lattice. Each octahedron shares a common vertex with its neighbour. The A cations are located in the cavities between the octahedra; they may also be absent as in the structure of ReO$_3$.

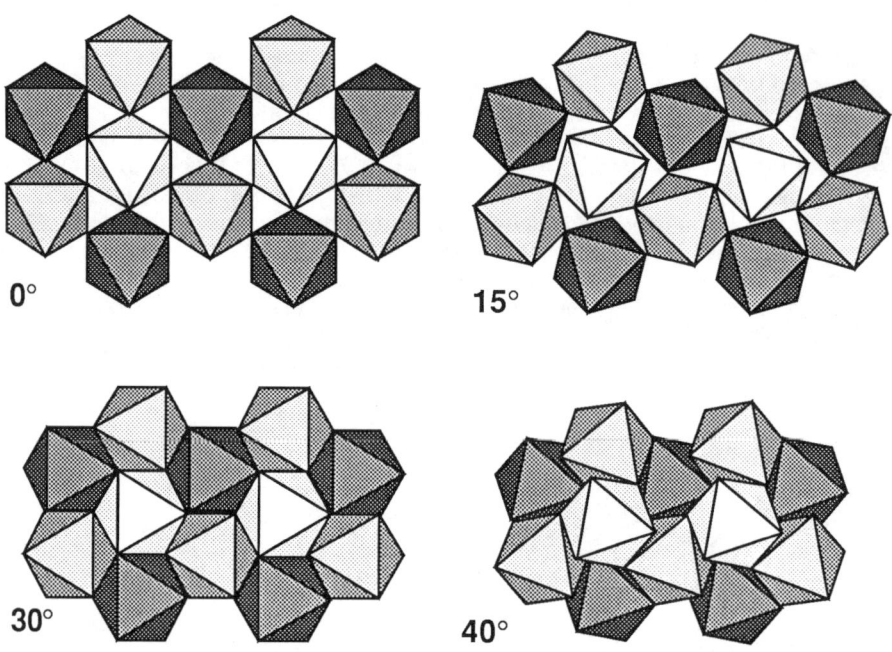

Fig 7. Possible structures obtained from deformations of the perowskite aristotype by rotating the octahedra about the [111] direction. The patterns of the octahedra are characteristic of their respective height.

Many deformations of the ideal perowskite structure are possible. In this context, one single type of deformation will be considered, namely a rotation of the octahedra along the [111] direction. Fig. 7 illustrates various ranges of ω values for which many real structures have been identified. They are listed on table 2 by increasing value of ω and fig. 8 shows their group-subgroup relationships.

Similar to the deformation obtained from the rutile structure, the hcp structure appears in this family after a rotation of the octahedra by 30°. After a 40° rotation, the structure of CaCO$_3$ is obtained. In this process, some of the bonds have been rearranged owing to the vicinity in the relative position of the octahedra. The inter-octahedral distances have become closer than the intra-octahedral distances.

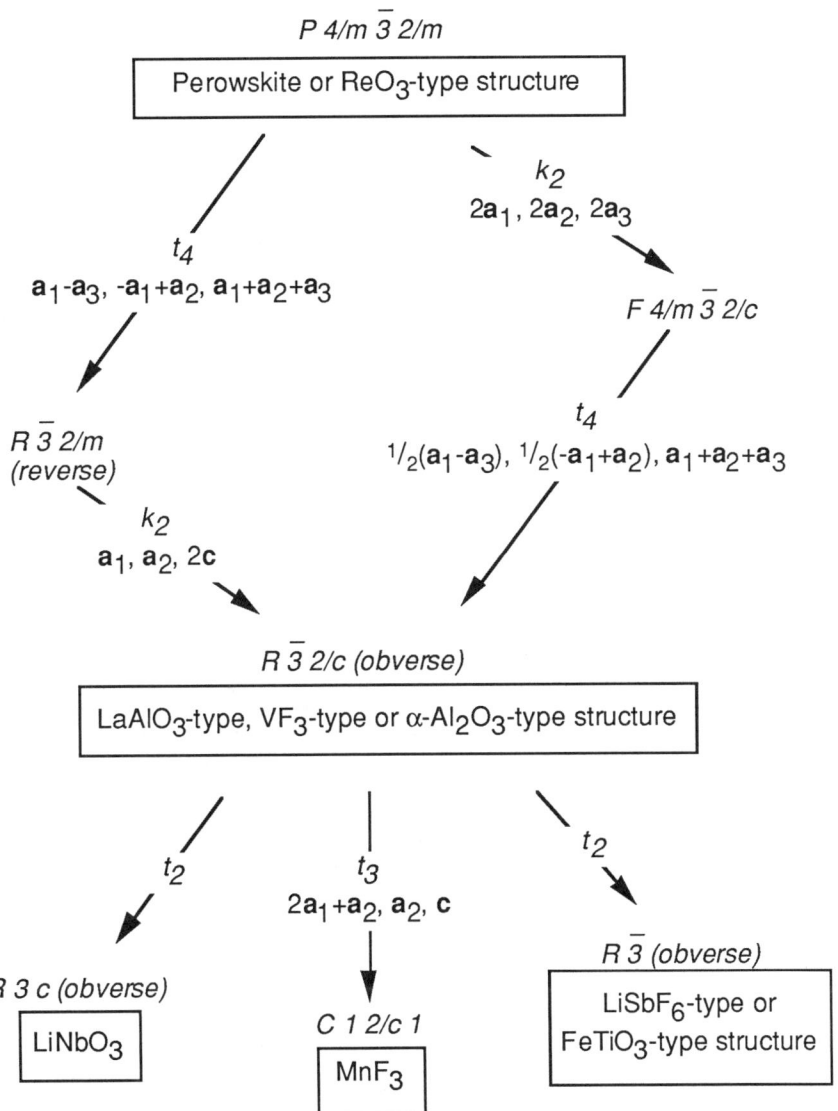

Fig 8. Group-subgroup relations of the perowskite family. Only structures resulting from a rotation of the octahedra about the [111] direction have been considered.

Table 2. List of structures derived from the perowskite aristotype ordered by increasing values of ω.

Compound	Comments	$\omega(°)$
ReO_3	Aristotype	0
$SrTiO_3$	Ideal perowskite type	0
$LaAlO_3$	Example of soft modes	5.8
$LaCoO_3$		10.3
$NaNbO_3$	at 123 K	12.1
FeF_3		17
MnF_3	Jahn-Teller deformation	≈20
$LiSbF_6$		21.3
VF_3		22
$LiNbO_3$	Important ferroelectrics	23.1
$TmCl_3$	High pressure modification	29.4
hcp structure		30
$Pd^{II}Pd^{IV}F_6$	PdF_3-type	31.3
$FeTiO_3$	Mineral ilmenite	33.8
Al_2O_3	Mineral corundum	33.9
$CaCO_3$	Mineral calcite	40.1

4. Conclusion

The number of examples relating many families of parent compounds can be multiplied at will. Many can be found in the literature derived e.g. from the fluorite or antifluorite aristotype or silicates. The group-subgoup relations is a powerful tool not only to classify structures. It also permits to introduce a hierarchical component in the classification of similar structures. This tool is also very convenient to describe the symmetry changes that occur in phase transitions of crystals. Many properties like order of transition or the formation of twin or anti-domain phase boundaries can be inferred from the group-subgroups diagrams relating the space groups of the various crystalline phases. This method can without doubt be extended towards a more quantitative tool to relate the degree of similarity between various structures.

5. References

Anselment, B. (1985). Die Dynamik der Phasenumwandlung vom Rutil- in den $CaCl_2$-Typ am Beispiel des $CaBr_2$ und zur Polymorphie des $CaCl_2$. Thesis. Universität Karlsruhe.

Bärnighausen, H. (1980). Group-subgroup relations between space groups: A useful tool in Crystal Chemistry. Match, **9**, 139-175.

Beister. H.J., Strössner, K. & Syassen, K. (1990). Rhomboedral to simple-cubic phase transition in arsenic under pressure. Phys. Rev. B, **41**, 5535-5543

Chapuis, G. & Zuñiga, F. (1988). The High temperature Phase of $(CH_3)_3NHCdCl_3$: or How to Minimize Order-Disorder Effects in Phase Transitions. Acta Cryst. **B44**, 243-249.

International Tables for X-ray Crystallography (1983). Dordrecht: Reidel.

Megaw, H. (1973). "Crystal Structures: A Working Approach." 563 pages. Philadelphia: W. B. Saunders.

6. Acknowledgement

I would like to thank warmly Prof. H. Bärnighausen for his advice and for providing an important part of this material.

7. Problems.

7.1 STRUCTURE OF $CoReO_4$

The structure of $CoReO_4$ has been recently determined (Bauer, W. H., Joswig, W., Pieper, G. and Kassner, D. (1992) $CoReO_4$, a new rutile type derivative with ordering of two cations. To appear in J. Solid State Chem.). The lattice constants are a=6.51, b=6.74 and c=2.89 Å. Space group Cmmm; Z=2. The atomic positions are given below.

Atom	multiplicity (Wyckoff pos.)	site symmetry	coordinates	parameters
Co	2(a)	mmm	0,0,0 ; $1/2,1/2,0$	
Re	2(c)	mmm	$1/2,0,1/2$; $0,1/2,1/2$	
O(1)	4(i)	m2m	0,y,0; 0,-y,0; $1/2,1/2+y,0$; $1/2, 1/2-y,0$	y=.30
O(2)	4(h)	2mm	$x,0,1/2$; $-x,0,1/2$; $1/2+x,1/2,1/2$; $1/2-x,1/2,1/2$	x=.22

This structure can be directly derived from the rutile aristotype with space group symmetry $P4_2/mnm$. From the data given above and in the appendix, sketch briefly the structure of $CoReO_4$ and establish the group-subgroup diagram relating the two structures.

7.2 STRUCTURES OF α- AND γ-Te

Both structures can be derived from the α-Po aristotype. Their relationships can best be seen from a rhomboedral description of α-Po with a cell volume three times the cubic cell and lattice parameters $a_r=4.75$ and $c_r=5.82$ Å. From the c/a ratio, it appears that the rhomboedral cell of γ-Te is strongly compressed along the space diagonal whereas for α-Te it is slightly elongated relatively to the rhomboedral cell of α-Po.

Structure	Space group	Z	a (Å)	c(Å)	multiplicity (Wyckoff pos.)	site symmetry	coordinates
α-Po	$P4/m\bar{3}2/m$	1	3.36		1(a)	$m\bar{3}m$	0,0,0
γ-Te	$R\bar{3}2/m$	3	4.71	3.82	3(a)	3m	0,0,0; $1/3,2/3,2/3$; $2/3,1/3,1/3$
α-Te	$P3_121$	3	4.45	5.93	3(a)	2	$x,0,1/3$; $0,x,2/3$; $-x,-x,0$; x=.26

The problem is to establish the group-subgroup diagram relating the three structures. Special attention should be given to the cell transformations, obverse and reverse choice of the rhomboedral cells and position of origins.

8. Appendix

SELECTION OF MAXIMAL SUBGROUPS OF SPACE GROUPS RELEVANT FOR THE EXAMPLES PRESENTED IN THE TEXT.

The data are a small selection from Volume A of the "International Tables for Crystallography". Type I represents *translationengleiche* or *t* subgroups, type II *klassengleiche* or *k* subgroups.
The subdivision of type II discriminates between the following categories:
IIa   stands for *k* subgroups, obtained by decentring of conventional Bravais-type cells.
IIb   stands for *k* subgroups, obtained by enlarging of the conventional unit cells.
IIc   stands for *k* subgroups (called *i* subgroups later on) that are even *isomorphic* with the group, i.e. they have the same standard Hermann-Mauguin symbol.

Note that there is an infinite number of maximal subgroups of this kind for each space group. Therefore only subgroups of the lowest index are given, but others may be quite important (see for instance space group *C 2/m*). Generally, the indexes of *isomorphic* subgroups are prime numbers. In addition prime numbers to the second or third power are possible, if the symmetry is trigonal, hexagonal, tetragonal or cubic, respectively (see space group $R\,3\,2$ or $P\,4/m\,\bar{3}\,2/m$).

The index of the subgroup is given in brackets in front of the space group symbol. Often this symbol is not the conventional one. In such cases the conventional space group symbol is shown in parentheses.

C 12/m 1 (No. 12)

I   [2] $C\,1\,2\,1$; [2] $C\,\bar{1}\,(P\bar{1})$; [2] $C\,1\,m\,1$
IIa [2] $P\,1\,2/m\,1$; [2] $P\,1\,2/a\,1\,(P\,2/c)$; [2] $P\,1\,2_1/m\,1$;
     [2] $P\,1\,2_1/a\,1\,(P2_1/c)$
IIb [2] $C\,1\,2/c\,1\,(c' = 2c)$; [2] $I1\,2/c\,1\,(c' = 2c)\,(C\,2/c)$
IIc [2] $C\,1\,2/m\,1\,(c' = 2c$ or $a' = a+2c, c' = 2c)$; [3] $C\,1\,2/m\,1\,(b' = 3b)$;
     [3] $C\,1\,2/m\,1\,(c' = 3c$ or $a' = a-2c, c' = a+c)$
     Note that the last case is not given in "International Tables, Vol. A", because index 3 is not the *lowest* one for cell enlargements in the a-c plane.

C 2/m 2/m 2/m (No. 65)

I   [2] $C\,2\,2\,2$; [2] $C\,1\,1\,2/m\,(P\,2/m)$; [2] $C\,1\,2/m\,1\,(C\,2/m)$;
    [2] $C\,2/m1\,1\,(C\,2/m)$; [2] $C\,m\,m\,2$; [2] $C\,m\,2\,m\,(A\,m\,m\,2)$
    [2] $C\,2\,m\,m\,(A\,m\,m\,2)$
IIa [2] $P\,m\,m\,m$; [2] $P\,b\,a\,n$; [2] $P\,b\,a\,m$; [2] $P\,b\,m\,n\,(P\,m\,n\,a)$
     [2] $P\,m\,a\,n\,(P\,m\,n\,a)$; [2] $P\,m\,m\,n$; [2] $P\,m\,a\,m\,(P\,m\,m\,a)$
     [2] $P\,b\,m\,m\,(P\,m\,m\,a)$
IIb [2] $C\,c\,c\,m\,(c' = 2c)$; [2] $C\,c\,m\,m\,(c' = 2c)\,(C\,m\,c\,m)$;
     [2]$C\,m\,c\,m\,(c' = 2c)$; [2] $I\,m\,m\,m\,(c' = 2c)$; [2] $I\,b\,a\,m\,(c' = 2c)$;
     [2] $I\,b\,m\,m\,(c' = 2c)\,(I\,m\,m\,a)$; [2] $Im\,a\,m\,(c' = 2c)\,(I\,m\,m\,a)$
IIc [3] $C\,m\,m\,m\,(a' = 3a$ or $b' = 3b)$; [2] $C\,m\,m\,m\,(c' = 2c)$

P 4₂/m 2₁/n 2/m (No. 136)

I [2] $P\,4_2\,2_1\,2$; [2] $P\,4_2\,n\,m$; [2] $P\,\bar{4}\,2_1\,m$; [2] $P\,\bar{4}\,n\,2$;
[2] $P\,4_2/m\,1\,1\,(P\,4_2/m)$; [2] $P\,2/m\,2_1/n\,1\,(P\,2_1/n\,2_1/n\,2/m)$;
[2] $P\,2/m\,1\,2/m\,(C\,2/m\,2/m\,2/m)$

IIa None
IIb None
IIc [3] $P\,4_2/m\,2_1/n\,2/m$ (**c'** = 3**c**) ; [9] $P\,4_2/m\,2_1/n\,2/m$ (**a'** = 3**a**, **b'** = 3**b**)

R 3 2 (No. 155) (hexagonal axes)

I [2] $R\,3\,1\,(R\,3)$; [3] $R\,1\,2\,(C\,2)$
IIa [3] $P\,3\,2\,1$; [3] $P\,3_1\,2\,1$; [3] $P\,3_2\,2\,1$
IIb None
IIc [3] $R\,3\,2$ (**a'** = -**a**, **b'** = -**b**, **c'** = 2**c**) ; [4] $R\,3\,2$ (**a'** = -2**a**, **b'** = -2**b**)

P $\bar{3}$ 2/m 1 (No. 164)

I [2] $P\,3\,2\,1$; [2] $P\,\bar{3}\,1\,1\,(P\,\bar{3})$; [2] $P\,3\,m\,1$; [3] $P\,1\,2/m\,1\,(C\,2/m)$
IIa None
IIb [3] $H\,\bar{3}\,2/m\,1$ (**a'** = 3**a**, **b'** = 3**b**) $(P\,\bar{3}\,1\,2/m)$; [2] $P\bar{3}\,2/c\,1$ (**c'** = 2**c**)
IIc [2] $P\,\bar{3}\,2/m\,1$ (**c'** = 2**c**) ; [4] $P\,\bar{3}\,2/m\,1$ (**a'** = 2**a**, **b'** = 2**b**)

R $\bar{3}$ 2/m (No. 166) (hexagonal axes)

I [2] $R\,3\,2$; [2] $R\,\bar{3}\,1\,(R\,\bar{3})$; [2] $R\,3\,m$; [3] $R\,1\,2/m\,(C\,2/m)$
IIa [3] $P\,\bar{3}\,2/m\,1$
IIb [2] $R\,\bar{3}\,2/c$ (**a'** = -**a**, **b'** = -**b**, **c'** = 2**c**)
IIc [2] $R\,\bar{3}\,2/m$ (**a'** = -**a**, **b'** = -**b**, **c'** = 2**c**) ; [4] $R\,\bar{3}\,2/m$ (**a'** = -2**a**, **b'** = -2**b**)

R $\bar{3}$ 2/c (No. 167) (hexagonal axes)

I [2] $R\,3\,2$; [2] $R\,\bar{3}\,1\,(R\,\bar{3})$; [2] $R\,3\,c$; [3] $R\,1\,2/c\,(C\,2/c)$
IIa [3] $P\,\bar{3}\,2/c\,1$
IIb None
IIc [4] $R\,\bar{3}\,2/c$ (**a'** = -2**a**, **b'** = -2**b**) ; [5] $R\,\bar{3}\,2/c$ (**a'** = -**a**, **b'** = -**b**, **c'** = 5**c**)

P 4/m $\bar{3}$ 2/m (No. 221)

I [2] $P\,2/m\,\bar{3}\,1\,(P\,2/m\,\bar{3})$; [2] $P\,4\,3\,2$; [2] $P\,\bar{4}\,3\,m$;
[3] $P\,4/m\,1\,2/m\,(P\,4/m\,2/m\,2/m)$; [4] $P\,1\,\bar{3}\,2/m\,(R\,\bar{3}\,2/m)$

IIa None
IIb [2] $F\,4/m\,\bar{3}\,2/m$ (**a'** = 2**a**, **b'** = 2**b**, **c'** = 2**c**) ;
[2] $F\,4/m\,\bar{3}\,2/c$ (**a'** = 2**a**, **b'** = 2**b**, **c'** = 2**c**) ;
[4] $I\,4/m\,\bar{3}\,2/m$ (**a'** = 2**a**, **b'** = 2**b**, **c'** = 2**c**)

IIc [27] $P\,4/m\,\bar{3}\,2/m$ (**a'** = 3**a**, **b'** = 3**b**, **c'** = 3**c**)

F 4/m $\bar{3}$ 2/m (No. 225)

I [2] $F\,2/m\,\bar{3}\,1\,(F\,2/m\,\bar{3})$; [2] $F\,4\,3\,2$; [2] $F\,\bar{4}\,3\,m$;
[3] $F\,4/m\,1\,2/m\,(I\,4/m\,2/m\,2/m)$; [4] $F\,1\,\bar{3}\,2/m\,(R\,\bar{3}\,2/m)$

IIa [4] $P\,4/m\,\bar{3}\,2/m$; [4] $P\,4_2/n\,\bar{3}\,2/m$

IIb	None
IIc	[27] $F\,4/m\,\bar{3}\,2/m$ (**a'** = 3**a**, **b'** = 3**b**, **c'** = 3**c**)

$F\,4/m\,\bar{3}\,2/c$ (No. 226)

I	[2] $F\,2/m\,\bar{3}\,1\,(F\,2/m\,\bar{3})$; [2] $F\,4\,3\,2$; [2] $F\,\bar{4}\,3\,c$; [3] $F\,4_2/m\,1\,2/n\,(I\,4/m\,2/c\,2/m)$; [4] $F\,1\,\bar{3}\,2/n\,(R\,\bar{3}\,2/c)$
IIa	[4] $P\,4_2/m\,\bar{3}\,2/n$; [4] $P\,4/n\,\bar{3}\,2/n$
IIb	None
IIc	[27] $F\,4/m\,\bar{3}\,2/c$ (**a'** = 3**a**, **b'** = 3**b**, **c'** = 3**c**)

9. Solutions of the problems.

9.1. STRUCTURE OF CoReO$_4$

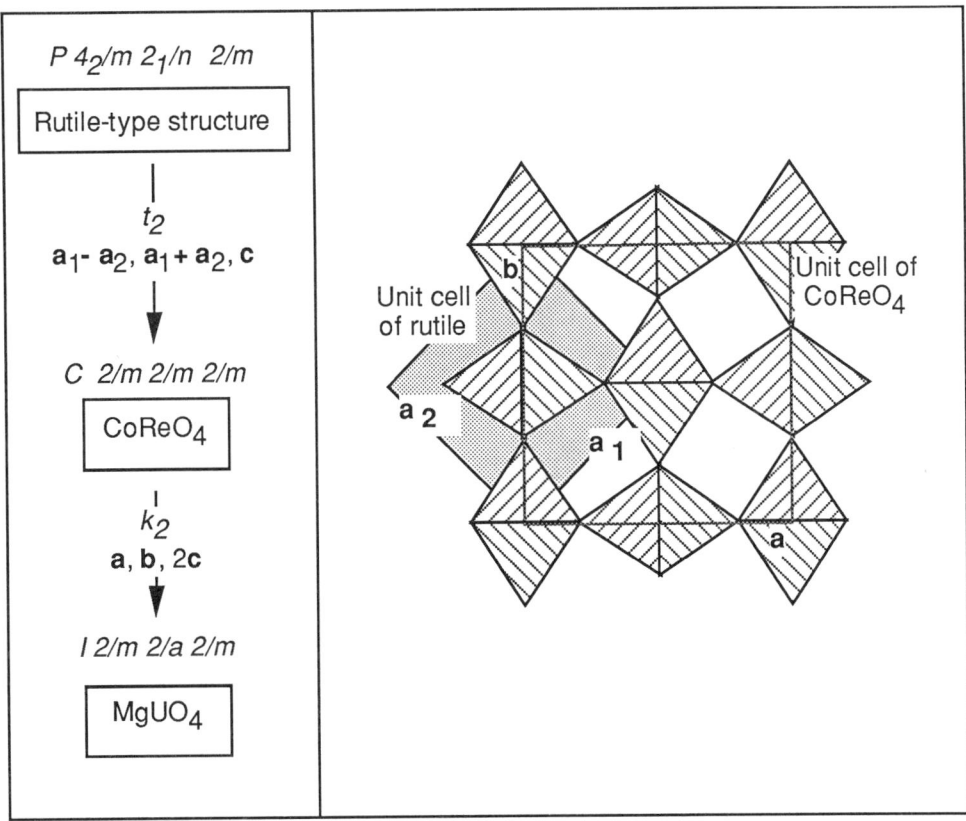

$P\ 4_2/m\ 2_1/n\ 2/m$

Rutile-type structure

$\downarrow\ t_2$
$a_1 - a_2, a_1 + a_2, c$

$C\ 2/m\ 2/m\ 2/m$

CoReO$_4$

$\downarrow\ k_2$
$a, b, 2c$

$I\ 2/m\ 2/a\ 2/m$

MgUO$_4$

In the above figure, the unit cells of CoReO$_4$ and rutile are represented. The c-axis is approximately the same for both. The index of the subgroup is obviously 2 as the volume of the CoReO$_4$ is twice the volume of rutile. The crystal class of the two space groups are different: 4/mmm (4/m2/m2/m) resp. mmm (2/m2/m2/m). The translation periods are however identical in both cases (remember the C centering!). the space group C2/m2/m2/m is a t_2 subgroup of P4$_2$/mnm which is confirmed by the list given in the appendix (type I subgroup).

The structure of MgUO$_4$ can also be described from the rutile aristotype. The space group Imam is a k_2 subgroup of Cmmm, the space group of CoReO$_4$. The I centering is a consequence of doubling the c-axis of a C-centered cell.

9.2 STRUCTURES OF α- AND γ-Te

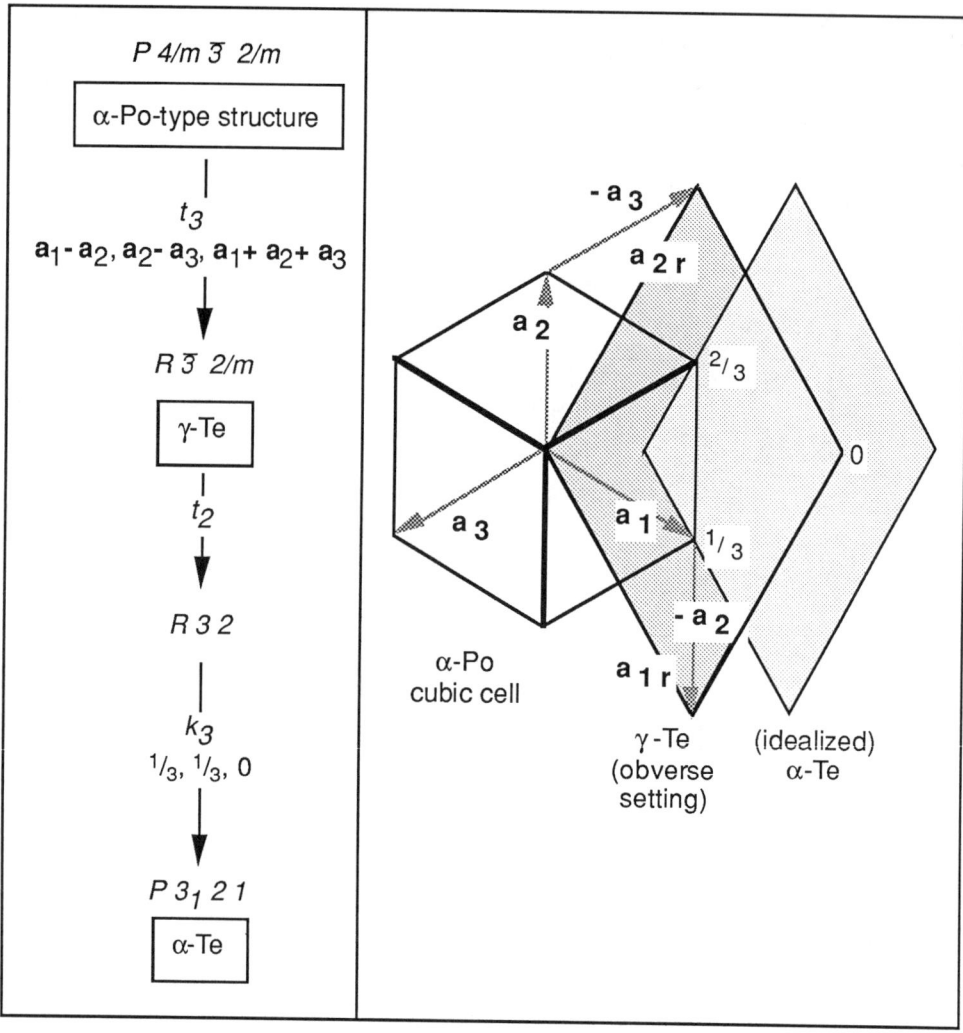

The cubic unit cell of α-Po is represented along the space diagonal. The same structure can be described in a (obverse) rhomboedral cell with a volume three times the cubic cell. This is also the structure of γ-Te with space group R$\bar{3}$2/m (t_3 subgroup of Pm$\bar{3}$m). From this space group, the structure of α-Te can be obtained in two steps: reduction of the crystal class ($\bar{3}$m to 32, t_2 subgroup) and decentering the rhomboedral cell (k_3 subgroup). In this process, a shift of the cell by (1/3, 1/3, 0) must be added to comply with the standard setting of space group P$3_1$21.

CLUSTER SOLID STATE CHEMISTRY : A FRONTIER DISCIPLINE BETWEEN METALLURGY AND MOLECULAR CHEMISTRY

ROGER CHEVREL
Laboratoire de Chimie du Solide et Inorganique Moléculaire
University of Rennes 1
U.R.A., C.N.R.S No. 1495
F - 35042, Rennes, France

ABSTRACT. Many compounds with early transition metals (Ti, Zr, V, Nb, Ta, Mo, W, Re...) in low oxidation states are known which are characterized by metal - metal bonds. These metals form aggregates, called clusters, such as pairs, triangles or regular shaped polyhedra such as tetrahedra, octahedra or polyoctahedra. These clusters are surrounded by ligands such as oxygen, halogen or chalogen atoms. In the case of suboxides, subhalides or chalcogenides where the formal oxidation state of the metal atoms is between 3 and 4, these clusters consist of pairs, triangles, rhomboids, zigzag chains or tetrahedra and are surrounded by ligands stacked in a close packed manner. On the other hand, the cluster compounds where the metal atoms have a formal oxidation state between 1 and 3 are built up of octahedral or polyoctahedral metal clusters which, surrounded by ligands, form "cluster units". The structures containing these cluster units are often arranged as the atoms in NaCl, CsCl or ReO_3 structure types. The number of valence electrons available for the metal-metal bonds is related to the transition metal/non-metal ratio and increases with the degree of condensation of the cluster (or nuclearity = number of metal atoms in the cluster).

1. Introduction

The understanding of the chemistry of compounds with metal - metal bonding has a great scientific and technological importance related, for instance, to problems in thin film technology, in catalysis, soil chemistry, corrosion, adhesion, electrochemical processes and in biological catalysis of some complex enzymes, in material science and metallurgy. During the last twenty years a major new area of solid state chemistry, based on compounds with metal-metal bonds, has emerged and has been developed by different teams : Chevrel & Sergent (1981, 1986), Corbett (1981, 1984), Cotton (1981), Cotton & Walton (1982), Fenske, Ohmer, Hachgenei & Merzweiler (1988), Kepert (1972), McCarley (1983), Mingos & Wales (1990), Perrin & Sergent (1988), Rogel, Zhang, Payne & Corbett (1990), von Schnering (1981), Simon, (1981, 1988). The development of more efficient analysis techniques (IR, NMR and X-ray diffraction) have greatly speeded up the characterization of solid state cluster compounds. In addition, an understanding of the metal - metal bonding in these compounds has arisen from molecular orbital theory and band structure calculations. Indeed, the field has reached an exciting point in its development because the sizes of new clusters, now being characterized, are sufficiently large to enable meaningful comparisons of physical and chemical properties to be made between both the cluster and the surface of the corresponding metal in bulk form.

2. Metal - metal bonding

At least two kinds of classes of transition metal compounds with metal - metal bonds can be distinguished :
- The clusters in organometallic compounds are formed by late transition metals. The surrounding ligands such as CO (carbonyl) are σ - dative bonding donor ligands and at the same time π - back bonding acceptor ligands. Example : $Os_6(CO)_{16}$.
- The clusters we want to discuss occur with early transition metals and have π - donor ligands such as halogen, oxygen, chalcogen and alkoxy (OR where R = alkyl group).

The chemistry of the compounds of the early transition metals in low oxidation states is full of examples of the occurrence of metal - metal bonds. These compounds show unusual compositions in terms of the traditional valence rules : metal - rich compounds contain more metal atoms than one expects for a normal valence compound (example : Nb_6I_{11} instead of NbI_5). The transition metal elements on the left in the periodic system, and of those mainly the 4d and 5d elements, are capable of using the excess valence electrons, not needed to complete the octets of the anions, to form metal - metal bonds.

The transition metals in their bulk state have maximum bonding energies at the centre of the transition metal series corresponding to a half filling of the d and s bands. In the cluster compounds the ligands modify the electronic configurations of the metals for them to obtain a more stable configuration : In the case of the cluster being formed by early transition metal atoms, the π - donor ligands contribute extra electrons. In the case of the cluster being formed by late transition metals, π - acceptor ligands remove some of the excess electrons.

For metal - metal bonds between transition elements to occur, the major conditions are :
1) The presence of relatively expanded d orbitals on the metal atoms to allow overlap with each other.
2) The availability of electrons that can be used for metal - metal bonding
3) The absence of unfavourable steric effects.
Expanded d orbitals are normally present with the early transition metals.

A metal cluster or better a "cluster" is defined here as a group or aggregate of two or more metal atoms where direct and substantial metal - metal bonding is present. It is a definition different from the one proposed by Cotton & Wilkinson (1988) who do not consider pairs of bonded metal atoms as clusters.

Often the term "cluster" is used to mean the complete geometric unit including also the surrounding ligands. Of course, the observed metal - metal distance does not always give a clear indication of the kind of metal - metal bonding. In certain cases metal - metal distances which exceed the sum of the covalent radii by 15% have formally been considered as a bond, although antiferromagnetic spin interactions rather than covalent bonding may actually provide a better description of such a "bond".

Some words must be said concerning the criteria for the existence of a real metal - metal bond as opposed to a weak metal - metal interaction which is due alone to a pairing of electron spins, and which may lessen or even quench the Curie - Weiss paramagnetism. At room - temperature and below, the interaction energy necessary to quench the paramagnetism, is only 2 to 3 kcal/mole. The spin coupling leads to an energy difference between the singlet and the triplet state of 2 to 3 times kT. That energy is too weak to constitute a genuine chemical bond (~ 20 kcal/mole).

Indeed, the atomization energy of the transition elements is the most important measure of the strength of the metal - metal bonds of the transition elements. The metal - metal bonding is more pronounced in the 4d and 5d transition metal compounds, being particulary strong with niobium, tantalum, molybdenum, tungsten and rhenium. It is just these transition elements that have the strongest metal - metal bonding in the metals themselves as indicated by their high melting points, boiling points, hardness and atomization energies.

Cluster compounds of the earlier transition metals are thus generally associated with π - donor ligands such as O^{2-}, S^{2-}, Cl^-, Br^-, I^- and the metals generally have formal oxidation states ranging from 2+ to 4+. This is in contrast with the cluster compounds having carbonyl ligands where the formal oxidation states of the transition metals are zero or even negative. That difference is essential in understanding the difference in bonding and in the properties of both classes of cluster compounds. Effectively, the valence shell d orbitals of the early transition elements are sufficiently separated in energy with respect to the outer s and p orbitals that is sufficient to consider only the five 4d or 5d orbitals once the valence electron requirements of the ligands have been fulfilled.

Interesting criteria for the occurrence of a metal - metal bond can be obtained from a study of the electronic structure of the compound (molecular orbital and band structure calculations) and from the number of electrons available for metal - metal bonds, expressed by the so - called VEC (Valence Electron Concentration). In this counting of electrons, no attempt is made to differentiate localized electron - pair bonds from bonding in terms of molecular orbitals delocalized over the entire cluster.

3. Valence electron concentration (VEC)

Let us assume a compound with composition $C_c M_m A_a A'_{a'}$, where :
  M are the transition metal atoms which form the cluster
  A and A' are two kinds of anions
  C are cations and
  c, m, a and a' are composition parameters.

Two kinds of valence electron concentration parameters are used : $VEC_{cluster}$ and VEC_{metal}. According to Chevrel & Sergent (1986), the first parameter can be calculated by means of :

$$VEC_{cluster} = m \cdot e_M + c \cdot e_C - a \cdot (8 - e_A) - a' \cdot (8 - e_{A'}) \tag{1}$$

where e_M, e_C, e_A and $e_{A'}$ are the valence electron numbers of the M, C, A and A' atoms in the non - ionized state which corresponds in the normal case to the group number of the element in the periodic system. Some of the exceptions for e_C are : $e_{In} = e_{Tl} = 1$, $e_{Sn} = e_{Pb} = 2$, $e_{Fe} = 2$ or 3.

The second VEC parameter is related to the first according to

$$VEC_{metal} = VEC_{cluster} / m \qquad (2)$$

Examples of compounds with Nb and Mo clusters are listed in Table 1 together with calculated $VEC_{cluster}$ and VEC_{metal} values. A cluster unit formula is given, provided cluster units can be recognized in the structure. Cluster units consist of the metal cluster polyhedron and the ligands bonded to the cluster.

Table 1 : Examples of Nb and Mo cluster compounds and their different VEC values.

Compound formula	Cluster unit formula *	$VEC_{cluster}$	VEC_{metal}	Comment
Nb_3Cl_8		7	2.33	c = 0; a' = 0
$GaNb_4S_8$	$[Nb_4S_4]^{5+}$	7	1.75	a' = 0
$CsNb_4Cl_{11}$		10	2.50	a' = 0
Nb_6Cl_{14}	$[Nb_6Cl_{12}]^{2+}$	16	2.67	c = 0; a' = 0
Nb_6I_{11}	$[Nb_6I_8]^{3+}$	19	3.17	c = 0; a' = 0
$Zn_2Mo_3O_8$		6	2	a' = 0
$GaMo_4S_8$	$[Mo_4S_4]^{5+}$	11	2.75	a' = 0
$Mo_4S_4Br_4$	$[Mo_4S_4]^{4+}$	12	3	c = 0
$Pb^{2+}Mo_6Cl_{14}$	$[Mo_6Cl_8]^{4+}$	24	4	a' = 0
$AgMo_6Cl_{13}$	$[Mo_6Cl_8]^{4+}$	24	4	a' = 0
Mo_6Cl_{12}	$[Mo_6Cl_8]^{4+}$	24	4	c = 0; a' = 0
$Mo_6Br_{10}S$	$[Mo_6(Br_7S)]^{3+}$	24	4	c = 0
Mo_6S_8	$[Mo_6S_8]^{0}$	20	3.33	c = 0; a' = 0
$Pb^{2+}Mo_6S_8$	$[Mo_6S_8]^{2-}$	22	3.67	a' = 0

* Formula given only if a pseudomolecular cluster unit can be recognized.

The VEC_{metal} parameter is plotted in Figure 1 versus the non-metal/transition metal ratio (a + a')/m in order to show the influence of VEC_{metal} on the number of metal atoms (nuclearity) of the clusters in halogenides, oxide or chalcogenide compounds.

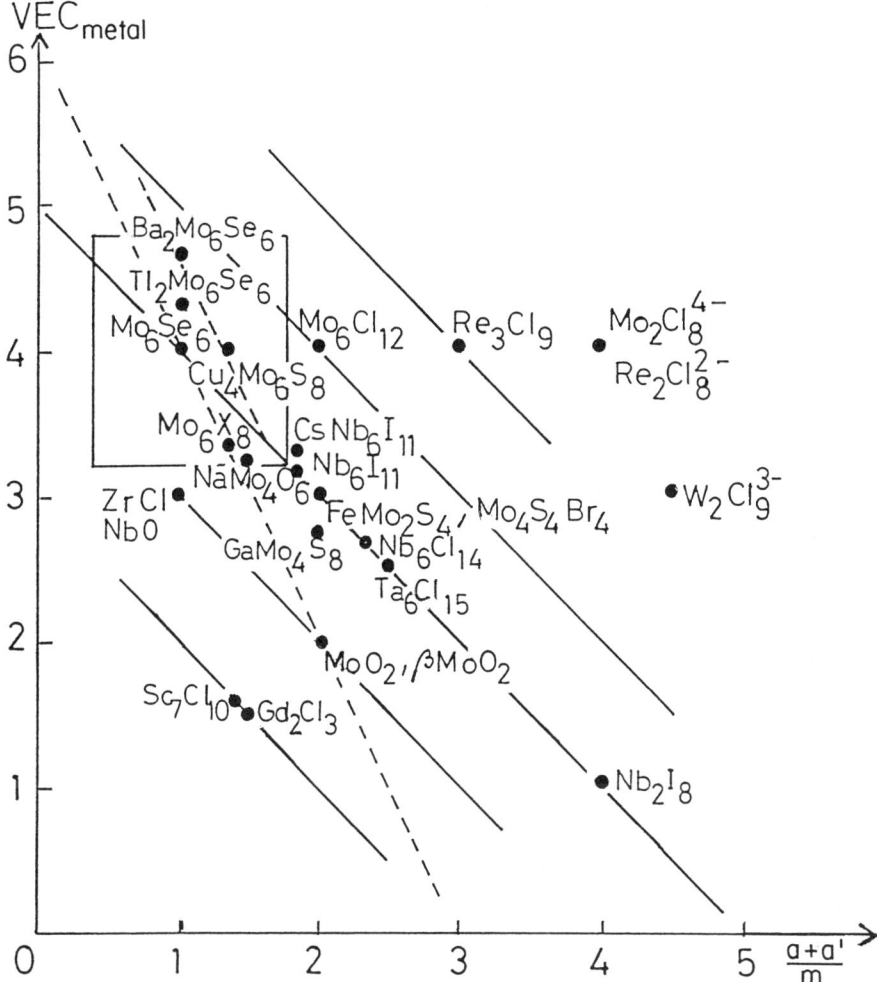

Figure 1 : VEC_{metal} versus the non-metal / transition metal ratio for $C_cM_mA_aA'_{a'}$ cluster compounds. Full lines connect data points of corresponding halides and dashed lines those of corresponding oxides and chalcogenides. The number of transition metal atoms in the cluster (nuclearity) increases in general with the value of VEC_{metal}.

It can be noticed that a relationship exists between VEC_{metal} and the formal oxidation state of the transition metals. Its value is given simply by the difference : $e_M - VEC_{metal}$

It is preferable to use the VEC_{metal} rather than the formal oxidation state in solid state cluster chemistry. From Figure 1 the following conclusions can be drawn :
Lowering the oxidation state or increasing the VEC_{metal} value leads to an increase of the nuclearity of the clusters.

	d^0	$\sim d^1$	$\sim d^2$	$\sim d^3$	$\sim d^4$	d^5	d^6	d^7
Ti, Zr, Hf	4	3	2	1	0			
	TiO_2 ZrO_2 no M-M $TiCl_4$	Ti_2O_3 Ti_2 pair β–$TiCl_3$ Ti_2 pair	TiO (NaCl metallic) ZrI_2 (Zr_6)	ZrCl ZrBr	Zr Ti_{metal} (d^2s^2)			
V, Nb, Ta	5	4	~ 3	2	1	0		
	$K_2Nb_2O_7$ Nb_2O_5	α-$NbCl_4$ Nb_2 pair	Nb_3Cl_8 (Nb_3) (2.67+) $CsNb_4Cl_{11}$ (Nb_4 plane) (2.50+)	NbO (condensed Nb_6) Nb_6F_{15}(Nb_6) (2.50+) Nb_6Cl_{14} (Nb_6) (2.33+)	Nb_6I_{11} (Nb_6) (1.83+) $CsNb_6I_{11}$ (Nb_6) (1.66+)	Nb (d^4s^1)		
Mo, W	6		4	~ 3	~ 2	1	0	
	Na_2MoO_4 MoO_3		MoO_2 (Mo_2 pair) $Zn_2Mo_3O_8$ (Mo_3) β-$MoTe_2$ (Mo_4 zigzag chain)	Na Mo_4O_6 (Mo_6 chain) $Cs_3Mo_2Cl_9$ (Mo_2 pair) Fe Mo_2S_4 (Mo_4 plane) Mo_2S_3 (Mo_4 plane) Ga Mo_4S_8 (Mo_4 tetrahedron)	$[Mo_2Cl_8]^{4-}$ (Mo_2) Mo_6Cl_{12} (Mo_6) $PbMo_6Cl_{14}$ (Mo_6) Mo_6S_8 (Mo_6) $Ag_{3.6}Mo_9Se_{11}$ (Mo_9) Mo_2S_6 ($Mo_6/2$)$_\infty$		Mo (d^5s^1)	
Re	7		5	4	3	2	1	0
	$KReO_4$			$ReCl_4$ (Re_2 pair) $ReSe_2$ (Re_4 plane) $M(Mo_2Re_2)S_8$ (Mo_2Re_2 tetrahedron)	$ReCl_3$ (Re_3) $[Re_3Cl_{12}]^{3-}$ (Re_3) $Re_6Se_8Cl_2$ (Re_6 octahedron)			Re (d^5s^2)

Table 2 : Examples of cluster compounds with transition metal atoms in different electron configurations together with their formal oxidation states.

Basic differences exist in respect to other procedures for counting electrons as for example, the "Wade & Mingos" rules (see Mingos & Wales, 1990), used for counting electrons in the cluster units in organometallic and inorganic chemistry.

According to Wade & Mingos, all the valence electrons are counted and summed up : the valence electrons on the metal - metal bonds plus the electrons on the metal - ligand bonds (dative and covalent bonds). Of importance here is the eighteen electrons per metal atom rule, the so - called "EAN" formalism (Effective Atomic Number) which assigns one electron pair per bond between two metal atoms. This localized scheme is largely used for the organometallic complexes where it permits the making of accurate predictions on the shape of the clusters. But this electronic description in terms of localized bondings between two centres is inadequate in e^- rich cluster compounds (as they have recently been found with some organometallic compounds) or e^- deficient cluster compounds (as they occur with early transition metal clusters).

The other drawback of the Wade & Mingos electron counting procedure is the necessity to know the molecular structure of the cluster beforehand in the case of the early transition metal clusters of interest here. In organometallic compounds, the carbonyl ligand functions always as a two - electron π - donating ligand in terminal, edge or face - sharing positions. Unfortunately with the early transition metal clusters, the chlorine ligand changes from a one (terminal) to a three (edge - bridging) to a five (face - capping) electron donor. Equally, the sulfur ligand changes according to the Wade & Mingos electron counting procedure from a two (edge - bridging) to a four (face - sharing) electron donor.

The chemistry of cluster compounds with early transition metals in low oxidation states is full of examples with often unusual compositions. Some of these compositions are reported in Table 2.

4. References

    Chevrel, R. & Sergent, M. (1982). 'Chemistry and Structure of Ternary Molybdenum Chalcogenides.' in Topics in Current Physics "Superconductivity in Ternary Compounds 1": Vol. **32**, 5 - 86. Editors : Ø. Fischer and M.B. Maple. Berlin : Springer.

    Chevrel, R. & Sergent, M. (1986). 'From Three Dimensional to One Dimensional Cluster Mo_6 Chalcogenides' in "Crystal Chemistry and Properties of Materials with Quasi-One-Dimensional Structures." Editor : J. Rouxel. 315 - 373. Dortrecht : Reidel.

    Corbett, J.D. (1981). 'Extended Metal-Metal Bonding in Halides of the Early Transition Metals.' Acc.Chem.Res. **14**, 239 - 246.

    Corbett, J.D. (1984). 'New Materials from High Temperature Synthesis.' Pure & Appl.Chem. **36** (11). 1527. New York : Pergamon.

    Cotton, F.A. (1981). "Metal-Metal Multiple Bonds and Metal Clusters." A.C.S. Symposium Series. **155**. Reactivity of Metal-Metal Bonds. Editor : M.H. Chisholm. 1 - 16.

Cotton, F.A. & Walton, R.A. (1982). "Multiple Bonds Between Metal Atoms." New York : Wiley.

Cotton, F.A. & Wilkinson, G. (1988). "Advanced Inorganic Chemistry." 5th Edition. New York : Wiley.

Fenske, D., Ohmer, J., Hachgenei, J. & Merzweiler, K. (1988). 'New Transition Metal Clusters with Ligands from Main Groups Five and Six.' Angew.Chem. Int.Ed.Engl. **27**, 1277 - 1296.

Kepert, D.L. (1972). "The Early Transition Metals." New York : Academic.

McCarley, R.E. (1983). 'Metal Clusters and Extended Metal-Metal Bonding in Metal Oxide Systems.' A.C.S. Symposium Series No. **211**. "Inorganic Chemistry. Toward the 21st Century." Editor : M.H. Chisholm. 273 - 290.

Mingos, D.M.P. & Wales, J.D. (1990). "Introduction to Cluster Chemistry ' 318 pages. Englewood Cliffs, New Jersey : Prentice-Hall.

Perrin, A. & Sergent, M. (1988). 'Rhenium Clusters in Inorganic Chemistry : Structures and Metal-Metal Bonding.' New J.Chem. **12**, 337- 356.

Rogel, F., Zhang, J., Payne, M.W. & Corbett J.D. (1990). 'Centered Cluster Halides for Group Three and Group Four Transition Metals. A Versatile Solid State and Solution Chemistry.' Adv.Chem.Ser. "Inorganic Compounds with Unusual Properties." **229**, 369 - 389.

Schnering, H.G. von (1981). 'Homoatomic Bonding of Main Group Elements.' Angew.Chem. Int.Ed.Engl **20**, 33 - 51.

Simon, A. (1981). 'Condensed Metal Clusters.' Angew.Chem. Int.Ed.Engl. **20**, 1 - 22.

Simon, A. (1988). 'Clusters of Valence - Electron Poor Metals. Structure, Bonding and Properties.' Angew.Chem. Int.Ed.Engl. **27**, 160 - 183.

5. Problems

Problem 1 : Sketch the molecular structure and calculate the values of $VEC_{cluster}$ and VEC_{metal} for the following compounds :

- Nb_3Cl_8, $Zn_2Mo_3O_8$, $GaMo_4S_8$, $Mo_4S_4Br_4$.

What can you say about the magnetic properties of these compounds.

Problem 2 : Draw and describe the two cubic $[M_6A_8]A'_6$ and $[M_6A_{12}]A'_6$ cluster units. Find for each cluster unit a geometrical relation between the ideal intra-cluster distances d_{A-A} and d_{M-M} (assuming an undistorted cluster unit).

Based on your results and the below listed experimental intra-cluster distances find out which of the two cluster units occur with $MoCl_2$ and $PbMo_6S_8$?

- $MoCl_2$: $d_{Cl^- - Cl^-} = 3.62$ Å; $d_{Mo - Mo} = 2.6$ Å.
- $PbMo_6S_8$: $d_{S^{2-} - S^{2-}} = 3.68$ Å; $d_{Mo - Mo} = 2.7$ Å.

Calculate for both compounds the two VEC parameters.

5.1 SOLUTIONS OF THE PROBLEMS

Problem 1 :
- Nb_3Cl_8 and $Zn_2Mo_3O_8$: The two structures are based on a distorted hexagonal close packed anion lattice. In both structures the metal - metal bonding is between groups of three metal atoms which form triangular clusters. Only 3/4 of the octahedral sites in every alternate layer are occupied. The tetrahedral sites in the other layer can be occupied by Zn^{2+} ions.

Nb_3Cl_8 : $\quad VEC_{cluster} = (3 \cdot 5) - (8 \cdot 1) = 7$ e⁻ per Nb_3 cluster.
$\quad\quad\quad\quad\quad VEC_{metal} = 7/3 = 2.33$ e⁻ per Nb atom.
$\quad\quad\quad\quad\quad\quad\quad\quad\quad \rightarrow$ Curie - Weiss paramagnetism (1 e-)

$Zn_2[Mo_3O_8]^{4-}$: $VEC_{cluster} = (3 \cdot 6) + (2 \cdot 2) - (8 \cdot 2) = 6$ e⁻ per Mo_3 cluster.
$\quad\quad\quad\quad\quad VEC_{metal} = 6/3 = 2$ e⁻ per Mo atom.
$\quad\quad\quad\quad\quad\quad\quad\quad\quad \rightarrow$ diamagnetism

- $GaMo_4S_8$ and $Mo_4S_4Br_4$: Both structures can be derived from a defect spinel structure with a cubic close packing of sulphur and bromine atoms.
Normal and defect spinels :
$\quad CM_2A_4 \rightarrow C\square M_4A_8 \rightarrow \square_2M_4A_8$: $Ga\square Mo_4S_8$ and $\square\square Mo_4S_4Br_4$.
Mo atoms are in octahedral sites and gallium atoms in tetrahedral sites. In both structures occur cubic $[Mo_4S_4]$ cluster units (as in the cubane-type) which contain tetrahedral Mo_4 clusters ($d_{Mo-Mo} = 2.8$ Å).

$GaMo_4S_8$: Here occurs a $[Mo_4S_4]^{5+}$ ionic cluster unit.
$\quad\quad\quad VEC_{cluster} = (4 \cdot 6) + (1 \cdot 3) - (4 \cdot 2) - (4 \cdot 2) = 11$ e⁻ per Mo_4 cluster.
$\quad\quad\quad VEC_{metal} = 11/4 = 2.75$ e⁻ per Mo atom.

In a model of localized M - M bonds, there are two e⁻ per metal - metal bond. In a tetrahedron there are six bonds which need alltogether 12 e⁻. $GaMo_4S_8$ is thus characterized by electron - deficient metal - metal bonding. A paramagnetic Curie - Weiss law is observed (1 e⁻ unpaired). This was the first found example of long range magnetic order in cluster compounds. Below $T_{Curie} = 20$ K, the compound becomes ferromagnetic. However, this compound has a semiconducting behaviour because the cluster units are isolated.

$Mo_4S_4Br_4$: Here is found a $[Mo_4S_4]^{4+}$ ionic cluster unit.
$\quad\quad\quad VEC_{cluster} = (4 \cdot 6) - (4 \cdot 2) - (4 \cdot 1) = 12$ e⁻ per Mo_4 cluster.
$\quad\quad\quad VEC_{metal} = 12/4 = 3$ e⁻ per Mo atom.

A diamagnetic behaviour is observed (2 e⁻ per Mo - Mo bond in the tetrahedral cluster).

Problem 2 :
The $[M_6A_8]$ cubic cluster unit consists of a M_6 octahedron with the faces capped by eight "inner" A^i ligands (denoted by $\mu_3 - A^i$ in inorganic chemistry).

The $[M_6A_{12}]$ cubic cluster unit is built up of a M_6 octahedron with its edges bridged by twelve "inner" A^i ligands (denoted by $\mu_2 - A^i$).

Each M atom is then bound to an additional ligand (A') in a terminal (ended) or "apical" position (labelled A^a) to form a $[M_6A^i_8]A'^a_6$ or a $[M_6A^i_{12}]A'^a_6$ cluster unit, respectively.

$[M_6A_8]$ cluster unit : In the ideal case the A - A distance is equal to a_{cubic}, thus $a_{cubic} = 2 \cdot r_A = d_{A-A}$; further $d_{M-M} = 2 \cdot r_M = a_{cubic} \cdot \sqrt{2}/2$. Substituting for a_{cubic} one obtains finally : $d_{A-A} = \sqrt{2} \cdot d_{M-M}$. Experimentally one finds $d_{A-A} \approx \sqrt{2} \cdot d_{M-M}$.

$[M_6A_{12}]$ cluster unit : In the ideal case $d_{A-A} = d_{M-M} = a_{cubic} \cdot \sqrt{2}/2$. Experimentally one finds $d_{A-A} \approx d_{M-M}$.

$MoCl_2$: Calculated with $d_{Mo-Mo} = 2.62$ Å (intra-cluster distance) the expected A - A distance, assuming a cubic $[Mo_6Cl_8]$ cluster unit, is $d_{A-A} = 2.62 \cdot \sqrt{2} = 3.70$ Å which agrees approximately with the observed distance $d_{Cl^- - Cl^-} = 3.62$ Å.
For $[Mo_6Cl_8]^{4+}Cl_{4/2}Cl_2$
$VEC_{cluster} = (6 \cdot 6) - (8 \cdot 1) - (4 \cdot 1) = 24$ e$^-$ per Mo_6 cluster.
$VEC_{metal} = 24/6 = 4$ e$^-$ per Mo atom.

$PbMo_6S_8$: $d_{Mo-Mo} = 2.70$ Å (intra-cluster distance), thus $d_{A-A} \approx 2.7 \cdot \sqrt{2} = 3.82$ Å and the observed distance is $d_{S^{2-} - S^{2-}} = 3.68$ Å. In $PbMo_6S_8$ there are the same kind of cubic cluster units as in $MoCl_2$.
For $Pb^{2+}[Mo_6S_8]^{2-}$
$VEC_{cluster} = (6 \cdot 6) + (1 \cdot 2) - (8 \cdot 2) = 22$ e$^-$ per Mo_6 cluster.
$VEC_{metal} = 22/6 = 3.67$ e$^-$ per Mo atom.

STRUCTURAL AND BONDING PRINCIPLES IN METAL HALIDE CLUSTER CHEMISTRY

JOHN D. CORBETT
Department of Chemistry
Iowa State University
Ames, Iowa 50011
USA

ABSTRACT. Metal cluster halides M_6X_{12} and M'_6X_8 and their alkali metal derivatives exhibit a variety of structure types because of the evident need to bond terminal or shared halide at all metal vertices. The versatility of systems with group 3 and 4 transition metals is greatly increased by the presence of interstitial heteroatoms (Z) within all thermodynamically stable clusters. The 24 examples of Z provide a wide range of valence electrons and include many that are themselves metals. Structural and bonding principles are described, including the novel characteristics of some cation sites, the regular effects that cations have on cluster bond distances, matrix effects that arise because of characteristic X···X or M - X distances and their influence on metal - metal bonding and electronic configurations, and the use of metal - metal bond orders to characterize matrix effects. The condensation products of octahedral clusters, particularly chains centered by heterometals and double-metal sheet products, are also noted.

1. Condensation to Clusters

The clustering of two to six (or more) metals within a nonmetal matrix is a frequent event among reduced compounds of the early transition (d) elements. The formation of such clusters is probably best known among the chalcogenides, while only a relatively few halide examples were known for Nb, Ta and Mo 10 – 15 years ago. The relative proportion of anions for a fixed oxidation state of a metal obviously increases in the series pnictide, chalcogenide, halide (e.g., VP, V_2S_3, VCl_3), and so the number of possible metal – metal interactions decrease in parallel. Thus, highly reduced pnictides and chalcogenides often exhibit condensed, complex, and relatively isotropic structures, whereas halides more often show limited and lower dimensional interactions between the valence-active metals, i.e., in isolated clusters and chains. The latter characteristics not only provide very novel results but also make more evident many of the bonding and structural principles that go into the determination of cluster and condensed cluster structures. Much of the background to this area can be found in Schäfer and Schnering (1964) and Wells (1984).

The beginning of metal – metal bonding is often seen structurally whenever the stoichiometric number of nonmetals (halides) per metal in a reduced compound is less than the preferred coordination number so that coordination polyhedra must be shared. Thus, ZrI_4 achieves CN6 by sharing two edges of each ZrI_6 octahedron, viz, $ZrI_2I_{4/2}$. Only when valence electrons are available does this allow metal – metal bonding, as with the dimers in α-NbI$_4$. Fewer halogen atoms and a d^1 configuration afford shorter M–M separations when polyhedral faces are shared, as in $^1_\infty$[ZrI$_{6/2}$] chains where metal dimers also form. More valence electrons may produce edge sharing of three MX$_6$ octahedra, as may be described by M$_3$X^i$_3$X^{i-i-i}X^a$_9$ (=MX^i$_{2/2}$X^{i-i-i}$_{1/3}$X^a$_3$)$_3$ where outer (exo or aussen) halide are distinguished by a and inner edge-bridging (or face capping) halide, by i (Schäfer & Schnering, 1964). The last two types of shared clusters are illustrated at the top of Figure 1.

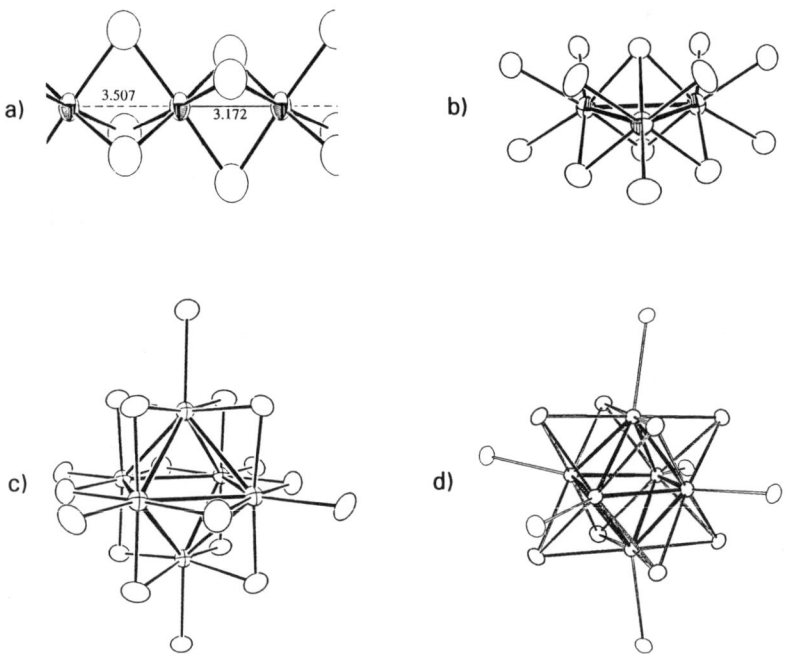

Figure 1. a) Confacial MX$_{6/2}$ chain with M–M bonding; b) M$_3$X$_{13}$ trimer from shared octahedra; c) M$_6$X^i$_{12}$X^a$_6$ cluster with edge-bridging halide; d) M'$_6$X^i$_8$X^a$_6$ analogue.

A common means of further clustering with fewer nonmetal atoms involves six MX$_6$ polyhedra, each of which shares edges with four others to give M$_6$X^i$_{12}$X^a$_6$ units ([MX^i$_{4/2}$X^a]$_6$), the one halide that would be a member of all six polyhedra and lie in the center of the cluster being eliminated. Figure I(c) shows this example for a cluster like Nb$_6$Cl^i$_{12}$Cl^a$_6$$^{4-}$. A less common halide cluster type has just eight inner halide atoms that cap faces and are each shared by three octahedra (M'X^i$_{4/3}$X^a)$_6$ = M'$_6$X^i$_8$X^a$_6$, Figure I(d). The lower proportion of halide in the last occurs with more cluster bonding electrons and, where it matters, larger X. At this stage it is useful to focus on the octahedral metal cores of the clusters and to forget about the MX_6 units with which we started.

These last two cluster types are often referred to as M_6X_{12} and M'_6X_8 to emphasize the differences in the cores, but it should be remembered that the six exo positions are evidently always occupied by halide or another Lewis base (Smith & Corbett, 1985). To do otherwise, leaving a substantial fraction of the bonding sphere about a transition metal empty, would be unusual.

Even at this stage, the condensation process is clearly headed toward the metal itself with its intrinsically higher coordination number, often 12. The described formation of $M_6X_{12}X_6$ or $M'_6X_8X_6$ units from MX_6 converts the coordination about each M to X_5M_4, one halide being replaced by four M. Later on we will see further condensation of clusters into infinite chains and then into double metal sheets in which the local metal coordination is 4–5 X plus 4–7 M and then 3 X plus 9 M, respectively. The dominance of metal octahedra throughout rather than other polyhedra may reflect the important contribution of polar M–X bonding to stability, each metal atom retaining four or five halide neighbors until double metal sheets form. (Cluster condensation via shared metal vertices is also known but not in the systems considered here; see Simon, 1981, 1988.)

Delocalization within the metal network increases in the above series as well, characteristic of a metal, and the M–M bonded unit eventually becomes structurally dominant. The metal cluster structure is in all cases found to be well-sheathed by halide, a condition that can be imagined as necessary in order to prevent further condensation. It is clear that the effectiveness of M–M bonding may be limited at least in the early stages where edges or faces of MX_6 polyhedra are shared. The fact that any M displacement in response to M–M bonding may be resisted by a necessary deformation and crowding of the nonmetal array (that is, the elastic energy of the lattice) was recognized early (Schäfer & Schnering, 1964). These constitute what have later been generalized as "matrix effects" (see 7).

2. Structural Networks via Halogen-Bridged Clusters

Isolated $M_6X^i_{12}X^a_6$ (or $M'_6X^i_8X^a_6$) clusters, usually anions, are the limiting compositions under the most basic (halide-rich) conditions, but a large range of stoichiometries, some different connectivities, and a rich variety of structures are possible with lower X:M ratios. This occurs through the introduction of bifunctional X^a atoms in $M_6X^i_{12}X^a_n$ compositions, $0 \leq n < 6$. These are illustrated at a single vertex in the following cartoon. The first figure represents one edge of a cluster and its X^i in isolation, and thereafter the modes of exo bonding starting with X^a.

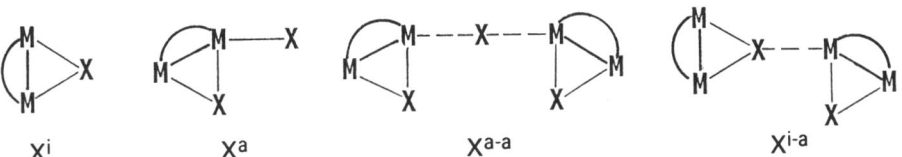

X^i \qquad X^a \qquad X^{a-a} \qquad X^{i-a}

With decreasing n, single atoms first bridge between clusters, X^{a-a}, and then inner X^i become exo to vertices in other clusters and vice versa, X^{i-a} and X^{a-i}, respectively. Compositions with uniform exo bonding modes are thus M_6X_{18},

M_6X_{15} (= $M_6X^i{}_{12}X^{a-a}{}_{6/2}$) and M_6X_{12} (= $M_6X^i{}_6X^{i-a}{}_{6/2}X^{a-i}{}_{6/2}$ for a simple proportioning), and intermediate examples are in principle achieved through anisotropic mixed connectivities via neighboring bridging functions X^a, X^{a-a}, X^{i-a}. Two common examples are the layered M_6X_{16} (= $M_6X^i{}_{12}X^{a-a}{}_{4/2}X^a{}_2$) and the three-dimensional M_6X_{14} (= $M_6X^i{}_{10}X^{i-a}{}_{2/2}X^{a-i}{}_{2/2}X^{a-a}{}_{4/2}$). Note that the sums of the i and a functions in all of these are 12 and 6, respectively. The same ideas work for M'_6X_8 cores as well, and the reader can verify that the well-known Mo_6Cl_{12} and Mo_6S_8 must for the simplest distributions have the connectivities $Mo_6Cl^i{}_8Cl^{a-a}{}_{4/2}Cl^a{}_2$ and $Mo_6S^i{}_2S^{i-a}{}_{6/2}S^{a-i}{}_{6/2}$, respectively, as observed.

Some exceptions to the foregoing regularities are more complex but also novel and informative. The first has two new aspects, anion chains of clusters joined by shared Cl^i, $Zr_6(Be)Cl^i{}_{10}Cl^{i-i}{}_{2/2}$, that are interbridged by three-coordinate $Cl^{a-a-a}{}_{6/3}$ atoms to give a $Zr_6Cl_{13}Be$ composition. The presence of three different types of chlorine in one phase presumably arises because of an advantageous three-dimensional structure (Ziebarth & Corbett, 1985). The first function is also found in the hypostoichiometric phases $Y_6I_{10}Ru$ [= $Y_6(Ru)I^i{}_2I^{i-i}{}_{4/2}I^{i-a}{}_{6/2}I^{a-i}{}_{6/2}$] and the related $Sc_6I_{11}C_2$ [$Sc_6(C_2)I^i{}_4I^{i-i}{}_{2/2}I^{i-a}{}_{6/2}I^{a-i}{}_{6/2}$] (Hughbanks & Corbett, 1989; Dudis & Corbett, 1987). The triply bridging X^a also occurs in the novel $Nb_6Cl_{12}I_2$ ($Nb_6Cl^i{}_{12}I^{a-a-a}{}_{6/3}$), a cubic relative of the Nb_6Cl_{14} structure (see 4) (Sägebarth & Simon, 1990).

Before these structures can be given detailed considerations, two other variables need to be introduced, those provided by *simple cations* and *interstitials*. Even the bridged networks described above have some flexibility and therefore an ability to accommodate particular numbers of cations of different sizes within intercluster cavities defined by X^i and X^a atoms. Nonetheless, binary Nb, Ta examples of the above types as well as their ternary derivatives with electropositive cations are relatively few, evidently because of the clusters' rather narrow valence electron requirements (see 5). Niobium chloride clusters, for example, appear limited to Nb_3Cl_8 (trimers), Nb_6Cl_{14}, $A^I{}_4Nb_6Cl_{18}$ and $CsNb_6Cl_{15}$ the last three all containing 16-electron units. Only the first and third types are known for the bromide, while the iodide provides only Nb_6I_8-type clusters in Nb_6I_{11} and $CsNb_6I_{11}$. Interstitial atoms with a range of valence electrons that are bound within all discrete halide clusters of group 3 (Sc,Y,La...Lu) and 4 (Zr,Hf) metals are the added feature that affords such great versatility in these compounds and their structures.

3. Interstitials in Clusters

The evident requirement of an interstitial atom (Z) within all thermodynamically stable 6-12 clusters of group 3 and 4 transition metals involves two obvious effects. The easier to assess (see 5) is the contribution of the interstitial's valence electrons to a bonding manifold that is intrinsically electron-poor. A second aspect of the stability must be the apparently strong, central M−Z bonding introduced, although this is difficult to quantify. Comparable examples of interstitials in clusters of group 5 metals seem to be limited to hydrogen in Nb_6I_{11} and $CsNb_6I_{11}$.

The first examples of zirconium and rare-earth-metal (R) clusters were largely synthetic mysteries until it was established that the clusters in these scarce but well-formed crystals were all centered and stabilized by foreign adventitious atoms, largely the ubiquitous C, N, or H. Inclusion of the correct ingredient then gave high yields of the same phases. Subsequent exploration of Z prospects over much of the periodic table has resulted in the encapsulation of 24 different elements in Zr and R examples. Remarkably, two-thirds of that number are metals in the pure state. All of the compounds discussed are obtained from high temperature reactions and therefore represent thermodynamically stable materials.

By far the greater number of discrete clusters and structure types and therefore systematics originate with zirconium systems, particularly for the chlorides (Ziebarth & Corbett, 1985, 1989a,b; Zhang & Corbett, 1991a) and iodides (Smith & Corbett, 1985, 1986; Hughbanks, Rosenthal & Corbett, 1988). The rare-earth elements are in contrast the only sources of condensed cluster chain phases while both group 3 and group 4 metals afford infinite sheet products (section 8).

Nearly all of these cluster phases fall in the family $A^I_x[M_6X_{12}(Z)]X^a_n$. Zirconium clusters occur in structures for all n between 0 and 6, as well as with x alkali-metal cations (A^I) per cluster over the same range. (A few phases are also known with Ba^{2+}.) Accepting for the moment the condition that 14 cluster-based electrons are optimal when Z has s and p valence orbitals, we can devise diverse ways in which 14-e counts can be achieved via the three variables x, n and the number of electrons brought by Z. Some sample categories are given in Table 1 for a few of the many dozen compositions known with $Zr_6Cl_{12}Z$-type clusters.

The first compound in the first series, $Zr_6Cl_{12}Be$, has $6 \cdot 4 - 12 \cdot 1 + 2 = 14$ e's available for cluster bonding after the low-lying chlorine 3p levels are filled. The

Table 1. The Versatility of Cluster Phases in the Family $A^I_x[Zr_6Cl_{12}(Z)]Cl_n$ for the Indicated Variables

1. Z, n:	$Zr_6Cl_{12}Be$	$Zr_6Cl_{13}B$	$Zr_6Cl_{14}C$	$Zr_6Cl_{15}N$
2. Z, n:	$Sc(Sc_6Cl_{12}N)$	$KZr_6Cl_{13}Be$	$KZr_6Cl_{14}B$	$NaZr_6Cl_{15}C$
3. Z, x:	$Zr_6Cl_{15}N$	$KZr_6Cl_{15}C$	$K_2Zr_6Cl_{15}B$	$K_3Zr_6Cl_{15}Be$
4. n, x:	$RbZr_6Cl_{14}B$	$Rb_2Zr_6Cl_{15}B$	$Rb_3Zr_6Cl_{16}B$	$Rb_5Zr_6Cl_{18}B$

rest of the isoelectronic members in this series are obtained by providing progressively electron-richer interstitials B, C, N which force a parallel stepwise oxidation of the cluster via n. The zirconium examples in the second series illustrate how a diagonal relationship with the first is accomplished by insertion of a single cation countered by a unit oxidation, e.g., $Zr_6Cl_{12}Be \rightarrow KZr_6Cl_{13}Be$. The compounds arranged vertically in series 1 and 2 are isostructural pairs if we allow this meaning to include cation insertion in a pre-existing interstice without a change in space group. In many cases, the question of structure is in fact dependent on the number and size of the cations accommodated and, sometimes, on the size of Z as well (see 4). The six fewer valence electrons available in a $Sc_6Cl_{12}Be$ cluster relative to $Zr_6Cl_{12}Be$ is compensated by three more electrons each from the interstitial (Be → N) and from an isolated Sc^{III} cation in a suitable cavity.

Series 3 shows how related changes can be effected by a parallel but opposite variations of x and the electron count of Z, while the compounds in 4 are achieved through successive additions of RbCl. Greater detail on the stoichiometries and structures possible for $Zr_6Cl_{12}Z$-type clusters is given by Ziebarth and Corbett (1989a).

The change to zirconium iodide clusters expands the possible interstitials with Al, Si, P and Ge. Surprisingly, most of the 3d elements Cr − Ni can be incorporated into either chloride or iodide clusters. Extension of the studies to rare-earth-metal iodide hosts (R = Sc, Y, Pr, Gd particularly) reveals an even more amazing chemistry — not only the 3d but many of the analogous 4d and 5d metals may be encapsulated in either $R_7I_{12}Z$ or $R_6I_{10}Z$ phases (Hughbanks & Corbett, 1988, 1989; Payne & Corbett, 1990). The optimal requirement for a cluster when the interstitial utilizes d-orbitals is 18 electrons (see **5**).

4. Structural Examples and Varieties

The progression of structures associated with n exo chlorine atoms about the basic $M_6X^i_{12}(Z)$ cores follows the principles laid out earlier. These will be illustrated for n = 0, 2, 3 and 4 in figures that emphasize cluster interconnectivities, but defer for a time any representation of the strong central Zr–Z bonding. Figure 2 shows the

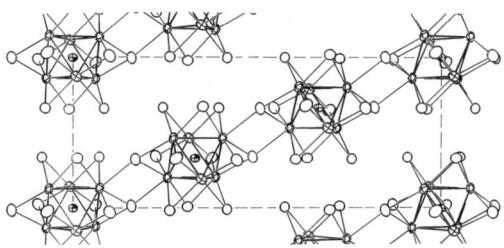

Figure 2. The [110] section of the rhombohedral structure of $Zr_6Cl_{12}Be$ with Zr and Be as crossed, chlorine as open ellipsoids.

[110] section of the rhombohedral $Zr_6Cl_{12}Be$ ($Zr_6I_{12}C$-type). With no exo chlorines, complementary Cl^{i-a} and Cl^{a-i} bridges are used to bond at zirconium vertices in adjoining clusters, as seen here as rhomboids at the upper right and lower left of each cluster. Vertical three-fold axes through the cluster "octahedra" with $\bar{1}$ centers at Be generate such linkages to six neighboring (c.c.p) clusters, while the six chlorine atoms about the two ends of each cluster remain Cl^i. (The antiprismatic site between clusters defined by these Cl^i (0,0,1/2) is the locale of the countercation in the $R(R_6X_{12}Z)$ structure; Hwu & Corbett, 1986.)

Starting with this Zr_6Cl_{12} structure, addition of chloride progressively converts Cl^{a-i} to Cl^{a-a} and then to Cl^a. The fairly regular way this first step is accomplished at n = 2 in the orthorhombic $Zr_6Cl_{14}C$ (Nb_6Cl_{14} type except for C) is shown in Figure 3. (For clarity, only the intercluster bridging halides will be shown in this and later cluster pictures, with all Cl^i omitted.) The additional chlorines have generated 4/2 Cl^{a-a} atoms, with the remaining pairs of Cl^{a-i} and Cl^{i-a} functions lying on a mirror plane perpendicular to $\bar{a}$.

A characteristic of this structure is a substantial compression of the clusters along the X^{a-i}–Zr–Z–Zr–X^{a-i} axes, these Zr–C distances in $Zr_6Cl_{14}C$ being 0.064 (2) Å less than those trans to the shorter Zr–Cl^{a-a} bridges. This doubtlessly arises from the lower basicity and coordinating ability of the three-bonded Cl^{i-a} atoms, as will be generalized later. Inclusion of a cation in such a structure can often be forced by inclusion of only an electron-poorer interstitial in the synthesis, as in $AZr_6Cl_{14}B$ for A = Li–Cs, Tl. Which of two cation sites is utilized depends on the size of A (Zhang, 1990).

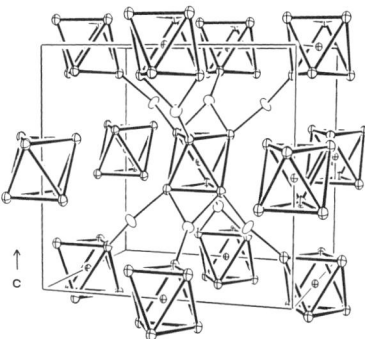

Figure 3. The C-centered orthorhombic structure of $Zr_6Cl_{14}C$. The intercluster bridging Cl^{a-a} and Cl^{i-a} are shown as open ellipsoids, while the Cl^i atoms are omitted. (90% ellipsoids here and later unless noted.)

The symmetric bridging arrangement in $(Zr_6Cl^i{}_{12}Z)Cl^{a-a}{}_{6/2}$ ($n = 3$) provides a beautiful series of four distinctive structures that are unrelated in the sense that they cannot be interconverted by twisting, only by breaking and rearranging the bridge bonds. The distinguishing features are listed in Table 2, and the first and last members are shown in Figure 4. The progression $Zr_6Cl_{15}N$, $KZr_6Cl_{15}C$, $K_2Zr_6Cl_{15}B$ (summarized by Ziebarth & Corbett, 1988) and $Zr_6Cl_{15}Co$ (Rogel, Zhang, Payne & Corbett, 1990; Zhang & Corbett, 1991a) parallels the size of Z, the angle at bridging Cl^{a-a} and, approximately, the number (and the size) of the cations. Thus, the first member can be derivatized using cations no larger than sodium (Figure 4, left). A single K, Rb, or Cs or two such cations with combinations of K plus Cs or larger give the second more open structure as the carbide or the boride, respectively. (Two cation sites of different sizes are available in the network.) Two or three potassium ions produce a still more open structure with one-third linear bridges. (Some of the cation sites utilized here are not very optimal — see 6.1.) Finally, the combination of a large interstitial (Mn – Ni), the relatively small Cl^{a-a}, and the small cation Li^+ (if any) allows a remarkable result: the two interpenetrating but

Table 2. Distribution of Cl^{a-a} bridges in $Zr_6Cl_{12}(Z)Cl_{6/2}$-type structures

Structure Type	Space Group	X^{a-a} Bridge Type, %		Criteria	
		Linear	Bent	Ring Size[a]	No. of FBS[b] Clusters so Linked
$Zr_6Cl_{15}N$	Ia3d	0	100	4	2
$KZr_6Cl_{15}C$	Pmma	17	83	3	–
$K_2Zr_6Cl_{15}B$[c]	Cccm	33	67	4	2,3
$Zr_6Cl_{15}Co$	Im3m	100	0	4	4

[a] y in $(-Zr_6-Cl^{a-a})_y$. [b] First bonding sphere (FBS) to a central cluster. [c] A distorted version $K_3Zr_6Cl_{15}B$ accommodates an additional cation (s.g. C2/c).

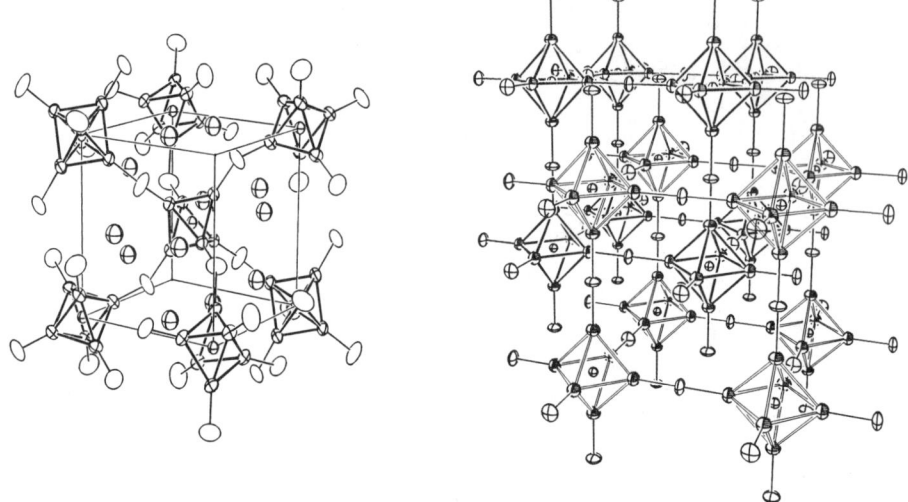

Figure 4. Left: The structure of $Na_{0.5}Zr_6Cl_{15}C$ ($Zr_6Cl_{15}N/Ta_6Cl_{15}$ type) with only cluster-bridging chlorine shown. Right: Cubic $Zr_6(Co)Cl^i{}_{12}Cl_{6/2}$ with Cl^i omitted. Note the two interpenetrating but not interconnected bcc networks (open and solid clusters).

not interconnected cubic networks with only linear bridges shown in Figure 4 (right) for $Zr_6Cl_{15}Co$. (The parent structure type, Nb_6F_{15}, had previously been a singularity; Schäfer, von Schnering, Niehues, & Nieder-Vahrenholz, 1965.) Each of the two networks is equivalent to the ReO_3 structure with a cluster replacing the rhenium atom. There is room only for lithium ions within chlorine-defined cavities, and their presence in the synthesis allows one to "drive" the encapsulation of earlier transition metals in the isoelectronic clusters $LiZr_6Cl_{15}Fe$ and $Li_2Zr_6Cl_{15}Mn$. Space for the lithium atoms in the series Z = Co, Fe, Mn is created by small contractions in d(Zr–Z) plus appreciably greater increases in d(Zr–Cl^{a-a}), both of which are consistent with the effects that such small cations are observed to have on dimensions in cluster structures (below). The structure has also been realized for $Th_6Br_{15}Z$ with Z = Mn–Ni (Böttcher et al., 1991).

The result of one more chlorine per cluster in opening up the 6-15 structure is illustrated in Figure 5 (left) for the anion structure of $Na_4[Zr_6(Be)Cl_{12}]Cl^{a-a}{}_{4/2}Cl^a{}_2$ (= $Na_4Zr_6Cl_{16}Be$). The puckered layers of bridged clusters are lined by Cl^a atoms, with the cations (not shown) lying both within and between the layers (Ziebarth & Corbett, 1989c).

A comparable series of bridged structure types is found in rhenium 6-8 clusters, e.g., $Ba_2(Re_6S_8S^{a-a}{}_{6/2})$ (Bronger & Miessen, 1982). Further versatility develops with the use of both chalcogenide and halide which provide the novel layered $Re_6Se_8Cl_2$ ($Re_6Se^i{}_4Se^{i-a}{}_{4/2}Se^{a-i}{}_{4/2}Cl^a{}_2$) and $Re_6Se_6Cl_6$ ($Re_6Se^i{}_6Cl^i{}_2Cl^{a-a}{}_{4/2}Cl^a{}_2$) and the 1-D $Re_6Se_5Cl_8$ (Leduc, Perrin & Sergent, 1983; Perrin, Leduc & Sergent, 1991).

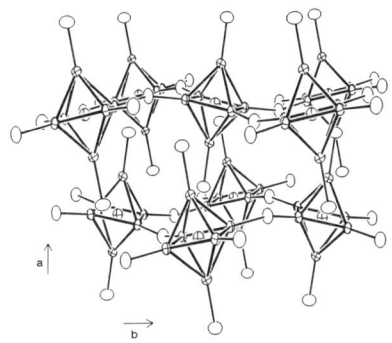 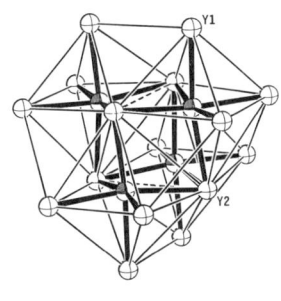

Figure 5. Left: Portions of two anion layers in $Na_4Zr_6Cl_{16}Be$. Right: The metal portion of the oligomeric $Y_{16}I_{20}Ru_4$ (Ru is shaded). The drawing emphasizes the strong central Y–Ru bonding.

Finally, a rare example of an oligomeric product rather than condensed cluster chains is $Y_{16}(Ru_4)I_{20}$, Figure 5 (right), where four $Y_6I_{12}Ru$ type clusters have been condensed two-by-two (pseudo tetrahedral) through edge sharing (Payne, Ebihara & Corbett, 1991). This new 16-atom polyhedron can be generated by capping the four hexagon faces of a truncated tetrahedron and can also be recognized in a more condensed array in certain Laves phases. The strength of the central Y–Ru bonding is evident in both distances and the M.O. results, while Ru–Ru interactions are very small. The motifs of cluster sheathing and intercluster bridging by iodine are of the types described earlier.

5. Bonding in Clusters

Some general observations on the bonding features and regularities in these isolated cluster examples are useful before we pass on to other structural systematic as well as to extended examples where such bonding generalities are not as easy to come by. Greater details and discussion are available in Hughbanks (1989). The extended-Hückel M.O. description that accounts for the 14-electron optimum for cluster-based electrons is relatively straightforward – Figure 6 (Ziebarth & Corbett, 1989b). Starting with the hypothetical empty $(Zr_6Cl_{12})Cl_6$ unit is important as the exo Cl^a atoms are trans to, and compete with, Z for the same orbitals on Zr. Occupied Cl 3p states, both nonbonding pairs and those associated with Zr–Cl bonding, lie below the interesting zirconium-based and cluster-bonding d orbital combinations a_{1g}, t_{1u}, t_{2g}, a_{2u} (in the O_h limit) in the order of increasing energy. The result of insertion of a main-group element like B is easy to judge since its s (a_{1g}) and p (t_{1u}) orbitals interact only with the metal-based states of the same symmetry. Since the a_{2u} M.O. is somewhat antibonding, clusters with 14 cluster-based electrons and t_{2g}^6 as the HOMO are plausible and are realized in a great many structures and compositions (Table 1). Note that boron contributes three electrons and four orbitals to the bonding manifold, but no new bonding orbitals appear; rather high-lying a_{1g}^* and t_{1u}^* states (not shown) are generated in the

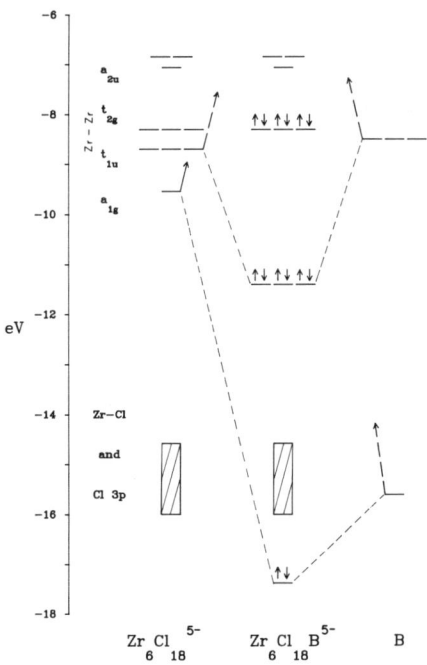

Figure 6. The metal and interstitial-based region of the extended-Hückel M.O. result for $Zr_6(B)Cl_{12}Cl_6^{5-}$.

process. The parameters of distance, overlap, valence energy and so forth require some care and consideration, but a decrease in energy of the system is evident. This is by no means sufficient to conclude that stability is likely; the process neglects not just the obvious heat of atomization of boron, etc., but also the more general and very real possibility that alternate products may be more stable, ZrZ_x, Zr and $ZrCl_3$, for example.

Eighty or so $Zr_6Cl_{12}Z$-type clusters distributed over 18 different structure types have been prepared with main group Z (H, Be, B, C, N), many to test the electronic boundaries, and only seven are exceptions to the I4-e expectation. Three are I3-e hydrides, suggesting that something in the cluster is better than nothing at all. Most of the others can be understood as extremes in "matrix effects". Zirconium iodides afford only $Zr_6I_{12}Z$ and $Zr_6I_{14}Z$ structure types but new examples of Al, Si, Ge or P interstitials and a few 15- and 16-electron clusters. This last change will be considered later (sec. 7). Tetragonally elongated clusters of the rare-earth elements are also known with dicarbon (acetylide) interstitials. These are more interesting for their bonding than for structural principles. For details see Simon (1985), Hughbanks (1989), or Simon et al. (1991).

The surprising existence of zirconium cluster phases centered by the 3d metals Cr, Mn, Fe, Co or Ni requires further consideration of both M.O. expectations and the nature of the Zr–Z bonding. The former are relatively straightforward, as shown in Figure 7 for the charge-consistent result of iron in an idealized octahedral

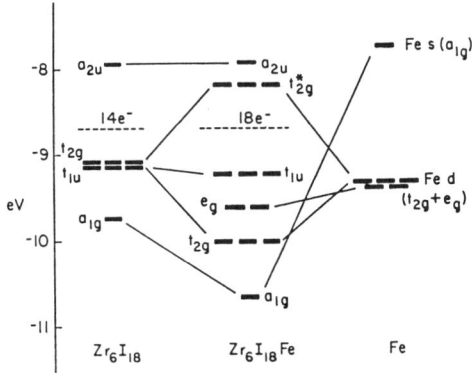

Figure 7. The M.O. diagram for the metal-based states in $Zr_6(Fe)I_{12}I_6$

Zr_6I_{18} cluster of typical dimensions (Hughbanks, Rosenthal & Corbett, 1988). Symmetry properties alone mean that the t_{1u} and t_{2g} cluster-based states exchange functions relative to main-group Z, the t_{2g} set now becoming Zr–Z bonding via Fe 3d orbitals while t_{1u}^6 is the probable HOMO. (Fe 4p states lie too high to be important in bonding to the cluster framework.) An important difference is the presence of the essentially nonbonding e_g^4 on the interstitial, the reason the closed shell count is raised from 14 to 18 with d elements. In this case, only four of the eight valence electrons of iron contribute to bonding within the cluster. In a valence sense, the iron is thus "carbon-like" and can also be ascribed an oxidation state of −IV since all 3d and 4s states are filled in the diamagnetic product. A number of lesser details depend on parameters such as cluster size and the relative valence energies of the host metal and the interstitial. Experimentally, the zirconium examples with 3d metals are all 18- or 19-electron clusters, the iodides representing most of the exceptions.

A remarkable stability and interstitial diversity occurs in the rare-earth-metal cluster iodides which encapsulate not just 3d but 4d and 5d metals as well. As summarized by Payne & Corbett (1990), any of the elements

Mn	Fe	Co	Ni	Cu
	Ru	Rh	Pd	
Re	Os	Ir	Pt	Au

can be found in $R_7I_{12}Z$ or $R_6I_{10}Z$ cluster phases for R = Pr or Gd, while R = Y allows the incorporation of about two-thirds as many. (Only these three host elements R have been investigated at all systematically.) Again, Z elements not listed represent systems in which other products are more stable. Some of these R–I–Z combinations also form a variety of extended, condensed cluster phases in addition to the $R_7I_{12}Z$ and $R_6I_{10}Z$ examples with isolated clusters (see **8**).

In contrast to the zirconium examples, clusters in $R(R_6I_{12}Z)$ phases centered by many late transition metals show ±2-electron divergences from the 18-e optimum. Two other factors now seem particularly important. The larger transition metals expand the cluster framework appreciably, raising the energy of the R−R bonding t_{1u} HOMO and making it more nearly nonbonding. A greater difference between the valence energies H_{ii} of R and Z, especially for the platinum metals, also lowers the relative energies of all of R−Z bonding levels, the $t_{2u}{}^*$ LUMO included. Evidently, the former t_{1u} level can be less than fully occupied in 16- and 17-electron cases, or the nominal $t_{2u}{}^*$ levels partially filled in 19- and 20-electron examples with larger R−Z polarities.

Calculations as well as qualitative expectations suggest that substantially all of the interstitials become somewhat negative with respect to the host cluster metals Zr, Y, etc. but certainly not to the degree of the enveloping halides. EHMO calculations do not quantify the degree of electron transfer very well, but XPS core data clearly indicate that carbon is somewhat negative. The carbon 1s levels are shifted −2.2 to −3.0 eV in $Zr_6Cl_{14}C$ and in the condensed phases $Sc_7Cl_{10}C_2$ and M_2Cl_2C, M = Sc, Zr (Hwu et al., 1986; Ziebarth & Corbett, 1989b). The "little bits of intermetallics" sampled in, e.g., $Zr_6Cl_{12}Fe$ or $Y_7I_{12}Ru$ are certainly novel. Their existence is certainly consistent with the exceptional stability known for binary phases formed between early and late transition metals (Brewer & Wengert, 1973). It is worth noting that M−Z separations within these are usually 0.1 to 0.3 Å less than exhibited in the analogous M_xZ intermetallics with similar local environments. The process of cluster formation involves oxidizing a bit of this reference and then wrapping it with halogen, which appears to remove electrons that are approximately nonbonding and M−Z screening. Alternatively, the observations may be a reflection of the much more anisotropic M−Z interactions that develop on removal of "backside" M−Z interactions in the intermetallic when the polyhedron is extracted, oxidized and enclosed.

6. Dimensionality Regularities

Phases containing discrete clusters afford many informative regularities regarding various dimensions, particularly among the zirconium chloride members where so many different compositions and structures have been achieved. The Zr−Cl and A−Cl distances naturally depend on halogen coordination number or basicity. Thus d(Zr−Cl^{i-a}) in $Zr_6Cl_{14}C$ (Figure 3) is 0.08 Å greater than for Zr−Cl^i; likewise, d(Zr−Cl^{a-i}) is longer than for Zr−Cl^{a-a}, while the last is in turn 0.10 Å longer than Zr−Cl^a in $Na_4Zr_6Cl_{16}Be$. However, neither these distances nor their differences are particularly transferable, and detailed comparisons show that we must also pay attention to the alkali-metal contributions, especially the effect that their simple, more-or-less ionic interactions with chlorine have on zirconium bonded to the same atoms. The phenomena turn out to be easily recognized, repeatedly observed, and quite instructional.

6.1. CATION SITES

The local cation structures and their differences originate with the clusters. The Zr−Cl bonding appears to be the most structurally demanding, particularly as expressed in the many and diverse networks with intercluster bridging chlorine. In contrast, the cations, particularly the larger and lower field members, are repeatedly found in relatively irregular, asymmetric, or even poorly defined sites compared with those in simple salts. These circumstances give the impression that the structural elements involving the countercations, which are required for electronic reasons by virtue of Z and the chlorine content, have the last "say" in what constitutes a "good" structure. The refinement characteristics of some of these may on first encounter engender considerable concern that crystallographic errors are instead involved, particularly regarding the correct local symmetry.

The earliest example encountered, one of two cesium sites in $CsKZr_6Cl_{15}B$, is shown in Figure 8(a) (Ziebarth & Corbett, 1987). Six of the eight neighboring Cl^i atoms about the cesium are rigorously planar, while the other two lie on axis inclined 55° thereto, all in the range of 3.44 – 3.57 Å. There are no other chlorine neighbors within 4.8 Å, and the "thermal" distribution found reflects this. The other half of Figure 8 shows the somewhat oversized site that is occupied by 20% of the cations in $Rb_5Zr_6Cl_{18}B$ where only packing of ions, not a cluster network, is involved (Ziebarth & Corbett, 1989b). What is illustrated is one-half of the split cation site that was refined about a $\overline{1}$ position coplanar with the Cl5 atoms. The rubidium distances to $Cl5^i$, $Cl3^a$ and $Cl3^a$ (dashed), 3.25, 3.92 and 4.37 Å, respectively, illustrate the problem. (The sum of crystal radii for CN6 is 3.44 Å.) The use of boron within the cluster and a comparatively large amount of RbCl have forced the lattice to accommodate some additional cations for which there are only relatively inferior sites. Many other examples of this type of problem are known in varying degrees.

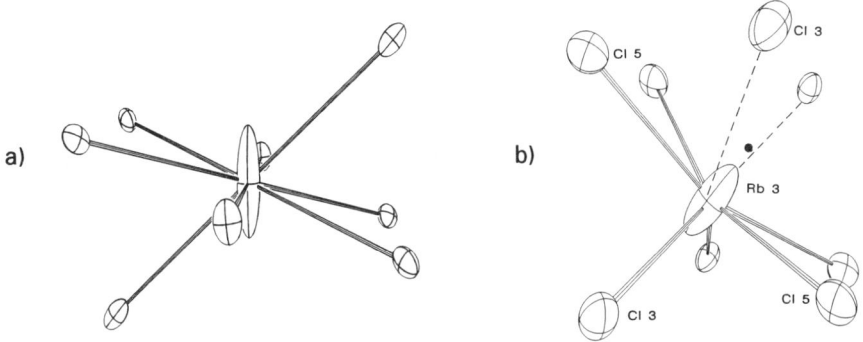

Figure 8. a) The Cl^i neighbors about the Cs2 site in $CsKZr_6Cl_{15}B$ (50% ellipsoids); b) One of the split sites refined for Rb3 in $Rb_5Zr_6Cl_{18}B$. The inversion-related member is at •.

6.2. CATION EFFECTS ON $d(Zr-Cl^a)$

Not surprising, the better-behaved cations occupy positions with neighboring anions that fall higher in the expected basicity ranking $Cl^a > Cl^{a-a} > Cl^i > Cl^{a-i}$, and these ligands lie distinctly closer as well. Second neighbor effects resulting from such cations also produce logical variations in $Zr-Cl^a$ and $Zr-Cl^{a-a}$ distances (Zhang, 1990; Zhang & Corbett, 1991b; Zhang, Ziebarth & Corbett, 1992). Thus, observed $Zr-Cl^a$ distances range from 2.56 Å and 2.60 Å in $Rb_4Zr_6Cl_{18}C$ and $Cs_3Zr_6Cl_{16}C$ to 2.68 and 2.69 Å in $Ba_3Zr_6Cl_{18}Be$ and $Li_6Zr_6Cl_{18}H$, respectively. Cations in the former pair have relatively few Cl^a neighbors and environments more of the character illustrated above, while the longer $Zr-Cl^a$ distances occur where distinctly higher field cations lie in more regular cavities and have more and distinctly closer Cl^a atoms. In $Ba_3Zr_6Cl_{18}Be$, the Cl^a atoms are all members of the coordination sphere of two cations, and the $Ba-Cl^a$ distances, 3.09 – 3.11 Å, compare with 3.16 Å for the sum of the six-coordinate crystal radii. An extreme in second neighbor effects occurs in several derivatives of the closely related $K_2(ZrCl_6)\cdot Zr_6Cl_{12}H$ where the exo chlorines (d = 2.77 Å) are more obviously members of a $ZrCl_6^{2-}$ (or similar) anion and actually lie closer to the isolated zirconium. Returning to the novel $Li_xZr_6Cl_{15}M$ structure, M = Mn, Fe, Co (Figure 4), the insertion of one or two lithium atoms per formula unit that affords compounds with the electron-poorer Fe and Mn, respectively, is similar. One-third of the lithium neighbors are Cl^{a-a}, and these cations lengthen the $Zr-Cl^{a-a}$ bridges by ~0.06 Å per lithium, enabling the cations to "dig their own hole" in the structure.

6.3. INTERSTITIAL RADII

Effective interstitial radii in clusters appear to be quite reproducible among different structural types (baring matrix effects – see 7), the clusters contracting or expanding as needed to accommodate Z elements of different sizes. This is nicely illustrated in Figure 9 as $\bar{d}(Zr-Z)$ in chloride clusters as a function of the number of electrons from (or the group of) Z for the second period elements Be – N plus H. (Note that d(Zr–Zr) is geometrically determined by d(Zr–Z).) Even the spread of d(Zr–Z) within the boron group in Figure 9 can be readily understood in detail in terms of opposed variations in $d(Zr-Cl^a)$ brought about by cation effects of the sort described in 6.2. Effective Z radii are not very predictable, however, based on standard radii of some sort or on Zr–Z separations in binary phases (see 5). In addition, the d(Zr–Z) values shown appear to reach an intrinsic minimum value with the small H and N, and hydrogen is in fact known to be "rattling" in $Zr_6Cl_{12}H$ (Chu et al., 1988).

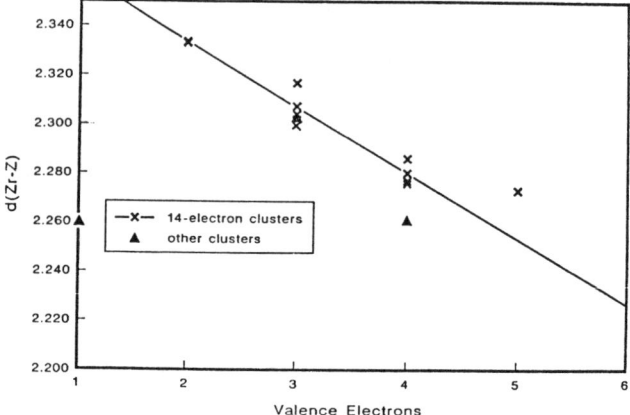

Figure 9. Average Zr–Z distances in $Zr_6Cl_{12}Z$ clusters as a function of interstitial H, Be, B, C, N.

6.4. M−X DISTANCES IN GENERAL

A widespread regularity among the shorter M−X distances in virtually all reduced compounds of these elements, metallic or clustered or not, warrants a brief comment. This property really arises from the absence of localized, reduced states on individual metal atoms in these (and many other) phases. The too-familiar behavior of the 3d metals in simple compounds is otherwise; six-coordinate radii increments of 0.14 to 0.19 Å accompany oxidation state changes of $+3 \rightarrow +2$ (high spin) and 0.06 to 0.11 Å with $+4 \rightarrow +3$ reductions (Shannon, 1976). No such differences appear on reduction of normal-valent compounds in many other situations; basically the differentiating (reduction) electrons in the latter are delocalized over a collection of metal cores and do not screen the metal−nonmetal interactions, but rather are in effect repelled by the latter. Accordingly, the shorter metal-nonmetal separations in all of these compounds can be well represented by the sum of crystal radii for the *normal-valent* components, Nb^V, Zr^{IV}, Y^{III}, etc. In other words, no distance increment is observed to accompany the onset of metallicity in a series like SrS, YS, ZrS. Some examples are collected in Table 3 to

Table 3. Shorter metal−nonmetal (M−NM) distances[a] in some reduced phases vs. sums of crystal radii (CN6) (Å)

	d(M−NM)[a]	Σ C.R.
Clusters		
$Sc_7Cl_{12}N$	2.55 – 2.60	2.55
KTi_4Cl_{11}	2.39 (ave)	2.41
$Zr_6Cl_{15}Co$	2.54	2.53
$Zr_6Cl_{14}C$	2.50 – 2.52	2.53
$Rb_4Zr_6Cl_{18}B$	2.56[b]	2.53
$Na_4Zr_6Cl_{16}Be$	2.54 – 2.60[b]	2.53
$Nb_6I_{11}S$	2.82 – 2.90	2.84
$K_4Nb_6Br_{18}$	2.58 – 2.60	2.60
Cluster Chains		
Gd_2Cl_3	2.73 – 2.79	2.75
Sc_5Cl_8C	2.54 – 2.58	2.55
Dimers in Confacial Chains		
$ZrCl_3$	2.54	2.53
ZrI_3	2.87 – 2.93	2.92
Three-Dimensional		
Zr_9S_2	2.59	2.56
Nb_2Se	2.62 – 2.64	2.62
YS	2.73	2.74
ZrS	2.58	2.56

[a] To X^i in clusters. [b] Cl^i also has alkali-metal neighbors.

illustrate the breadth. Since the effect of coordination environment, if any, is difficult to sort out in complex structures, a CN = 6 has been used throughout. Self-consistent atom radii of 1.46 and 1.40 Å for carbide and nitride, respectively, can be similarly derived from the many NaCl-type examples.

7. Matrix Effects in Diverse Structures

The foregoing comparisons avoided the cluster iodides because the observed M−M and M−I distances in these often appear to be strongly affected by closed-shell contacts between the large iodine atoms. This kind of limitation is more serious with the larger nonmetals, or where the NM:M ratios are relatively large, as with coordination octahedra that share edges or faces like in $NbCl_4$, M_3X_8, etc. One would also expect this to be a problem in layered structures such as CdX_2, BiI_3, MoS_2 and related types where the M−M repeats within the slabs are usually determined mainly by the greater nonmetal contact separations within their parallel and commensurate layers. The same pertains to M−M bonding in high symmetry NaCl, CaF_2, etc.-type lattices; clearly d(M−M) in the rock salt structure is fixed at $\sqrt{2}$ d(M−X). Since nonmetal radii appear to be fairly transferable, d(M−M) values in structures with such "matrix effects" are generally not meaningful measures of metal−metal bond strengths (Corbett, 1981a,b).

7.1. MATRIX EFFECTS IN CLUSTERS

The ideal M_6X_{12} cluster probably consists of cuboctahedron of halide with the metals more-or-less centered on the square faces. In practice, all clusters with X = Cl, Br, I deviate from this norm to an increasing degree. Because of relative sizes, the $X^i \cdots X^i$ distances about the cluster are often intrinsically greater than optimal M−M distances in most bonded M_6 octahedra (centered or not). Therefore, the metal atoms are often significantly withdrawn from the X^i_4 planes about them, as can be seen in the vertices at the top and the right of Figures 1(c) and 2, respectively. This drive to achieve more appropriate M−M bonding pulls the halides toward each other and into increasingly repulsive interactions, limiting near 3.4 − 3.6 Å in chlorides and ~ 4.00 Å in iodides. The $M-X^i$ distances do not seem to elongate substantially in problem cases, although $M-X^i$ π bonding is lost on distortion (see 7.3); rather the M−M bonding is compromised and forced to be less than optimal. Interstitials act as props within the M_6 units and independently determine d(M−M), but the competing effects may still seen with small Z and large X. In the isotypic $Zr_6Cl_{14}C$ and $Zr_6I_{14}C$, $\overline{d}$(Zr−Zr) and $\overline{d}$(Zr−Z) are 0.08 and 0.05 Å greater, respectively, in the iodide, and displacement of the zirconium from the plane of its four X^i atoms increases from 0.25 to nearly 0.5 Å. M'_6X_8-type clusters are naturally less hindered in this respect, but problems still arise with smaller metals. Of course, large Z makes the clusters larger and more nearly regular but at some expense to M−M bonding in the framework. (This occurs not only via the t_{2g} or t_{1u} HOMO's for normal clusters, Figures 6 and 7, but in metal−metal bonding components within other M.O.s as well.)

These cluster distortions can be generalized better in terms of the trans $X^i - M - X^i$ angles across the square faces of the clusters. Some meaningful ranges are: $(Zr_6I_{12}C)I_2$, 156−163°; $(Nb_6Cl_{12})Cl_6^{2-}$, 161−167°; $(Zr_6Cl_{12}C)Cl_n$, 167−171°; $(Zr_6Cl_{12}Co)Cl_3$, 176.4°. The last provides a good reference since $Cl^i - Cl^a$ and $Cl^i - Cl^i$ (intra- and inter-network) distances are 3.62, 3.58 and 3.60 Å, respectively, and $d(Zr-Cl^a)/d(Zr-Cl^i)$ (all for two-coordinate chloride) is only 1.041. Caution is necessary in dealing with the effects of changing Z in terms of these angles, however, as our interpretation has the implicit assumption that the opposed $M-X^a$ and $M-Z$ bonding are without size problems equally effective in competing for the same orbitals on M. Although $Zr_6Cl_{15}Co$ seems to qualify, phases like $Y_6I_{10}Os$ are not crowded by a $d(I\cdots I)$ criterion, yet $I-Y-I$ angles of $162-167°$ on the faces indicate that Os, Ir, Pt, etc. are "winning" the competition with I^a in bonding to the host metal atoms. Pi bonding with d-element interstitials seems to be responsible.

The magnitude of the matrix effect in clusters also appears to play a major role in determining $M-X^a$ distances and bond strengths, particularly with the larger nonmetals I, Te, and so forth. When the metal vertex is well withdrawn, an approaching X^a may encounter significant closed-shell repulsions from the four surrounding X^i long before an optimal $M-X^a$ bond is achieved. Although $M-X^a$ bond lengths do appear to be intrinsically somewhat longer than for $M-X^i$ in M_6X_{12} clusters, a relative elongation of $M-X^a$ bonds in iodides with smaller Z is clear. The particularly large family of isomorphous $Zr_6I_{14}Z$ (and $CsZr_6I_{14}Z$) compounds illustrate this. These compounds contain a sizable range of average $Zr-Z$ distances, from 2.32 Å with C through the intervening B, Al, P ~ Fe ~ Mn to 2.53 Å with Si, a range of 0.21 Å. The role of Z in expanding the clusters is seen particularly well at the two more withdrawn metal vertices with I^{a-i} bonding (Figure 3). Here the opposed $Zr-I^{a-i}$ distances *decrease* by 0.26 Å as $d(Zr-Z)$ increases by 0.21 Å over the same series! Thus, the Zr vertex can be thought of as moving within a relatively fixed iodine lattice in response to the push (or pull) supplied by Z, with $d(Z\cdots I^{a-i})$ remaining relatively constant (Hughbanks, Rosenthal & Corbett, 1988).

Trends observed for empty clusters over the periodic table relate to the natural decrease in metal bonding radii with increasing group number and also testify qualitatively to the same problems. Nb_6Cl_{12}-type clusters give way to (electron-richer) Mo_6Cl_8-type units (although $(W_6Cl_{12})Cl_6$ exists), while niobium bromide analogues are known only for the $Nb_6Br_{18}^{4-}$ anion, and the iodide forms only Nb_6I_8 cluster products. The anion effect in empty clusters is well demonstrated by average $Nb-Nb$ distances, which are at a minimum of 2.80 Å in Nb_6F_{15} and 2.80 – 2.83 Å in condensed Nb_6O_{12} systems and increase to 2.89 – 2.95 Å in Nb_6Cl_{14} and 2.97 – 2.98 Å in $K_4Nb_6Br_{18}$ (Schäfer & Schnering, 1964; Burnus, Köhler & Simon, 1987; Köhler, Tischtau & Simon, 1991; Ueno & Simon, 1985). Attempts to use interstitials to expand the niobium bromide or iodide clusters and thus to produce new $Nb_6X_{12}Z$ examples have been unsuccessful, perhaps because the cluster electron counts are already high relative to zirconium analogues.

These matrix ideas are also useful for understanding differences in optimal electronic configurations as a function of matrix effects and distortion. Before considering these, it is worthwhile to note a contrasting and yet similar example in M_6X_8-type cluster, namely with Mo_6S_8, etc. (Corbett, 1981b). This excursion will be kept brief since another chapter in this Volume is concerned with these "Chevrel phases".

The Mo_6S_8 cluster, like $Zr_6X_{12}Z$, lacks any additional ligands for exo bonding at the metal vertices. Furthermore, the lower nonmetal content in a face-capping role leaves the metal vertices more exposed than before. Two views of a portion of the structure of Mo_6S_8 in Figure 10 illustrate a mode of intercluster bridging that is comparable to that in $Zr_6Cl_{12}Be$, Figure 2, which is also rhombohedral. Pairs of complimentary Mo−$S^{a\text{-}i}$ bonds between neighboring clusters (top view) again fulfill the exo bonding needs, but these bonds are now as short as any between Mo and the face-capping $S^{i\text{-}a}$ and S^i atoms. This bridging also affords the short Mo−Mo intercluster interactions important to the phase's metallic conduction. A 3-fold inversion axis centered in each cluster again generates a c.c.p. array of bridged clusters as shown at the bottom of the Figure in a view along $\bar{c}$.

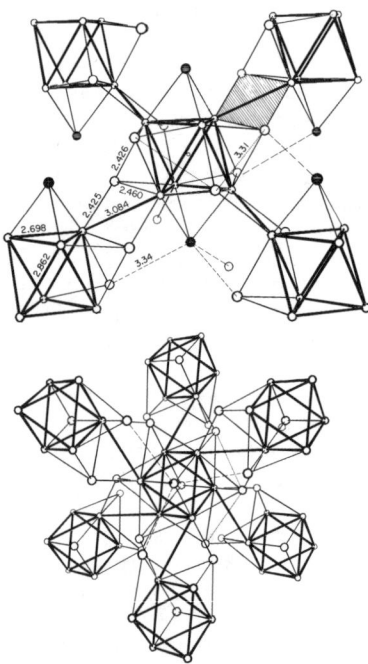

Figure 10. Two views of the rhombohedral Mo_6S_8. The top section with $\bar{c}$ vertical illustrates the intercluster bridging and the shorter intracluster (dotted) and intercluster (dashed) S⋯S contacts that serve to elongate the cluster. One cluster bridged to its six neighbors is shown at the bottom along $\bar{c}$.

A remarkable feature of the structure is that strong matrix effects − close intra- and intercluster S⋯S contacts − again give rise to cluster distortion. Examples of the shorter distances (3.31 Å and 3.34 Å) are shown as dotted and dashed lines, respectively, in the upper portion; these are judged to represent "tight" packing relative to the smallest values observed in other metal-rich sulfides. It will be noted that these repulsions all act across the waists of the bridged clusters, strongly suggesting that these are responsible for cluster elongation along $\bar{c}$ that produces a 0.16 Å difference in the Mo−Mo edges within. Reduction of the cluster to $M_xMo_6S_8$ phases lengthens the intercluster Mo−$S^{a\text{-}i}$ bridges, and (with some other changes) all of the S⋯S contacts and allows a 0.15 − 0.18 Å decrease in Mo−Mo separations around the waist of the cluster. Elusive "electronic effects" that otherwise were thought to be responsible for the cluster dimensions and their changes do not seem necessary.

7.2. METAL–METAL BOND ORDERS

A very noteworthy increase in effective Mo–Mo bonding (a decrease in bond lengths) relative to the foregoing example occurs when the necessity of intercluster bridging is completely eliminated through the formation of either ligated clusters $Mo_6X_8L_6$ or the condensed chain phases $A^IMo_3S_3$. This observation prompted a further inquiry as to whether a useful quantification of the degree of metal–metal bonding in diverse compounds relative to that in the parent metals themselves might be achieved when allowance is made for the diminished number of electrons available for metal–metal bonding in the compounds (Corbett, 1981a,b).

The very insightful equation $D_n = D_1 - 0.60 \log n$ was developed by Pauling to describe a metal–metal bond order n as a function of distance D_n relative to a derived single bond D_1 for $n = 1$ (Pauling, 1960). Its application to metals is particular straightforward: a 12-coordinate (h.c.p. or c.c.p.) example with an observed D_{12} and v metal–metal bonding valence electrons gives $D_1 = D_{12} + 0.60 \log (v/12)$. (Each neighbor also has v electrons so there are v bonding pairs distributed over 12 neighbors.) The question in more specific terms is then: will the Pauling Bond Order sum (PBO) over all independent distances in an ideal case be near 1.0 per electron pair as it is in the metal reference? (The sum per electron for an average atom is equivalent since this counts all bonds twice.) Matrix effects will then be diagnosed by values of PBO/e significantly below unity. This approach assumes that there are 1) no localized states on M (see 6.4), 2) no significant participation of the nonmetal in M–M bonding states so as to reduce distances or, alternatively, to contribute electrons via X–M pi bonding (as seems to occur in some oxides and fluorides of Mo, W and perhaps Nb, often in relatively high oxidation states), and 3) that valence electrons can be clearly assigned to the nonmetal arrangement to leave an unambiguous count for metal–metal bonding. The model is otherwise taken to be independent of geometry, oxidation state, degree of delocalization, or nonmetal.

The apparent successes of this approach make it a useful means to ascertain apparent matrix (or other) problems and, in their absence, to assure us that some reduced phases are bonded "just like the metal". The Mo_6S_8 structure above is an excellent example. Its low PBO/e_2 value of 0.73 supports our notion that the Mo–Mo bonding is not as good as it could be without restraints imposed by the nonmetal. The PBO/e_2 value increases smoothly to ~ 0.85 in the most reduced ternary derivatives as the matrix effects relax somewhat. But it's clear that the latter structures are still restricted when we look at the results for *isolated* $Mo_6X_8L_6$ or similar clusters where problem contacts seem to be negligible. Typical examples are $Mo_6S_8(PEt_3)_6$ (Saito et al. 1990), PBO/e_2 = 1.07; $(Mo_6Cl_8)Cl^i_{4/2}Cl^a_2$, 1.05; $(Mo_6Cl_7Se)Cl_{6/2}$, 1.01; $Hg(Mo_6Cl_8)Cl_6$, 1.00. Interestingly, the phase $TlMo_3Se_3$ with confacial $Mo_{6/2}S_{6/2}$ anion chains has only large Se–Se separations and thence PBO/e_2 = 1.01. Other rather "metal-like" phases include Nb_6F_{15} ($\sim$1.12, π-bonding?), $(Mo_6Br_8)Br_4(OH_2)_2$ (0.94), $Nb_6I_8I_{6/2}$ (0.80, marginal), the chain phases Y_4Cl_6 (face-capped, 0.96) and $NaMo_4O_6$ (1.03), some metal-rich 3D phases like Zr_9S_2 (1.01), NbO (0.95, but a matrix effect can still be operative), and the layered ZrCl (0.81), Hf_2S (0.82), Ti_2O (CdI_2^{-1}, 1.00). It is pleasing to note that the 24-electron rhenium clusters $(Re_6Se_6Cl_2)Cl_4$ (like $MoCl_2$), $(Re_6Se_5Cl_3)Cl_5$ and $Na_4[(Re_6S_8)S_{4/2}(S_2)_{2/2}]$ also fit (Perrin, Leduc & Sergent, 1991; Bronger & Spangenberg, 1980). These lack inner atoms in a bridging role and give PBO/e_2

values of 0.98, 1.03 and 0.99, respectively, if *seven* valence electrons are assumed for Re metal, contrary to Pauling.

Equally useful are the many trends to lower values of PBO/e_2 that are in accord with expectations for certain atom or structure changes. A partial collection is given in Table 4. In addition, some familiar compounds with very serious matrix effects (reduced M−M bonding) are β-MoTe$_2$ (0.34), Ti$_7$Cl$_{16}$ (0.31, Figure 1b), ZrI$_3$ (0.38), Cs$_3$Zr$_2$I$_9$ (Cs$_3$Cr$_2$Cl$_9$, 0.44), NbCl$_4$ (dimers, 0.26). (In the opposite direction are Ca$_2$N (CdCl$_2^{-1}$, 22.4), and Ag$_2$F (CdI$_2^{-1}$, 1.34). These are electron-poor relative to the short M−M distances, which are largely determined by the sizes of N or F.) (All of the centered cluster examples noted earlier (except hydrides) have M−M dimensions determined mainly by the sizes of Z, a different matrix effect, not by the electron count. Some of the PBO/e_2 values originally reported were misleading because of the unrecognized presence of interstitials.)

Table 4. Trends in matrix effects evidenced by PBO/e_2 values[a]

Compound (type)	PBO/e_2	vs	Compound (type)	PBO/e_2
Mo$_6$S$_8$(PEt$_3$)$_6$	1.07		Mo$_6$Se$_8$(PEt$_3$)$_6$	0.83
Nb$_6$F$_{15}$	1.12		Nb$_6$Cl$_{14}$	0.66
			Ta$_6$Cl$_{15}$	0.69
			(Me$_4$N)$_2$Nb$_6$Cl$_{18}$	0.52
			Li$_6$Zr$_6$Cl$_{18}$H[b]	0.64
W$_6$Br$_8^{6+}$	1.07		W$_6$Cl$_{12}^{2+}$	0.44
Ti$_2$O (CdI$_2^{-1}$)	1.00		Ti$_2$S (Ta$_2$P)	0.77
ZrCl$_2$ (3R-MoS$_2$)	0.51		MoS$_2$ (3R)	0.38
ZrCl	0.81		ZrBr	0.66
PrI$_2$ (MoSi$_2$)	0.44		PrI$_2$ (MoS$_2$, CdCl$_2$)	0.15
Ti$_2$S	0.77		Ti$_2$Se	0.55
Zr$_2$Se	0.73			
Nb$_5$Se$_4$	0.8		Nb$_5$Te$_4$	0.5
Ta$_5$As$_4$	0.80			

[a] Corbett, 1981a. [b] The hydrogen is rattling, but it still provides central bonding.

7.3. CLUSTER ELECTRONIC CONFIGURATIONS

The operation of matrix effects that put the surrounding halide cuboctahedron outside the metal vertices in a M$_6$X$_{12}$-type cluster also provides the basis for a semiquantitative interpretation of different cluster electron counts (Ziebarth & Corbett, 1989b). Larger distortions correlate quite well with increased cluster electron counts. This arises because the a_{2u} M.O., the LUMO with 14 electrons and the HOMO with 16 in chlorides, is M−M bonding but antibonding for M(xy) − X^i(p) π interactions. (The X^i atoms lie on ±x, ±y, so the π^* orbital set is normal to M−Z.) An increasing matrix effect causes the metal atom to withdraw from the plane of the X^i neighbors and thereby to decrease the antibonding contribution to a_{2u}. Clusters with 16 (or 15) electrons should therefore be favored by small metal, large halide, or both, while the opposite trends, or a large Z, should enhance a t_{2g}^6 (or t_{1u}^6) HOMO.

These expectations correlate rather well with known halide systems. Niobium 6-12 examples prepared at high-temperature are limited to the 16-e members like Nb_6Cl_{14} and $Nb_6Br_{18}^{4-}$ and the 15-e Nb_6F_{15}, while Ta_6X_{14} (16-e) phases form for X = Br, I plus the 15-e Ta_6X_{15} for the three heavier halides. All are appropriate to a significant matrix effect. (The condensed niobium oxides don't appear to behave so simply as examples range between 13 and 15-e, but these are rather different systems, and there are some inconsistencies in properties; see Köhler, Tischtau & Simon, 1991.) The zirconium chloride clusters with a larger metal bonding radius and centered by main group Z (which also alters the bonding scheme to some degree) show a sharp stability peak at 14 electrons. Correspondingly, the trans Cl^i−M−Cl^i angles fall in the range of 167 − 171° in the zirconium chloride carbides, compared with 161 − 167° for the niobium chlorides where M−M distances are 0.2 to 0.4 Å less than with zirconium. The small nitrogen interstitial gives rise to a rare 15-e member $Zr_6Cl_{15}N$. The trans angles fall to 156 − 163° with the larger iodide in zirconium carbides, and 15-e $CsZr_6I_{14}C$ and 16-e Zr_6I_{12} are found. Other examples are $CsZr_6Br_{14}C$ and $Zr_6Br_{12}B$ (15-e).

Oxidized, metastable clusters are obtained for both families via solution processes near room temperature and, in parallel, these contain 14-e niobium and tantalum members, but 12-e (and 13-e) zirconium chloride clusters. Rare-earth-metal clusters don't provide such persuasive evidence; small main-group Z examples are rather scarce but similar (12−14-e for $R_7X_{12}Z$), while iodides centered by large metals cover a sizable range (16−20), as was rationalized earlier (see 5).

8. Cluster Condensation

The reduced transition metal halides so far considered provided no forecast or warning that condensed cluster phases could also be synthesized. Many are now known, based principally on the elimination of two edge-bridging halides on trans edges of each M_6X_{12} cluster followed by sharing those metal edges and the four neighboring X^i between them to give quasi-infinite chains. This process may continue to produce double-chain and even double-metal-layered structures. The structure types found were all without precedent, striking testimony to the versatility and wonders possible in the solid state (Corbett, 1980).

Condensed cluster halides can be generalized as follows:

1. Only a few are binary compounds and do not require interstitials (Gd_2Cl_3, Sc_7Cl_{10}, ZrCl and ZrBr types).

2. The only condensed zirconium and hafnium examples known at present are the double-metal-layered ZrCl, HfCl, etc. and their interstitial derivatives; the rest,

including all chain structures, presently utilize only group 3 elements, lanthanides included.

3. Centered boron, nitrogen and carbon (and dicarbon) give one group of infinite chain structures, while many centered heterometals afford another.

4. The range of stability that is possible in each chain structure type appears quite narrow in terms of R, X and Z members.

5. The electronic rules governing stability, and whether the phases are metallic or semiconducting, have not been discerned.

6. The degree of condensation appears to be controlled by the X:M ratio, not the electron count (Corbett & McCarley, 1986). In a general sense, condensation appears to take place until the unit is sheathed by halogen.

7. Face sharing of square antiprismatic metal units in halides is to date very rare, while chains constructed from face-sharing metal octahedra are unknown.

8. Halide bonding modes at metal vertices are similar to those for isolated cluster structures.

8.1. CLUSTER CHAINS

The variety and structural principles present in the eleven structural and compositional varieties that contain condensed cluster chains of the rare-earth-metals will be illustrated by four types: chains derived from empty 6-8 clusters in Y_4Cl_6 (Gd_4Cl_6-type); single chains based on centered 6-12 clusters in Pr_4I_5Ru (Y_4I_5C-type); double chains in Pr_3I_3Ru; square antiprismatic single chains in Pr_4I_4Os. Many of the group have been recently reviewed by Simon and coworkers (1991).

The very remarkable Gd_4Cl_6 structure, the first example of halide chains (Lokken & Corbett, 1973) is illustrated in two views in Figure 11 for the better determined

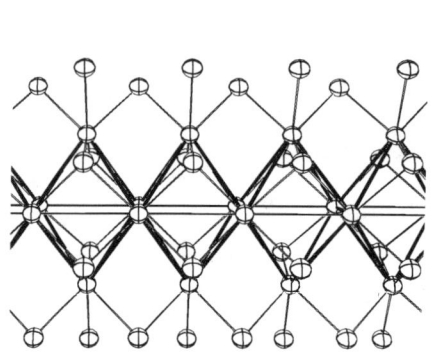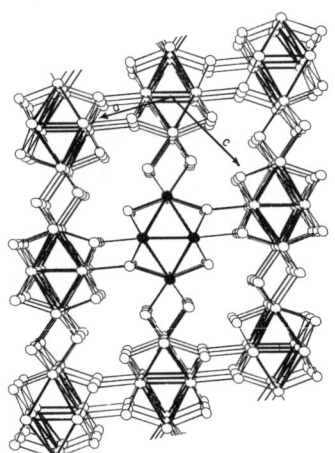

Figure 11. A side view of a single 6-8-type condensed cluster chain in Y_4Cl_6 (left) and a projection along the chains (down the short b-axis) showing the manner of intercluster bridging (right).

Y_4Cl_6 (Mattausch et al., 1980). The structure is a doubly rare example of condensed empty clusters of the M_6X_8 type in which the exposed (side) triangular faces are capped by halide ($Y_2Y_{4/2}Cl^i_4$). The two rows of chlorine at the top and bottom in the side view on the left are duplicates; zig-zag rows bridge the chains into vertical sheets, as seen in projection on the right. One-half the halogen atoms capping side faces are also exo to yttrium vertices on the sides of adjoining chains, and vice versa, completing the 3D character of the structure. Short shared edges (3.27 Å) and elongation along the chain repeat (3.82 Å) are characteristic, while distances across chlorine bridges represent quite tight packing of the anions (3.30 – 3.42 Å). Chlorides and bromides of only Y, Gd and Tb form this novel arrangement, and the results are semiconductors; in other words, the d band is well split with this structure and oxidation state (+1.5) (Simon et al., 1991). The only other chain structure with face-capping halide is Sc_7Cl_{10} with double metal chains; in contrast, centered units are all 6-12 based, including the closely related $Sc_7Cl_{10}C_2$, at least in part because $Z \cdots X^i$ distances in 6-8-Z types would probably be too short (Hwu, Corbett & Poeppelmeier, 1985).

Single chains of metal-edge-sharing and halogen-edge-bridged $R_{4/2}R_2$ octahedral repeat units occur in three basic types, Sc_4Cl_6B, Y_4I_5C and Sc_5Cl_8C. It is particularly noteworthy that $NaMo_4O_6$ is very closely related to the first in both the chain structure and bridging (Torardi & McCarley, 1979). The three halide types differ in the interchain bridging functionalities, those in Sc_5Cl_8C consisting of parallel chains of $Sc(III)Cl_{6/2}$ octahedra that also share edges. Figure 12 shows Pr_4I_5Ru (Y_4I_5C-type) in two views in order to illustrate some novel features (Payne, Dorhout & Corbett, 1991). Four I^{i-i} atoms per repeat simultaneously bridge edges of two adjoining octahedra (left), while the projection on the right illustrates how a fifth square planar I^{i-i} unit bridges side edges in parallel chains. The remaining

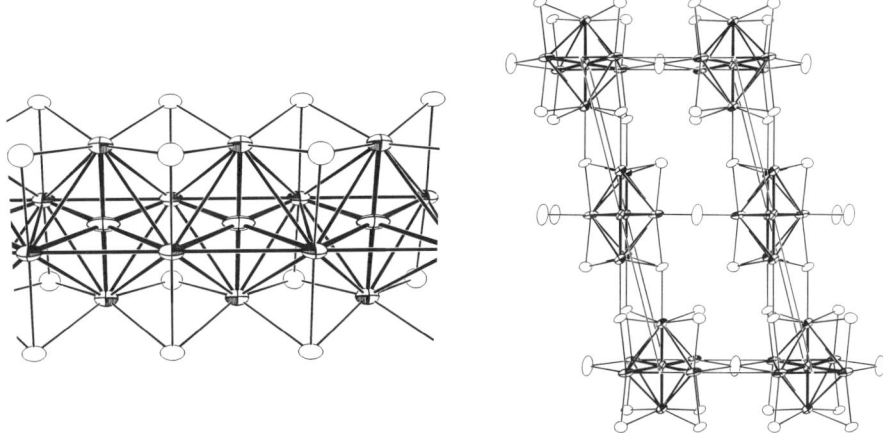

Figure 12. Pr_4I_5Ru: Left – a side view of one chain of condensed clusters with edge-bridging iodine. Line widths approximate bond populations. Right – a projection down the chains with the interchain bridging. Pr is shaded, Ru is crossed and iodine is open ellipsoids.

vertices are bonding in a complementary I^{i-a} fashion as with $Zr_6Cl_{12}Be$ clusters, Figure 2. There are no tight I··I contacts in this structure (4.17 Å between chains, 4.26 Å along the chain repeat), yet the reduced trans-angle I1 – Pr2 – I2 = 159° shows that the apical Pr2 – Ru bonds are clearly "winning" over the opposed $Pr2 - I2^{a-i}$. The dimensions of the Pr octahedra relative to those in Gd_4I_5Si (Nagaki, Simon & Borrmann, 1989) suggest, and band calculational evidence supports, the idea that $d\pi - d\pi$ bonding is especially important in the distinctly shorter apical Pr2 – Ru bonds.

The "bond" widths in the left part of the Figure are proportioned to reflect the relative magnitudes of atom pair bond populations, emphasizing the strong Pr – Ru and Pr – I interactions and the contrasting weak Pr – Pr with such a large Z (d ≥ 3.91 Å). These and the Pauli-like susceptibility of the isostructural La_4I_5Ru favor a description of the chain structure in terms of "heterometal wires" sheathed by iodine. Only Z = Co, Ru, Os appear to form Pr_4I_5Z.

Double-metal chains with edge-bridging halide and different stoichiometries occur in the $Sc_7Cl_{10}C_2$ (= $Sc_6Cl_7C_2 \cdot ScCl_3$), $Y_6I_7C_2$ and Pr_3I_3Ru types in order of increasing coordination of interbridging halogen. In all of these, two single chains of the R_4X_5 type, Figure 12, are further condensed by sharing metal side edges between pairs of chains displaced by half the repeat distance. This can be seen best in the projection down the chains shown on the right of Figure 13 for

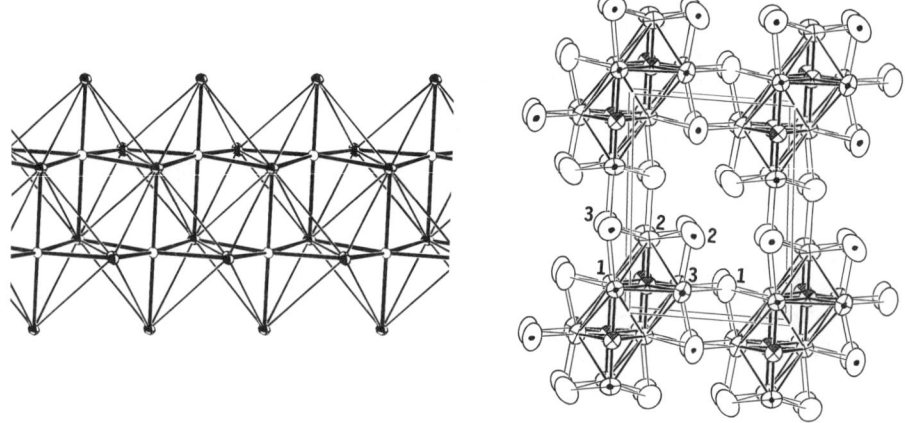

Figure 13. Pr_3I_3Ru – the double chains of condensed, Ru-centered clusters with iodine omitted (left); a projection along the chains with iodine added (right). Dotted atoms differ by b/2 from those without.

monoclinic Pr_3I_3Ru (Payne et al., 1992). The side view of the chain on the left omits the iodine atoms in order to make the condensed octahedra clearer. The interstitials are 0.3 Å closer to the apical Pr2, suggesting strong π bonding again pertains. Calculations suggest a quasi-filled band character. The structure can be generated from that of $Y_6I_7C_2$ by displacement of sheet of double chains bridged by square planar I^{i-i} (as in Figure 12) to eliminate half the bridging iodine and to generate the unusual five-coordinate I1 (Figure 13 right) which is in part face-capping. Smaller R elements, as in the Y_3I_3Ru and Y_3I_3Ir derivatives, cause a strange, continuous distortion in which the two single chain members move more nearly over one another so that the concept of octahedra is lost and more R−R, R−Z and even Z−Z bonding develops.

An aside regarding the variety of cluster phases that may be obtained in a single system is worthwhile. In the Pr−I−Os and Y−I−Ru systems for example, four different cluster products have been identified in each, $Y_6I_{10}Ru$ and $Y_7I_{12}Ru$ clusters, $Y_{16}I_{20}Ru_4$ oligomers and the Y_3I_3Ru chains in the latter. Significant temperature dependencies are accordingly observed in the mixed phase equilibria involved in the syntheses.

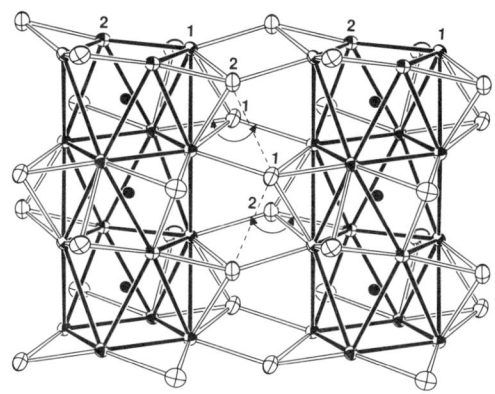

Figure 14. A pair of the quasi-infinite chains built from square-antiprismatic clusters in Y_4Br_4Os. The face capping and interchain bridging role of bromine is also shown.

A most novel structure develops with a smaller bromide and equally large Os, that for Y_4Br_4Os (as well as Er_4Br_4Os) shown in Figure 14 (Dorhout & Corbett, 1992). Here *square-antiprismatic* $Y_{8/2}Os$ units share square faces to generate infinite chains. The shared faces tilt slightly in alternate directions and what would nominally be edge-bridging bromine atoms bend out of the plane of the shared faces to cap metal faces. These changes open up the metal vertices sufficiently to allow strong interchain Br^{i-a} bridging that involves all Pr and Br atoms, as illustrated. The additional Pr−Br bonding is doubtlessly energetically favorable, and it also makes the crystals considerably more robust than without. The antiprismatic construction puts face-capping Br significantly further from Os than it would be in the analogous trigonal antiprismatic construction.

An interesting contrast is found in the only other example of square antiprismatic construction, Ta_4Te_4Si (Badding & DiSalvo, 1990). Here, the relatively smaller interstitial in the $Ta_{8/2}Si$ units gives them a more flattened disposition, and the large edge-bridging Te atoms remain coplanar with the shared faces and relatively tightly packed around the chain. In this case, there is clearly no room for an exo Te atom from another chain to bond at the Ta vertices.

8.2. SHEET STRUCTURES

The structure principles revealed by this class of compounds are less extensive and new, and the more dominant rare-earth-metal members have been reviewed recently (Simon et al., 1991). Accordingly, our presentation of this group will be less detailed and comprehensive.

The end member of side-by-side chain condensation is the double-metal-between-double-halogen layers first found in the metallic ZrCl and ZrBr, Figure 15.

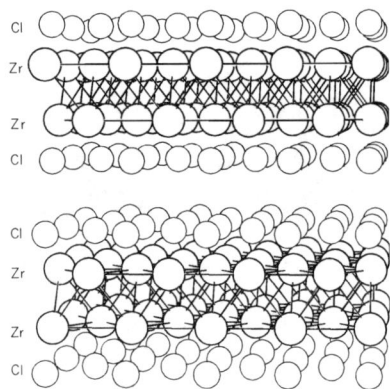

Figure 15. Portions of two of the four-layer slabs present in ZrCl and ZrBr. Interstitials in the intermetal region afford a diverse chemistry for a variety of host metals.

The c.p. layering sequence in these is AbcA, and these four-layer slabs stack ABC (ZrCl) or ACB (ZrBr). The Zr–Zr interlayer distances, 3.09 and 3.13 Å, reflect strong bonding, while the layer repeats, 3.42 and 3.50 Å, respectively, are more a result of halogen size (Adolphson & Corbett, 1976; Daake & Corbett, 1977). Not surprisingly, the face-capping halide motif in these changes to edge-bridging in M_2Cl_2Z, M = Sc, Y or Zr, Z = C or N, where Z is again in trigonal antiprismatic sites (Ab(γ)aB) (Hwu et al. 1986). The novel Gd_2XC version has single X layers between GdCGd triple layers (Bauhofer et al., 1990). A remarkable product with wavy sheets and partial carbon centering is also known for $Gd_6Cl_5C_3$ (Simon, Schwarz & Bauhofer, 1988).

The rare-earth-metal monohalide analogues of ZrX that were once thought to be stable all turned out to be nonstoichiometric hydrides RXH_x, $\sim 0.7 \leq x \leq 1.0$, and an extensive chemistry has been developed for these (Simon et al., 1991). The electron-richer $Zr(Cl,Br)H_x$ phases exist for x = 0.5 and 1.0. In ZrBrD, the heavy atom layers are stacked AbaB as in the carbide, but the deuterium is not in the tetrahedral cavities as suggested by the stoichiometry (and found for $YClH_x$) but at opposite ends of elongated trigonal antiprisms of metal, viz., Ab(γγ)aB (d(D–D) = 2.20 Å). The complex intermediate $Zr_2Br_2(H,D)$ is near the midpoint of the slab shear AbcA → AbaB with hydrogen in distorted tetrahedra. The contrast between these and the unstable RX and the RXH_x structures can be understood in terms of electronic effects (Wijeyesekera & Corbett, 1985, 1986).

9. Acknowledgements

The considerable number of excellent coworkers responsible for many of the results cited are evident in the references following. The early investigations were supported financially by the Ames Laboratory of the U.S. Department of Energy. This program has been supported since 1984 by the National Science Foundation – Solid State Chemistry – via grants DMR – 8318616 and – 8902954.

10. References

Adolphson, D. G. & Corbett, J. D. (1976). 'Crystal Structure of Zirconium Monochloride. A Novel Phase Containing Metal – Metal Bonded Sheets,' Inorg. Chem. **15**, 1820 – 1823.

Badding, M. E. & DiSalvo, F. J. (1990). 'Synthesis and Structure of Ta_4SiTe_4, a New Low-Dimensional Material,' Inorg. Chem. **29**, 3952 – 3954.

Bauhofer, C., Mattausch, Hj., Miller, G. J., Bauhofer, W., Kremer, R. K. & Simon A. (1990). 'Struktur und Eigenschaften von Gd_2XC (X ≡ Cl, Br, I),' J. Less-Common Met. **167**, 65 – 79.

Böttcher, F., Simon, A., Kremer, R. K., Buchkremer-Hermanns, H. & Cockcroft, J. K. (1991). 'Thoriumbromidcluster mit oktaedrischen Th_6-Einheiten,' Z. Anorg. Allg. Chem. **598/599**, 25 – 44.

Brewer, L. & Wengert, P. R. (1973). 'Transition Metal Alloys of Extraordinary Stability; An Example of Generalized Lewis-Acid-Base Interactions in Metallic Systems,' Met. Trans. **4**, 83 – 104.

Bronger, W. & Miessen, H.-J. (1982). 'Synthesis and Crystal Structures of $Ba_2Re_6S_{11}$ and $Sr_2Re_6S_{11}$, Compounds Containing $[Re_6S_8]$ Clusters,' J. Less-Common Met. **83**, 29 – 38.

Bronger, W. & Spangenberg, M. (1980). '$Na_2Re_3S_6$ und $K_2Re_3S_6$, Zwei Thiorhenate mit $[Re_6S_8]$-Clustern,' J. Less-Common Met. **76**, 73 – 79.

Burnus, R., Köhler, J. & Simon, A. (1987). '$Mn_3Nb_6O_{11}$ und $Mg_3Nb_6O_{11}$ – Darstellung von Einkristallen und Strukturverfeinerung,' Z. Naturforsch. **42b**, 536 – 538.

Chu, P. J., Ziebarth, R. P., Corbett, J. D. & Gerstein, B. C. (1988). 'Characterization of Interstitial Hydrogen Within Metal Clusters in $Zr_6Cl_{12}H$ and $ZrClO_xH_y$ by Solid-State NMR,' J. Am. Chem. Soc. **110**, 5324 – 5329.

Corbett, J. D. (1980). 'Extended Metal – Metal Interactions in Binary Halides of the Early Transition Metals: A New Structural Chemistry,' Adv. Chem. Ser. **186**, 329 – 347.

Corbett, J. D. (1981a). 'Correlation of Metal – Metal Bonding in Halides and Chalcides of the Early Transition Elements with That in the Metals,' J. Solid State Chem. **37**, 335 – 351.

Corbett, J. D. (1981b). 'Chevrel Phases: An Analysis of Their Metal – Metal Bonding and Crystal Chemistry,' J. Solid State Chem. **39**, 56 – 74.

Corbett, J. D. & McCarley, R. E. (1986). 'New Transition Metal Halides and Oxides with Extended Metal – Metal Bonding.' Crystal Chemistry and Properties of Materials with Quasi-One-Dimensional Structures, edited by J. Rouxel, D. Reidel, Dordrecht, pp. 179 – 204.

Daake, R. L. & Corbett, J. D. (1977). 'Zirconium Monobromide, a Second Double Metal Sheet Structure. Some Physical and Chemical Properties of the Metallic Zirconium Monochloride and Monobromide,' Inorg. Chem. **16**, 2029 – 2033.

Dorhout, P. K. & Corbett, J. D. (1992). 'A Novel Structure Type in Reduced Rare-Earth Metal Halides. One-Dimensional Confacial Chains Based on Centered Square Antiprismatic Metal Units: Y_4Br_4Os and Er_4Br_4Os,' J. Am. Chem. Soc. **114**, 1697–1701.

Dudis, D. S. & Corbett, J. D. (1987). 'Two Scandium Iodide Carbides Containing Dicarbon Units Within Scandium Clusters – $Sc_6I_{11}C_2$ and $Sc_4I_6C_2$. Synthesis, Structure and the Bonding of Dicarbon,' Inorg. Chem. **26**, 1933–1940.

Hughbanks, T. (1989). 'Bonding in Clusters and Condensed Cluster Compounds that Extend in One, Two and Three Dimensions,' Prog. Solid St. Chem. **19**, 329–372.

Hughbanks, T. & Corbett, J. D. (1988). 'Rare-Earth-Metal Iodide Clusters Centered by Transition Metals: Synthesis, Structure, and Bonding of $R_7I_{12}M$ Compounds (R = Sc, Y, Pr, Gd; M = Mn, Fe, Co, Ni),' Inorg. Chem. **27**, 2022–2026.

Hughbanks, T. & Corbett, J. D. (1989). 'Encapsulation of Heavy Transition Metals in Iodide Clusters. Synthesis, Structure, and Bonding of the Unusual Cluster Phase $Y_6I_{10}Ru$,' Inorg. Chem. **28**, 631–635.

Hughbanks, T., Rosenthal, G. & Corbett, J. D. (1988). 'Encapsulation of the Transition Metals Chromium Through Cobalt in Zirconium Cluster Iodides,' J. Am. Chem. Soc. **110**, 1511–1516.

Hwu, S.-J. & Corbett, J. D. (1986). 'Metal-Metal-Bonded Scandium Cluster ($Sc_7Cl_{12}Z$) and Infinite Chain (Sc_4Cl_6Z) Phases Stabilized by Interstitial Boron or Nitrogen (Z),' J. Solid State Chem. **64**, 331–346.

Hwu, S.-J., Corbett, J. D. & Poeppelmeier, K. R. (1985). 'Interstitial Atoms in Metal–Metal Bonded Arrays: The Synthesis and Characterization of Heptascandium Decachlorodicarbide, $Sc_7Cl_{10}C_2$, and Comparison with the Interstitial-Free Sc_7Cl_{10},' J. Solid State Chem. **57**, 43–58.

Hwu, S.-J., Ziebarth, R. P., von Winbush, S., Ford, J. E. & Corbett, J. D. (1986). 'Synthesis and Structure of Double-Metal-Layered Scandium, Yttrium, and Zirconium Chloride Carbides and Nitrides, M_2Cl_2C and M_2Cl_2N,' Inorg. Chem. **25**, 283–287.

Köhler, J., Tischtau, R. & Simon, A. (1991). 'Oxoniobates with 13 and 15 Electrons in $[Nb_6O_{12}]$ Clusters: the Structures of KNb_8O_{14} and $LaNb_8O_{14}$,' J. Chem. Soc. Dalton 829–832.

Leduc, L., Perrin, A. & Sergent, M. (1983). 'Chalcohalogénures et chalcogénures a "clusters" octáedriques dans la chimie de basse valence du rhenium,' Comp. Rend. Acad. Sci (Paris), Ser. II, **296**, 961–964.

Lokken, D. A. & Corbett, J. D. (1973). 'Rare Earth Metal–Metal Halide Systems. XV. Crystal Structure of Gadolinium Sesquichloride. A Phase with Unique Metal Chains,' Inorg. Chem. **12**, 556–559.

Mattausch, Hj., Hendricks, J. B., Eger, R., Corbett, J. D. & Simon, A. (1980). 'Reduced Halides of Yttrium with Strong Metal–Metal Bonding: Yttrium Monochloride, Monobromide, Sesquichloride, and Sesquibromide,' Inorg. Chem. **19**, 2128–2132.

Nagaki, D., Simon, A. & Borrmann, H. (1989). 'Synthesis and Structure of Gd_4I_4Si and Gd_3I_3Si: The First Examples of Reduced Rare Earth Halides Containing Second-Row Non-Metal Interstitial Atoms,' J. Less-Common Met. **156**, 193–205.

Payne, M. W. & Corbett, J. D. (1990). 'Encapsulation of the Platinum and Neighboring Metals within Cluster Iodides of Rare-Earth Elements,' Inorg. Chem. **29**, 2246–2251.

Payne, M. W., Dorhout, P. K. & Corbett, J. D. (1991). 'Heterometallic Condensed Cluster Compounds: Pr_4I_5Z (Z = Co, Ru, Os) and La_4I_5Ru. Synthesis, Structure, and Bonding,' Inorg. Chem. **30**, 1467–1472.

Payne, M. W., Dorhout, P. K., Kim, S.-J., Hughbanks, T. R. & Corbett, J. D. (1992). 'Chains of Centered Metal Clusters with a Novel Range of Distortions: Pr_3I_3Ru, Y_3I_3Ru and Y_3I_3Ir,' Inorg. Chem. **31**, in press.

Payne, M. W., Ebihara, M. & Corbett, J. D. (1991). 'A Novel Oligomer of Condensed Metal Atom Clusters in $[Y_{16}Ru_4I_{20}]$,' Angew. Chem. Int. Ed. Engl. **30**, 856–858.

Perrin, A., Leduc, L. & Sergent, M. (1991). 'Halogen bridged Re_6L_8 units in octahedral cluster rhenium chalcohalides,' Eur. J. Solid State Inorg. Chem. **28**, 919–931.

Pauling, L. (1960). "The Nature of the Chemical Bond." 3rd ed., Cornell University Press, Ithaca, N.Y., pp. 400–403.

Rogel, F., Zhang, J., Payne, M. W. & Corbett, J. D. (1990). 'Centered Cluster Halides for Group-Three and Group-Four Transition Metals,' Adv. Chem. Ser. **226**, 369–389.

Sägebarth, M. & Simon, A. (1990). '$Nb_6Cl_{12-x}I_{2+x}$ (x < 2) – ein neues Niobhalogenid,' Z. Anorg. Allg. Chem. **587**, 119–128.

Saito, T., Yamamoto, N., Nagase, T., Tsuboi, T., Kobayashi, K. Yamagata, T., Imoto, H. & Unoura, K. (1990). 'Molecular Models of the Superconducting Chevrel Phases: Syntheses and Structures of $[Mo_6X_8(PEt_3)_6]$ and $[PPN][Mo_6X_8(PEt_3)_6]$ (X = S, Se: PPN = $(Ph_3P)_2N$),' Inorg. Chem. **29**, 764–770.

Schäfer, H. & Schnering, H.-G. (1964). 'Metall-Metall-Bindungen bei niederen Halogeniden, Oxyden und Oxydhalogeniden schwerer Übergangsmetalle,' Angew. Chem. **76**, 833–849.

Schäfer, H., Schnering, H.-G., Niehues, K.-J. & Nieder-Vahrenholz, H. G. (1965). 'Beitrage zur Chemie der Elemente Niob und Tantal XVII. Niobfluoride,' J. Less-Common Met. **9**, 95–104.

Shannon, R. D. (1976). 'Revised Effective Ionic Radii and Systematic Studies of Interatomic Distances in Halides and Chalcogenides,' Acta Cryst. **A32**, 751–767.

Simon, A. (1981). 'Condensed Metal Clusters,' Angew. Chem. Intl. Ed. Engl., **20**, 1–22.

Simon, A. (1985). 'Empty, Filled, and Condensed Metal Clusters,' J. Solid State Chem. **57**, 2–16.

Simon, A. (1988). 'Clusters of Valence Electron Poor Metals – Structure, Bonding, and Properties,' Angew. Chem. Int. Ed. Engl. **27**, 159–183.

Simon, A., Mattausch, Hj., Miller, G. J., Bauhofer, W. & Kremer, R. K. (1991). 'Metal-Rich Halides.' Handbook on the Physics and Chemistry of Rare Earths, vol. 15, edited by K. A. Gschneidner and L. Eyring, North-Holland, Amsterdam, pp. 191–285.

Simon, A., Schwarz, C. & Bauhofer, W. (1988). 'Ein Neues Lanthanoidcarbidhalogenid, $Gd_6Cl_5C_3$', J. Less-Common Met. **137**, 343–351.

Smith, J. D. & Corbett, J. D. (1985). 'Stabilization of Clusters by Interstitial Atoms. Three Carbon-Centered Zirconium Iodide Clusters, $Zr_6I_{12}C$, $Zr_6I_{14}C$, and $MZr_6I_{14}C$ (M = K, Rb, or Cs),' J. Am. Chem. Soc. **107**, 5704–5711.

Smith, J. D. & Corbett, J. D. (1986). 'Four Zirconium Iodide Cluster Phases Centered by Boron, Aluminum, or Silicon,' J. Am. Chem. Soc. **108**, 1927–1934.

Torardi, C. C. & McCarley, R. E. (1979). 'Sodium Tetramolybdenum Hexoxide (NaMo$_4$O$_6$). A Metallic Infinite-Chain Polymer Derived by Condensation of Octahedral Clusters,' J. Am. Chem. Soc. **101**, 3963–3964.

Ueno, F. & Simon, A. (1985). 'Structure of Tetrapotassium Dodeca-μ-bromohexabromo-*octahedro*-hexaniobate(4-), K$_4$[(Nb$_6$Br$_{12}$)Br$_6$],' Acta Cryst. **C41**, 308–310.

Wells, A. F. (1984). "Structural Inorganic Chemistry." 5th ed., Clarendon Press, Oxford, Chap. 9.

Wijeyesekera, S. D. & Corbett, J. D. (1985). 'A Neutron Powder Diffraction Study of Zirconium Bromide Deuteride, ZrBrD. Two Hydrogen Atoms in the Same Metal Interstice,' Solid State Commun. **54**, 657–660.

Wijeyesekera, S. D. & Corbett, J. D. (1986). 'Structure of the Intermediate Zr$_2$Br$_2$H by Neutron Diffraction and Its Structural and Bonding Relationships to Other Phases,' Inorg. Chem. **25**, 4709–4714.

Zhang, J. (1990). 'Solid State Chemistry of Centered Zirconium Chloride Clusters. Synthesis and Characterization of New Compounds,' Ph.D. Dissertation, Iowa State University.

Zhang, J. & Corbett, J. D. (1991a). 'Zirconium Chloride Cluster Phases Centered by Transition Metals Mn–Ni. Examples of the Nb$_6$F$_{15}$ Structure,' Inorg. Chem. **30**, 431–435.

Zhang, J. & Corbett, J. D. (1991b). 'Synthesis and Structure of the Centered Cluster Compound Ba$_3$(Zr$_6$Cl$_{18}$Be),' Z. Anorg. Allg. Chem. **598/599**, 363–370.

Zhang, J., Ziebarth, R. P. & Corbett, J. D. (1992). 'Two Novel Phases Containing Centered, Isolated Zirconium Clusters, Rb$_4$Zr$_6$Cl$_{18}$ and Li$_6$Zr$_6$Cl$_{18}$H,' Inorg. Chem. **31**, 614–619.

Ziebarth, R. P. & Corbett, J. D. (1985). 'Zirconium Cluster Compounds Stabilized by Interstitial Atoms,' J. Am. Chem. Soc. **107**, 4571–4573.

Ziebarth, R. P. & Corbett, J. D. (1987). 'Cation Distribution within a Cluster Framework. Synthesis and Structure of the Carbon- and Boron-Centered Zirconium Cluster Compounds KZr$_6$Cl$_{15}$ and CsKZr$_6$Cl$_{15}$B,' J. Am. Chem. Soc. **109**, 4844–4850.

Ziebarth, R. P. & Corbett, J. D. (1988). 'Cluster Framework Structures and Relationships. Two Compounds with a New Connectivity, K$_2$Zr$_6$Cl$_{15}$B and K$_3$Zr$_6$Cl$_{15}$Be,' J. Am. Chem. Soc. **110**, 1132–1139.

Ziebarth, R. P. & Corbett, J. D. (1989a). 'Centered Zirconium Chloride Clusters. Synthetic and Structural Aspects of a Broad Solid-State Chemistry,' Acc. Chem. Res. **22**, 256–262.

Ziebarth, R. P. & Corbett, J. D. (1989b). 'Isolated Zirconium Chloride Clusters in the Phase Rb$_5$Zr$_6$Cl$_{18}$B. Generalities Regarding in Bonding, Stability, and Size of M$_6$X$_{12}$-Type Clusters with and without Interstitial Atoms,' J. Am. Chem. Soc. **111**, 3272–3280.

Ziebarth, R. P. & Corbett, J. D. (1989c). 'New Layered Phases Achieved with Centered Zirconium Chloride Clusters. The Stoichiometry Zr$_6$Cl$_{16}$,' Inorg. Chem. **28**, 626–631.

MODERN CONCEPTS OF ATOM COORDINATIONS, ATOM VOLUME AND CHARGE TRANSFER IN INTERMETALLIC COMPOUNDS

MARIA L. FORNASINI
Istituto di Chimica Fisica
Università di Genova
Corso Europa 26
I - 16132 Genoa
Italy

ABSTRACT. The crystal structure determination of a large number of binary intermetallic compounds has allowed to build up structural maps, by means of which atom coordinations or coordination polyhedra of a new phase can be predicted as a function of a few atomic properties of the two elements. The volume effect (generally a contraction) that accompanies the formation of a binary phase can be evaluated in several ways. Three methods are described, fulfilling the requirements of being applicable to a large number of metallic systems and easy to use. The semi-empirical Miedema's model, based on the concept of atomic cells, applies to transition-metal compounds and emphasizes charge transfer effects. In Hafner's approach the first - order perturbation theory and the pseudopotential concept are used to yield a simple rule for the variation of the mean atomic volume of sp binary alloys with concentration. In Merlo's approach a three - parameters empirical formula is proposed to calculate volume contractions in compounds formed by alkaline earths and bivalent rare earths with other elements.

1. Atomic Coordination

Intermetallic compounds form a particular class in the inorganic chemistry. They present different behaviours : large fields of solid solubility and simple structures as in the brass alloys, but also in most cases fixed composition and complicated structures. Anyway, the formula of an intermetallic compound is not in principle a priori known, as it does not follow the normal rules of the valence. Some formulae, for example AB, AB_2, A_5B_3, occur very frequently, but we have also exotic formulae as $CoZn_{13}$, $Ca_{31}Sn_{20}$, $Sm_{11}Cd_{45}$. Therefore, the determination of the crystal structure must be preliminary to any other measurement or property investigation.

From the crystal structure we can derive two important information : besides the exact formula of the compound, we can also know the atomic coordination, that means how many atoms surround a given atom, at what distances, and what is the shape of the coordination polyhedron.

Atomic coordination in intermetallics show characteristic features.
- 1) The coordination number (CN) is generally greater than in other classes of inorganic structures. For example, in the minerals we have tetrahedra or octahedra with CN4 or CN6. In the intermetallic compounds there is a large range of coordination numbers from CN4 to CN24.

- 2) A certain variability occurs in the bond distances. For example, in Ba_3Al_5 the Ba - Ba distances range from 3.65 to 3.95Å [$2r_{Ba}$ = 4.49Å] and the Al - Ba distances from 3.49 to 4.00Å [$r_{Al} + r_{Ba}$ = 3.67Å]. As a guide we can use the sum of the metallic radii of the two elements. Distances are generally contracted up to 10% of this sum, but sometimes they are slightly greater than the sum of the metallic radii and nevertheless belong to the coordination polyhedron. In several cases distances are spread in a range of values.
- 3) On the other hand, neither molecules (as in organic structures) nor groups of atoms (as in silicates or oxides) can be a priori recognized. So, the identification a posteriori of the coordination polyhedra is important for the classification of the structure.

1.1. HOW TO DETERMINE ATOM COORDINATIONS

A new problem arises, as no univocal determination of the coordination can be given. Several methods are proposed.

In the **maximum gap method** (Brunner & Schwarzenbach, 1971) a histogram of distances from the chosen atom to the surrounding atoms is plotted. It is convenient to express all distances relative to the shortest distance using d/d_{min} as abscissa. For most structures a large (or significantly larger) gap appears in the distance distribution. All atoms before the gap are considered as belonging to the coordination polyhedron.
A modification of this method (Bruzzone et al., 1970) takes into account the different sizes of the atoms. The histogram is plotted using normalized distances $d_n/(r_0 + r_n)$ where r_0 is the radius of the central atom and r_n is the radius of the atom at distance d_n. The radii used are those for CN12 of Teatum et al. (1968).

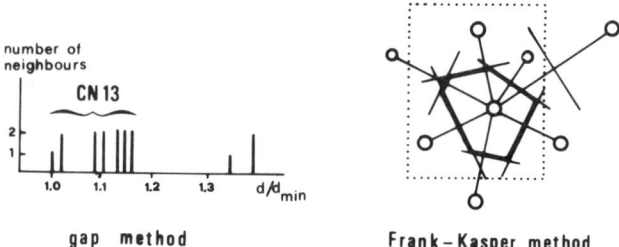

Figure 1. Examples of the maximum gap and the Frank - Kasper methods in evaluating atom coordinations.

The **Frank - Kasper method** (Frank & Kasper, 1958) defines as domain of an atom all points which are closer to the atom than to all other atoms. We draw lines joining the central atom to the surrounding atoms and then bisect these lines with perpendicular planes. The polyhedron limited by these planes is the domain of the atom (called also Voronoi polyhedron or Wigner-Seitz cell). All atoms emerging from the faces of the domain are coordinated with the central atom and define its coordination polyhedron.

A modification of this method (O'Keeffe, 1979) calculates a weighted CN, where the contribution of each face of the domain is weighted in proportion to the solid angle S_n subtended by that face at the centre : WCN = $\Sigma S_n/S_1 = 4\pi/S_1$ where S_1 is the largest solid angle.
Examples are given in Figure 1.

1.2. STRUCTURAL MAPS

An important simplification in the study of intermetallic compounds is provided by the concept of isotypism. Several compounds have the same composition, the same structure with the same coordination polyhedra. We say that they are isotypic and identify a "structure type". For example, nearly 280 CsCl-type and more than 150 Mn_5Si_3-type binary phases are reported in the literature. We can take advantage from this fact and, chosen a given composition, try a systematization of the structural types. In other words, we can build a map where are reported the representative points of all the isotypic phases. If the coordinates of the map are correctly chosen, we should be able to define a region for the occurrence of each structure type. It is clear that such a map could be a useful guide in synthesizing new materials with a desired structure and with desired physical properties, but it is also clear that the approach is phenomenological and based on already known experimental data.

In recent years an extensive application of this concept was made by several researchers. As an example one can mention the studies of Villars (1983) on binary equiatomic phases.

A more advanced step could be to use all phases together and to build maps for the coordination types. By means of these maps one can predict which polyhedra should occur for a given compound, and as a natural derivation also which structure types. This application was made by Villars et al. (1989) who considered nearly 5000 binary compounds A_xB_y, crystallizing with 147 structure types. All phases were divided into 5 groups as shown in Table 1.

TABLE 1. Distribution of the coordination polyhedra in the binary phases.

Presence of	No.of struct.types	No.of examples
1 polyhedron	25	1347
2 polyhedra	51	1952
3 polyhedra	34	879
4 polyhedra	28	543
more than 4 polyh.	9	113
	147	4834

Villars et al. recognized 26 most frequently occurring polyhedra ranging from CN3 to CN20 and 26 less frequent ones ranging from CN2 to CN24.

Neglecting structures with more than 4 polyhedra, tridimensional maps were constructed for the four cases (1, 2, 3 or 4 polyhedra) by plotting each binary compound A_xB_y (x ≤ y) against three coordinates :

- the mean number of valence electrons

$$\overline{VE} = [x/(x+y)] \cdot VE_A + [y/(x+y)] \cdot VE_B \qquad (1)$$

- the electronegativity difference (scale of Martynov - Batsanov)

$$\overline{\Delta X} = [2x/(x+y)] \cdot (X_A - X_B) \qquad (2)$$

- the difference of pseudo-potential radii sums (Zunger radii, a.u.)

$$\overline{\Delta R} = [2x/(x+y)] \cdot [(r_s + r_p)_A - (r_s + r_p)_B] \qquad (3)$$

In these maps can be identified regions where particular coordination types occur. In defining the coordination type three criteria are applied : 1) minimum number of different polyhedra; 2) maximum gap in the distance histogram; 3) maximum convexity of the coordination polyhedron. All structure types with the same or similar polyhedra are adjoined to the same coordination type. For example, the AuCu, $AuCu_3$ and $HfGa_2$ structures based on the same close packing have the CN12 cuboctahedron as coordination polyhedron. In the case of structures with 3 or 4 polyhedra only the two more symmetrical polyhedra are reported in the maps. For examples of the maps see Problem 1.

1.3. COORDINATION IN ALKALINE EARTH PHASES

We now restrict the study to the coordination types found in the binary alkaline earth phases. The problem is limited but still significant as nearly 100 structure types occurring in 350 phases are involved.

In each phase an electropositive element with large size (alkaline earth) is alloyed with a partner with smaller size. As a consequence, the polyhedra around the alkaline earths present high coordination numbers ranging from 14 to 24. They have shapes generally different in each structure type and cannot be easily classified. On the contrary, the polyhedra around the partner elements have a lower number of vertices and a limited number of shapes, and they will be now considered.

According to Bruzzone & Merlo (1982) the most frequent polyhedra can be grouped into two classes. In the first class (I) the polyhedra have only alkaline earth atoms at the vertices, but the lateral faces of the trigonal prism can be capped by other atoms, not necessarily alkaline earths (Figure 2). In the second class (II) the vertices are occupied by both kind of atoms, indifferently. Trigonal prisms in the first class and icosahedra in the second class are the most frequently found polyhedra. Moreover, a third type of coordination has to be added, according to the presence of an anionic sublattice.

The partner atoms can occupy either only one Wyckoff position or more than one. Examples are given in Table 2. Two empirical rules emerge : a) When the partner atom in a structure occupies more than one Wyckoff position, the coordination polyhedra belong to the same class. There were found only two exceptions to this rule in the 94 structure types examined. b) If several intermediate phases occur in a binary system, their crystal structures can show polyhedra of both classes or of one only. However, if both classes are present, a boundary composition exists which separates phases containing class I polyhedra from those containing class II polyhedra and no mixing of the two types of polyhedra occurs.

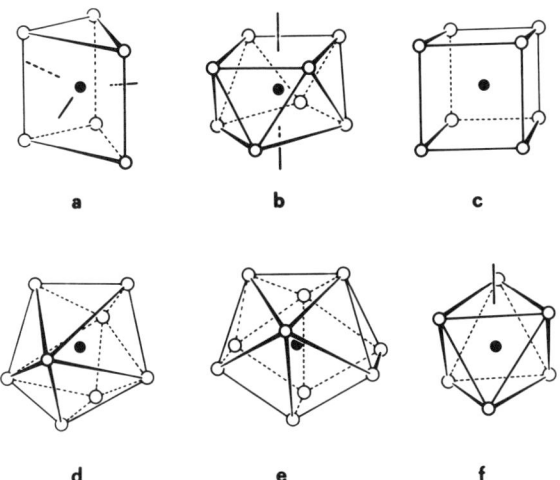

Figure 2. First class polyhedra : a) trigonal prism, b) square antiprism, c) cube, d) bisdisphenoid, e) cubicosahedron, f) trigonal antiprism.

TABLE 2. Different occupation of Wyckoff positions in two structures.

Structure type	Example	Space group	Pearson symbol	Atom	Wyckoff position
Fe_3C	Ca_3Pd	Pnma	oP16	Pd	4c
$Ho_{11}Ge_{10}$	$Ca_{11}Sb_{10}$	I4/mmm	tI84	Sb1	16m
				Sb2	8i
				Sb3	8h
				Sb4	4e
				Sb5	4d

The occurrence of each type of coordination is strictly connected with the composition of the phase and the kind of atom involved. Just with these two parameters we can build up a distribution map of the coordination types. The binary Ca phases are represented in Figure 3.

We can note that class I polyhedra (full squares), class II polyhedra (circles) and anionic sublattices (triangles) are found in definite regions of the graph. Class I polyhedra occur in an intermediate range of composition : the phase must be not too rich in alkaline earth nor in partner element. Beyond the 1 : 2 composition (< 33 at.% Ca) there is the region of the class II polyhedra, however anionic sublattices alternate with class II polyhedra for elements of groups 13, 14 and 15. The same trend is found with binary Sr and Ba alloys.

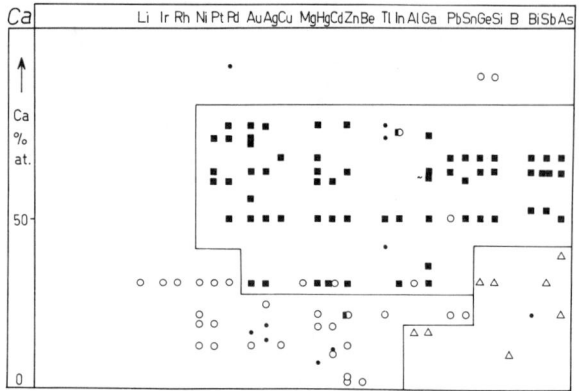

Figure 3. Distribution map of the coordination types of the partner atoms in the binary Ca phases. Vertically is reported the atomic percentage of the calcium atom in the phase, and horizontally the partner elements ordered according to the scale proposed by Pettifor (1984). Each symbol refers to an intermediate phase with known structure. Small full circles represent phases with unknown structure.

2. Atomic Volumes

The prediction of the variation of the mean volume per atom is a fundamental problem in the theory of alloys. In a simple way the volume per atom in a metallic alloy can be evaluated as an average of the atomic volumes of the components according to Vegard's or Zen's law. [Strictly speaking Vegard's law refers to the lattice constants, and Zen's law to the atomic volumes].

$$V_{Zen}(A_{1-x}B_x) = (1-x) \cdot V°_A + x \cdot V°_B \qquad (4)$$

Zen's law is fulfilled in several cases.
- 1) When two metals form a continuous series of solid solutions. E.g. Pt - Cu, Pd - Cu, Ni - Cu alloys. The mean atomic volume is calculated in terms of the atomic volumes of the two metals :

$$V(Pt_{1-x}Cu_x) = (1-x) \cdot V°_{Pt} + x \cdot V°_{Cu} \qquad (5)$$

- 2) When two compounds with the same structure form a solid solution. E.g. GaSb and InSb, both with sphalerite structure, form a solid solutions by rapid cooling. The mean atomic volume is calculated in terms of the atomic volume of the binary compounds :

$$V(Ga_{1-x}In_xSb) = (1-x) \cdot V^A_{GaSb} + x \cdot V^A_{InSb} \qquad (6)$$

- 3) When, starting from a binary compound, an element is substituted by another element obtaining a pseudo-binary phase with the same structure or closely related structure. E.g. for the $CaCu_{1-x}Ag_x$ alloys :

$$V(CaCu_{1-x}Ag_x) = (1 - x) \cdot V^A_{CaCu} + x \cdot V^A_{CaAg} \qquad (7)$$

Nevertheless, if we apply Zen's law to binary metallic systems where intermediate phases are formed, generally more or less large deviations occur. These deviations are normally negative; in other words, chemical bonding effects lead to a contraction :

$$\Delta V(A_{1-x}B_x) = V(A_{1-x}B_x) - V_{Zen}(A_{1-x}B_x) < 0 \qquad (8)$$

Examples are given in Figure 4.

Several authors have pointed out the influence of the volume contraction on the alloy cohesion. For instance, the volume contraction was linked to the glass forming ability of the alloys. It was found that deviations of 15% or more are required for glass formation.

Other authors like Kubaschewski & Alcock (1979) have correlated the enthalpies of formation with the volume contractions upon compound formation. For ionic compounds the more exothermic is the reaction the greater the contraction and with good confidence one can derive the enthalpy from crystallographic information. For intermetallic compounds the absolute values of the enthalpies of formation are very small and poorly correlated with volume contractions. Then, we can deduce in intermetallic compounds the volume effects do reflect a **more complex bond situation.**

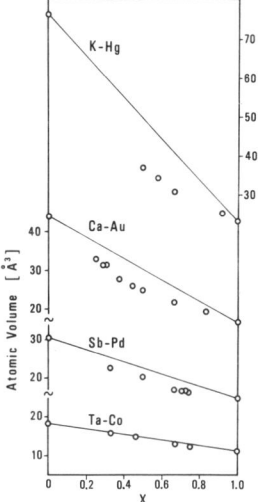

Figure 4. Atomic volumes versus composition in the systems Ta - Co, Sb - Pd, Ca - Au and K - Hg. Open circles indicate experimental values of the intermediate phases occurring in each system. Straight lines correspond to Zen's law.

2.1. FACTORS INFLUENCING VOLUME VARIATIONS

The main ideas about the volume variations can be summarized into three points : 1) elastic effects; 2) charge transfer effects; 3) structure effects. The elastic effects are due to the difference in compressibility of the two metals and can be related to their different electron densities. We can expect that in the alloy one metal is expanded and the other metal is compressed until their electron densities become equal. The charge transfer effects are due to the electronegativity difference between the two components. A charge transfer occurs from the less electronegative to the more electronegative atom. A contribution can come also from the change in crystal structure leading to a different atom coordination or to a different packing in the alloy compared with the elements. However, these effects are generally smaller than elastic or charge transfer effects and, as a first approximation, they can be neglected. Indeed, if we compare the atomic volumes of elemental allotropes we see the average difference be no more than 0.8%.

In conclusion, one can say that generally both elastic and charge transfer effects contribute to volume variations, but sometimes for some classes of alloys one of the two contributions could dominate.

Many authors were interested in studying this problem and many papers were published on the subject. Here only a few approaches will be presented for the volume calculation, fulfilling the requirements of being applicable to a large number of metallic systems and easy to use.

2.2 MIEDEMA'S APPROACH

One of the most useful models in evaluating both enthalpy of formation and volume effects in intermetallic compounds is Miedema's approach (Miedema & Niessen, 1982; de Boer et al., 1988). It uses an atomic cellular model of the alloy, based on the concept of Wigner-Seitz (W.S.) cell. The intermetallic compound AB_n is considered as being formed by the atomic W.S. cells of the pure metals A and B (see Figure 5).

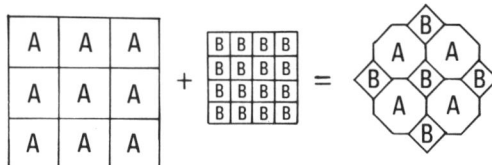

Figure 5. Formation of an alloy in the atomic cellular model.

This model is used for calculating the enthalpy of formation of intermetallic phases. Two properties of the pure metals are important in this calculation : n_{WS}, the electron density at the boundary of the W.S. cells and Φ^*, the electronegativity. On the whole:

$$\Delta H = f(c) \cdot [- P \cdot (\Delta \Phi^*)^2 + Q \cdot (\Delta n_{WS}^{1/3})^2] \tag{9}$$

where f(c) is a function of the composition of the alloy and P, Q are empirical constants.

When the two types of cells A and B approach each other two events take place. Firstly, the electron density in the alloy must be continuous across the boundary between dissimilar atomic cells. In other words, the difference between the electron densities of the two metals at the boundary must be smoothed out. This event requires an energy contribution proportional to $(\Delta n_{WS}^{1/3})^2$ and gives a positive term in the enthalpy of formation of the alloy. On the other hand, the electronegativities of the two metals are, in general, different. So, a small amount of electronic charge will be transferred from cells with the higher chemical potential to cells with the lower one, in order to equalize the chemical potential in the alloy. This event requires an energy contribution proportional to $-(\Delta \Phi^*)^2$ and gives rise to a negative term in the enthalpy of formation of the alloy.

n_{WS} [arbitrary electron density unit] is related to the compressibility of the metal and derived from the bulk modulus; Φ^* [volt] is derived from the work function of the metal (energy necessary to remove an electron from the metal).

Many concepts used by Miedema for the enthalpy evaluation hold also in the volume calculation, which is applied to transition-metal compounds. In these phases the observed volume contraction or expansion is principally due to ionic effects.

Let's consider again an intermetallic compound AB_n and assume B more electronegative than A. Charge transfer will occur from atoms A to atoms B, related to the electronegativity difference $(\Phi^*_B - \Phi^*_A)$. Moreover, the electrons are transferred from a region where the electron density is n_{WS}^A to a region with electron density n_{WS}^B. So, the volume effect is given by

$$\Delta V \propto (\Phi^*_B - \Phi^*_A)[(n_{WS}^B)^{-1} - (n_{WS}^A)^{-1}] \qquad (10)$$

where Δn_{WS}^{-1} is the difference between the volumes occupied by an electron in the two metals at the boundary. Generally, the more electronegative is the metal, the higher is the electron density and then smaller is the volume, so n_{WS}^{-1} is small for electronegative atoms. In our assumption $\Phi^*_B > \Phi^*_A$ and then $(n_{WS}^B)^{-1} < (n_{WS}^A)^{-1}$, so we expect a volume contraction. The complete formula is:

$$\Delta V \text{ (cm}^3\text{/mol A)} = \frac{P_o \cdot f^A_B \cdot V_A^{2/3}}{(n_{WS}^A)^{-1/3} + (n_{WS}^B)^{-1/3}} \cdot (\Phi^*_B - \Phi^*_A) \cdot [(n_{WS}^B)^{-1} - (n_{WS}^A)^{-1}] \qquad (11)$$

where $P_o = 1.3$ for phases containing Ca, Sr, Ba;
$P_o = 1.5$ for all other phases.

$$f^A_B = (1 - c^s_A) \cdot \{1 + 8[c^s_A(1 - c^s_A)]^2\} \qquad (12)$$

is a function of the composition of the alloy and represents the degree to which atoms A (**solute**) are surrounded by atoms B (**solvent**) in the binary ordered compound.

$$c^s_A = x_A V_A^{2/3} / [x_A V_A^{2/3} + (1 - x_A) V_B^{2/3}] \qquad (13)$$

is the surface concentration of metal A, and x_A is the atomic fraction of metal A.
By using the pure metal volumes one immediately derives all quantities. More correct values of the volumes should take into account the changes upon alloying due to charge transfer :

$$V_A^{2/3}(\text{alloy}) = V_A^{2/3}(\text{pure}) \cdot [1 + a_A \cdot f^A_B \cdot (\Phi^*_A - \Phi^*_B)] \tag{14}$$

where a_A is a proportionality constant derived from experimental volume contractions. Since f^A_B is a function of $V_A^{2/3}$, $V_B^{2/3}$ the calculation of these quantities requires an iterative procedure. However, a sufficient accuracy is obtained calculating c^s_A with pure metal volumes and substituting this value in the f^A_B expression. $V_A^{2/3}(\text{alloy})$, $V_B^{2/3}(\text{alloy})$ are then calculated and a new f^A_B value is obtained. On the average, the accuracy in the ΔV predictions is of the order of 3%. Examples are reported in Table 3.

TABLE 3. Experimental and by Miedema's formula calculated atomic volumes (Å^3/at) and volume contractions (%) in some transition-metal phases.

Phase	V_{exp}	$V_{Miedema}$	$\Delta V_{exp}\%$	$\Delta V_{Mied.}\%$
$CoTa_2$	15.49	15.59	-1.3	-0.7
$TaCo_2$	13.00	13.25	-2.9	-1.0
$LaCu_5$	15.34	15.38	-4.7	-4.5
$YCo_{8.5}$	12.92	12.56	-3.5	-6.2
$LaRu_2$	19.05	18.13	-11.6	-15.9
$IrLa_3$	31.30	29.61	-1.2	-6.6
$LaIr_2$	18.93	18.27	-13.7	-16.7
$LaIr_5$	17.68	16.00	-2.0	-11.3
$HfCr_2$	15.27	15.01	-1.1	-2.8

2.3. HAFNER'S APPROACH

Hafner (1985) applies to binary alloys formed by s and p elements the same approach already used by Heine & Weaire (1970) in the evaluation of the cohesion energy of simple metals (sp-bonded metals).

A simple metal can be characterized by a small ion core plus a "gas" of valence electrons distributed through the volume of the crystal. A total energy per atom is written in the hypothesis that the electron gas is weakly perturbed by the ion lattice, that is to say perturbed to first order. The perturbation is weak because the valence electrons undergo the combined effects of the attraction from the ionic positive charge and the Pauli repulsion arising from the filled electronic shells. So, a weak effective potential called "pseudopotential" acts on the valence electrons. The total energy per atom is:

$$E = Z \cdot (E_g + E_{ion}) \tag{15}$$

where Z is the number of electrons per atom, E_g the electron "gas" energy and E_{ion} the change in the energy to first order due to the presence of the ions.

$$E_g = 2.21/R_s^2 - 0.916/R_s - (0.115 - 0.031 \ln R_s) \tag{16}$$

where the first term is the electron kinetic energy, the second term the electron exchange, the third term the electron correlation (electrons are not completely independent).

$$E_{ion} = 1.2\, Z^{2/3} / R_s - 3\, Z^{2/3} / R_s + 3\, R_c^2 / R_s^3 \qquad (17)$$

where the first term is the electron - electron potential (Coulomb repulsion), the second term the electron - ion potential, the third term the electron repulsion due to orthogonality constraints.

R_s is the radius of a sphere which on average contains one electron, defined by $R_s = (3\, V/4\pi Z)^{1/3}$ where V is the atomic volume of the metal. It is a conventional parameter used by physicists to characterize the electron density in the metal. R_c is the core radius. Energy and radii are expressed in a.u..

We can impose the equilibrium condition, by differentiating the energy with respect to the atomic volume, or in other words by imposing zero pressure to the elctronic gas : p = - ($\delta E / \delta V$) = 0. Neglecting the very small contribution from the correlation energy, we obtain a relation between R_s and R_c :

$$R_c^2 = (0.2\, Z^{2/3} + 0.102)\, R_s^2 - 0.491\, R_s \qquad (18)$$

This relation can be used to fix the core radius.

Now, let's consider the formation of an $A_{1-x}B_x$ alloy. Using again the Wigner-Seitz approximation we think each atom be confined in a cell. If the electronic density is lower in A than in B cells, we compress the A cells and expand the B cells until each cell reaches a constant average electron density. As for the pure metals seen above, we can write the energy per atom in the alloy :

$$E^{AB} = Z \cdot E_g(R_s) + (1-x) \cdot Z_A \cdot E^A_{ion} + x \cdot Z_B \cdot E^B_{ion} \qquad (19)$$

as a function of R_s, the conventional parameter now representing the electron density in the alloy; R_{cA} and R_{cB}, the core radii of the A and B metals; Z_A and Z_B, the number of valence electrons of the two metals; x, the stoichiometric coefficient defining the composition of the alloy $A_{1-x}B_x$. An average valence can be also defined as $Z = (1-x) \cdot Z_A + x \cdot Z_B$. Again, as before, we can state the equilibrium condition by differentiating : p = - ($\delta E^{AB} / \delta V$) = 0.

We obtain a relation between R_s and the core radii R_{cA}, R_{cB}. The core radii can be expressed in terms of R_{sA}, R_{sB} using equation (18). Concluding R_s is now expressed as a function of R_{sA}, R_{sB}, Z_A, Z_B, x. The complete formulae are :

$$\bar{R}_s^2 - 1 = (1-x) \cdot D_A (\bar{R}_{sA}^2 - 1) + x \cdot D_B (\bar{R}_{sB}^2 - 1) \qquad (20)$$

with

$$\bar{R}_s = -1 + 0.452\, R_s \{0.916 + 1.8\, [(1-x)\, Z_A^{5/3} + x\, Z_B^{5/3}] / Z\} \qquad (21)$$

$$\bar{R}_{si} = -1 + 0.452\, R_{si}\, (0.916 + 1.8\, Z_i^{2/3}) \qquad i = A, B \qquad (22)$$

and

$$D_i = \frac{Z_i}{Z} \cdot \frac{0.916 + 1.8\, [(1-x) \cdot Z_A^{5/3} + x \cdot Z_B^{5/3}] / Z}{0.916 + 1.8\, Z_i^{2/3}} \qquad (23)$$

In equation (22) a misprint in the original paper -1/0.452 was corrected as -1 + 0.452.

This method works fairly well for compounds formed by alkali and alkaline earth metals, as can be seen in Table 4.

TABLE 4. Experimental and by Hafner's formulae calculated volume variations (%) in some sp alloys.

Phase	Structure type	$\Delta V_{exp}\%$	$\Delta V_{Hafner}\%$
$RbBi_2$	$MgCu_2$	-29.8	-37.5
CsK_2	$MgZn_2$	-0.7	-0.6
KPb_2	$MgZn_2$	-21.5	-33.0
CaCd	CsCl	-13.5	-1.5
BaZn	CsCl	-32.2	-5.2
LiZn	NaTl	-17.8	-12.8
NaTl	NaTl	-20.6	-21.8
KGe	KGe	-30.4	-46.8
$BaLi_4$	$BaLi_4$	+4.9	+3.4

For $BaLi_4$ is correctly predicted an expansion, because of the presence of a component (Li) with low valence and large electron density. The agreement is less good for compounds formed by alkaline earths with Zn, Cd, Hg. We can note that in this case both elements have the same number of valence electrons. The disagreement can be explained by the physical mechanism involved in these calculated volume changes. The energy change in alloy formation is purely given by the elastic energy necessary to expand or compress the individual atomic cells. No difference of electronegativity, no charge transfer is taken into account in this model. Concluding, a successful application of Hafner's model can be predicted in all cases where volume changes in the alloy formation are dominated by elastic effects.

2.4. MERLO'S APPROACH

Different from the other methods before examined, this approach (Merlo, 1988) is empirical. A formula is proposed to calculate the volume contraction for a family of binary compounds formed by the alkaline earths Ca, Sr, Ba and the bivalent rare earth elements Eu and Yb. To evaluate the volume per atom for a given $A_{1-x}B_x$ compound we use the expression

$$V = (1 - x) \cdot V_A + x \cdot V_B \qquad (24)$$

with A = Ca, Sr, Ba, Eu, Yb; B = element of group 10 to 15. V_A, V_B are the "effective" atomic volumes of the elements in the given compound.

Again we can think that atom A with a lower electron density is contracted, while atom B with a high electron density is expanded, in order to obtain the same electron density in the compound. However, as it was pointed out by several authors, the compression of the electropositive element is always much stronger than the expansion of the electronegative one.

As a first approximation then it is supposed that only V_A undergoes a contraction, while V_B maintains the elemental value. Moreover, the effective A volume appears to decrease as a function of the composition of the alloy, according to an empirical equation. So we have

$$V_A = V°_A - K x^n \quad \text{with K, n > 0 adjustable parameters} \tag{25}$$

$$V_B = V°_B \tag{26}$$

Substituting (25) and (26) in equation (24) we obtain

$$V = V_{Zen} - K x^n (1 - x) \tag{27}$$

Two hypotheses are now introduced : 1) the K parameter is related to the difference $|\Gamma_A - \Gamma_B|$ between the charge transfer parameters of the two elements; 2) the area ΔV_{int} is evaluated and put equal to $|\Gamma_A - \Gamma_B|$ (see Figure 6). The larger is the area the greater the charge transfer between the two atoms.

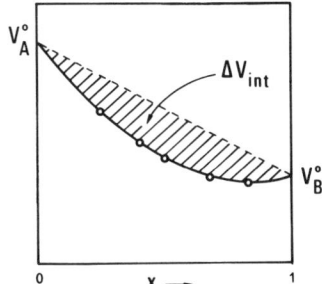

Figure 6. The quantity ΔV_{int} defined as the area comprised between the linear Zen-like trend and the experimental trend.

$$\Delta V_{int} = \int_0^1 (V_{Zen} - V) \, dx = \int_0^1 K x^n (1 - x) \, dx = K / [(n + 1) \cdot (n + 2)] \tag{28}$$

$$K / [(n + 1) \cdot (n + 2)] = |\Gamma_A - \Gamma_B| \tag{29}$$

Then

$$V = V_{Zen} - (n + 1) \cdot (n + 2) \cdot x^n \cdot (1 - x) \cdot |\Gamma_A - \Gamma_B| \tag{30}$$

It is easy to see that the parameter n is related to the composition of the phase AB_n with the maximum volume contraction in the A - B system :

$$n = x_{max} / (1 - x_{max}) \tag{31}$$

The next step is the determination of the Γ values for the 25 considered elements, and the n value for each system. On the whole, 100 binary systems are involved with

nearly 500 intermediate phases. An iterative process is applied and the parameters optimized until the difference between observed and calculated atomic volumes becomes minimum. The n parameter is the same for alloys with the same B element and ranges from 0.5 to 2. The difference $|\Gamma_A - \Gamma_B|$ ranges from 0.4 to 6.8.
As can be seen in Figure 7, where atomic volumes versus composition are plotted for five systems, the continuous curve calculated by the equation (30) approximates very well the experimental trend.

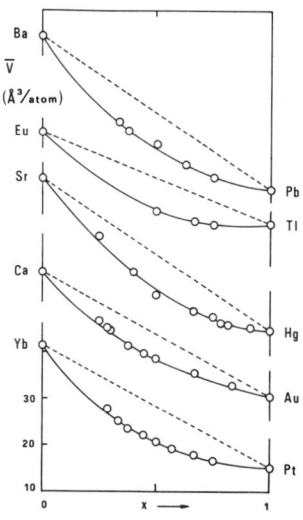

Figure 7. Atomic volumes versus composition in the systems Yb - Pt, Ca - Au, Sr - Hg, Eu - Tl and Ba - Pb. Open circles indicate experimental values for the intermediate phases occurring in each system. Dashed lines correspond to Zen's law; continuous curves are calculated with equation (30).

3. References

Boer, F.R. de, Boom, R., Mattens, W.C.M., Miedema, A.R. & Niessen, A.K. (1988). "Cohesion in Metals: Transition Metal Alloys." F.R. de Boer and D.G. Pettifor (series eds.), Vol.1. Amsterdam : North Holland.

Brunner, G.O. & Schwarzenbach, D. (1971). 'Zur Abgrenzung der Koordinationssphäre und Ermittlung der Koordinationszahl in Kristallstrukturen.' Z.Kristallogr. **133**, 127-133.

Bruzzone, G., Fornasini, M.L. & Merlo, F. (1970). 'Rare-Earth Intermediate Phases with Zinc.' J.Less - Common Met. **22**, 253-264.

Bruzzone, G. & Merlo, F. (1982). 'Crystal Chemical Remarks on the Alloying Behaviour of Calcium, Strontium and Barium.' J.Less - Common Met. **85**, 285-306.

Frank, F.C. & Kasper, J.S. (1958). 'Complex Alloy Structures Regarded as Sphere Packings. I. Definitions and Basic Principles.' Acta Crystallogr. **11**, 184-190.

Hafner, J. (1985). 'A Note on Vegard's and Zen's Laws.' J.Phys.F **15**, L43-L48.

Heine, V. & Weaire, D. (1970). 'Pseudopotential Theory of Cohesion and Structure.' In H. Ehrenreich, F. Seitz and D. Turnbull (eds.), "Solid State Physics" Vol. 24. New York : Academic Press, pp.249-463.

Kubaschewski, O. & Alcock, C.B. (1979) . "Metallurgical Thermochemistry." Oxford : Pergamon Press.

Merlo, F. (1988). 'Volume Effects in the Intermetallic Compounds Formed by Ca, Sr, Ba, Eu and Yb with Other Elements.' J.Phys. F **18**, 1905-1911.

Miedema, A.R. & Niessen, A.K. (1982). 'Volume Effects upon Alloying of Two Transition Metals.' Physica **114B**, 367-374.

O'Keeffe, M. (1979). 'A Proposed Rigorous Definition of Coordination Number.' Acta Crystallogr. **A35**, 772-775.

Pettifor, D.G. (1984). 'A Chemical Scale for Crystal-Structure Maps.' Solid State Commun. **51**, 31-34.

Teatum, E., Gschneidner, K.A. Jr. & Waber, J. (1968). Report LA-4003. National Technical Information Service, Springfield, VA 22151, USA.

Villars, P. (1983). 'A Three-Dimensional Structural Stability Diagram for 998 Binary AB Intermetallic Compounds.' J.Less - Common Met. **92**, 215-238.

Villars, P., Mathis, K. & Hulliger, F. (1989). 'Environment Classification and Structural Stability Maps' in F.R. de Boer and D.G. Pettifor (series eds.), Vol. 2, "The Structure of Binary Compounds." Amsterdam : North-Holland, pp.1-103.

4. Problems

Problem 1 : Using the structural maps for the single-environment - type and the two-environment-type compounds reported in Figure 8 (Villars et al., 1989), predict, if possible, coordination and structure of the compounds Mg_2Si, Ca_2Si, $ThMg_2$, ThSi and CaTl.

Element	VE	X (electronegativity)	$r_s + r_p$ (a.u.)
Ca	2	1.17	3.00
Mg	2	1.31	2.03
Si	4	1.98	1.42
Th	4	1.3	4.98
Tl	3	1.69	2.235

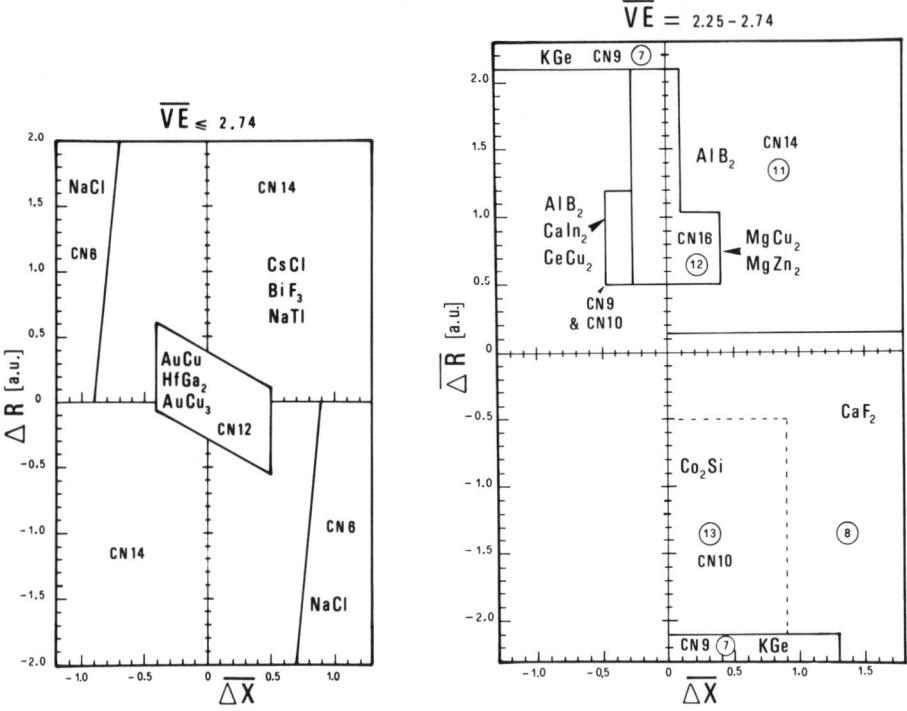

Figure 8. Single-environment (left) and two-environment (right) maps for mean number of valence electrons $\overline{VE} \leq 2.74$ and $2.25 \leq \overline{VE} \leq 2.74$, respectively. In the two-environment map the second coordination number is indicated by a circle around a number.

Problem 2 : In Figure 9 projections of three structures are shown. Recognize the coordination polyhedron around Si in ThSi$_2$, around Cu in CuAl$_2$ and around Th in Th$_3$P$_4$. Establish whether the atoms capping some faces of the main polyhedron around the Si and Cu atoms are to be counted or not.

ThSi$_2$ (tI12)	a = 4.134Å	c = 14.375	r(Th) = 1.798Å	r(Si) = 1.319
CuAl$_2$ (tI12)	a = 6.063Å	c = 4.872	r(Cu) = 1.278Å	r(Al) = 1.432
Th$_3$P$_4$ (cI28)	a = 8.617Å		r(Th) = 1.798Å	r(P) = 1.28

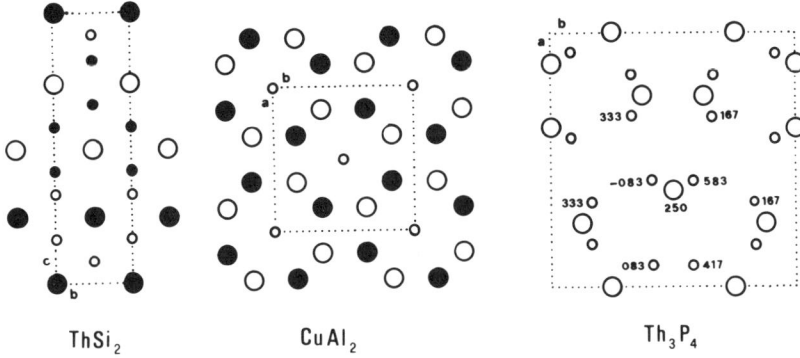

Figure 9. Projection of the three structures : $ThSi_2$ (full circles : $x = 0$; open circles : $x = 1/2$; small circles : Si), $CuAl_2$ (full large circles : Al at $z = 0$; open large circles : Al at $z = 1/2$; small circles : Cu at $z = 1/4$ and $3/4$) and Th_3P_4 (large circles : Th. The z coordinates of the atoms belonging to the Th coordination polyhedron have been multiplied with 10^3).

Problem 3 : Calculate the lattice constant of the compound CaTl (cubic CsCl-type, Ca in 0,0,0; Tl in 1/2,1/2,1/2) in the hypothesis that Ca and Tl maintain the same size as in the elemental state. Compare the volume per atom so obtained with the corresponding value from Zen's law and with the experimental value.

$r(Ca) = 1.974Å$ $r(Tl) = 1.716$
$V°_{Ca} = 26.274$ cm^3 mol^{-1} $V°_{Tl} = 17.214$
$a_{exp} = 3.855Å$

Problem 4 : Compare experimental and calculated volumes per atom for the six phases of the Ca - Pd system. To obtain calculated volumes apply Zen's law, Miedema's approach and Merlo's approach.

Ca_3Pd (oP16) $a = 7.699Å$ $b = 9.937$ $c = 6.691$
Ca_5Pd_2 (mS28) $a = 16.694Å$ $b = 6.708$ $c = 7.704$ $\beta = 97.3°$
Ca_3Pd_2 (hR45) $a = 8.939Å$ $c = 16.900$
$CaPd$ (cP2) $a = 3.518Å$
$CaPd_2$ (cF24) $a = 7.652Å$
$CaPd_5$ (hP6) $a = 5.147Å$ $c = 4.224$

Zen : $V°_{Ca} = 43.631$ Å^3/at $V°_{Pd} = 14.717$ Å^3/at

Miedema : $V_{Ca}^{2/3} = 8.82$ $\Phi^*_{Ca} = 2.55$ $a_{Ca} = 0.1$ $(n_{WS}^{Ca})^{1/3} = 0.91$
           $V_{Pd}^{2/3} = 4.29$ $\Phi^*_{Pd} = 5.45$ $a_{Pd} = 0.04$ $(n_{WS}^{Pd})^{1/3} = 1.67$

Merlo : $\Gamma_{Ca} = 2.15$ $\Gamma_{Pd} = 6.65$ $n = 1$

Problem 5 : Compare experimental and calculated volumes per atom for the five phases of the Ba - Pb system. To obtain calculated volumes apply Zen's law, Hafner's approach and Merlo's approach.

Ba_2Pb (oP12) $a = 8.78$Å $b = 5.75$ $c = 10.69$
Ba_5Pb_3 (tI32) $a = 9.040$Å $c = 16.816$
$BaPb$ (oS8) $a = 5.289$Å $b = 12.610$ $c = 4.822$
Ba_3Pb_5 (oS32) $a = 11.148$Å $b = 9.049$ $c = 11.368$
$BaPb_3$ (hR36) $a = 7.298$Å $c = 25.765$

Zen, Hafner : $V^*_{Ba} = 63.367$ Å^3/at $V^*_{Pb} = 30.326$ Å^3/at

Hafner : $Z_{Ba} = 2$ $Z_{Pb} = 4$

Merlo : $\Gamma_{Ba} = 1.45$ $\Gamma_{Pb} = 6.75$ $n = 0.8$

4.1. SOLUTIONS OF THE PROBLEMS

Problem 1 : The observed structure type is printed in bold faced characters.

$SiMg_2$: $\overline{VE} = 2.67$ $\overline{\Delta X} = 0.45$ $\overline{\Delta R} = -0.41$
CN12 $HfGa_2$-type or CN10 + CN8 **CaF_2-type**.

$SiCa_2$: $\overline{VE} = 2.67$ $\overline{\Delta X} = 0.54$ $\overline{\Delta R} = -1.05$
CN13 + CN10 **Co_2Si-type**.

$ThMg_2$: $\overline{VE} = 2.67$ $\overline{\Delta X} = 0$ $\overline{\Delta R} = 1.97$
CN16 + CN12 **$MgCu_2$**- or $MgZn_2$-type (Laves phases)

ThSi : $\overline{VE} = 4$ is outside the structural maps given in the problem.

CaTl : $\overline{VE} = 2.5$ $\overline{\Delta X} = -0.52$ $\overline{\Delta R} = 0.765$
CN14 **CsCl**- or NaTl-type.

Note that the composition of the phase automatically eliminates some possibilities.

Problem 2 : In $ThSi_2$ the Si atom is surrounded by a trigonal prism of 6 Th atoms with the lateral faces capped by 3 Si atoms (d ≈ 2.5Å) : CN9. Note that all Si atoms occupy the same atomic position. In the drawing trigonal prisms are found in two different orientations, with the prism axis perpendicular to or lying in the sheet, but they are equivalent.
In $CuAl_2$ the Cu atom is surrounded by a square antiprism of 8 Al with the square faces capped by 2 Cu atoms (d = 2.44 Å) : CN10.
In Th_3P_4 the Th atom is surrounded by a bisdisphenoid of 8 P atoms.

Problem 3: $d_{Ca-Tl} = a_{cal} \sqrt{3}/2 = r_{Ca} + r_{Tl}$

$a_{cal} = 2(r_{Ca} + r_{Tl})/\sqrt{3}$

$V_{cal} = a_{cal}^3/2 = 4(r_{Ca} + r_{Tl})^3/3^{3/2} = 38.68$ Å^3/at

$V^*_{Ca} = (26.274 \cdot 10^{24})/(6.022 \cdot 10^{23}) = 43.630$ Å^3/at

$V^*_{Tl} = 28.585$ Å^3/at

$V_{Zen} = 0.5(V^*_{Ca} + V^*_{Tl}) = 36.11$ Å^3/at

$V_{exp} = a_{exp}^3/2 = 28.64$ Å^3/at

The calculated volume per atom is greater than V_{Zen}, because the CsCl structure is less compact than the face-centred-cubic (Ca) or the hexagonal-close-packed (Tl) structures. The experimental volume is much smaller owing to a contraction in the compound formation.

Problem 4: All volumes are given in Å^3/at.

Phase	V_{exp}	V_{Zen}	$V_{Miedema}$	V_{Merlo}
Ca$_3$Pd	32.0	36.4	31.7	31.3
Ca$_5$Pd$_2$	30.6	35.4	30.0	29.9
Ca$_3$Pd$_2$	26.0	32.1	24.6	25.6
CaPd	21.8	29.2	20.4	22.4
CaPd$_2$	18.7	24.4	16.3	18.4
CaPd$_5$	16.2	19.5	15.3	15.8

A good agreement is observed between experimental and calculated values by both Miedema's and Merlo's approaches, confirming the validity of these methods. On the contrary, the values calculated by Zen's law are always higher, CaPd showing the maximum contraction.

Problem 5: All volumes are given in Å^3/at.

Phase	V_{exp}	V_{Zen}	V_{Hafner}	V_{Merlo}
Ba$_2$Pb	45.0	52.4	42.5	45.0
Ba$_5$Pb$_3$	42.9	51.0	41.2	43.4
BaPb	40.2	46.8	37.8	39.2
Ba$_3$Pb$_5$	35.8	42.7	35.2	35.8
BaPb$_3$	33.0	38.6	33.2	33.3

A good agreement is observed between experimental and calculated values by both Hafner's and Merlo's approaches. Note that Hafner's approach using only elemental volume and valence of the two metals reaches good results.

THE INTERGROWTH CONCEPT AS A USEFUL TOOL TO INTERPRET AND UNDERSTAND COMPLICATED INTERMETALLIC STRUCTURES

YURI N. GRIN'
Max-Planck-Institut für Metallforschung
Institut für Werkstoffwissenschaft
Seestrasse 75
D - 7000 Stuttgart 1
Germany

ABSTRACT. A great number of the structures of intermetallic compounds can be interpreted as an intergrowth of special determined structure segments. The crystal structures built up of the same segments belong to the same series of intergrowth structures. The homogeneous intergrowth structures are constructed of identical segments, in the case of inhomogeneous intergrowth structures the building segments are different. The identity or the difference of the constructing segments is determined by their composition and the coordinations of the atoms (coordination numbers and shape of the coordination polyhedra of the atoms in the parent structures). One distinguishes between an one-dimensional intergrowth (linear intergrowth-structures) where two-dimensional segments (slabs) are intergrown, a two-dimensional intergrowth where one-dimensional segments (columns) are intergrown and also a three-dimensional intergrowth of zero-dimensional segments (blocks). The use of the intergrowth concept allows one not only to interpret the known structures and to understand their construction, but also to predict the unit cells, the symmetry and the composition of all possible structures belonging to the same structure series.

1. Introduction

The idea of describing selected structures as stacking variants of more simple common structure slabs was born in the twenties and its evolution was shown in a detailed review article by Parthé, Chabot & Cenzual (1985). The use of the intergrowth concept for the interpretation of complicated intermetallic structures was first proposed by Kripyakevich & Gladyshevskii (1972). Kripyakevich (1976a, 1976b) described the first series of intergrowth structures, consisting of identical segments. The classification of linear (one-dimensional) homogeneous intergrowth structures was made by Kripyakevich & Grin' (1978, 1979). The systematic description of inhomogeneous intergrowth structure series was started by Grin', Yarmolyuk & Gladyshevskii (1982). The term "intergrowth" has been proposed first by Parthé, Chabot & Cenzual (1985) for the definition of such a structure relationship and was finally included by Lima-de-Faria et al. (1990) into the recommendations for the nomenclature of inorganic structure types. Pani & Fornasini (1990) presented a set of new intermetallic compounds interpreted by means of the intergrowth model. The present paper is an attempt at a systematics of structure interpretations based on this concept considering also symmetry relationships.

2. The Intergrowth structure series

Structures belonging to the same series can be described as being constructed from structure segments of more simple structures. These simple structures are named parent structures. Each structure series can be labelled by denoting the formulae of their parent structure types. The parent structure types sometimes have their representatives in the same chemical system. For example, the linear inhomogeneous series $BaAl_4$ - AlB_2 has both parent structure representatives and one intergrowth variant in the binary system Eu - Ga (for a description of the symbols, see 3.2.) :

$EuGa_4$ ($BaAl_4$ structure type) $\rightarrow$ Eu_3Ga_8 (own structure type, intergrowth structure with ($BaAl_4^3$ | $2AlB_2^4$)$_2$ as code) $\leftarrow$ $EuGa_2$ (AlB_2 structure type).

Often the parent structure types are represented in chemically analogous systems. For example, the linear homogeneous series α-ITl - FeB has the parent structure representatives among the light rare earth mononickelides (α-ITl structure type, also called CrB type) and the heavy rare earth mononickelides (FeB structure type), whereas a large number of intergrowth variants occurs in the ternary systems $R'_xR''_{1-x}Ni$, where R' and R" are different rare earth elements. Structure series are also known where one of the parent structure types has not yet been found in nature.

2.1. THE DIFFERENT KINDS OF STRUCTURAL SERIES

The structure series can be classified according to :
i)   the method of construction and
ii) the kind of fragments.

According to the method of construction, the structure series can be grouped into linear (one-dimensional), two- and three-dimensional series. In the linear structure series the two-dimensional segments (infinite slabs) of the parent structures are stacked one-dimensionally along one direction (stacking direction). In a two-dimensional structure series the member structures are mosaically constructed from one-dimensional fragments (infinite columns of different shape). The structures belonging to the three-dimensional structure series are built from zero-dimensional parent structure blocks (usually cubes or generally parallelohedra) stacked in three dimensions.

According to the kind of fragments, one distinguishes between homogeneous and inhomogeneous structure series. The homogeneous intergrowth structures consist of identical segments, while in the case of a inhomogeneous structure series the segments are from different parent structures. The identity or the difference in the constructing segments is determined by their composition and the atom coordinations (coordination numbers and shape of the coordination polyhedra of the atoms in the parent structures). The parent structure segments used for the construction of a homogeneous series, are equal in respect to both composition and coordination. In the case of an inhomogeneous series the segments differ in composition and/or atom coordination.

2.2. THE CHOICE OF BASIC SEGMENTS IN THE PARENT STRUCTURES

Slicing a parent structure into segments can be done in different ways. For a segment to be used in a particular structure series for which we want to make predictions concerning the composition and symmetry of its members it must conform to the four following requirements :

i) The segments selected from each parent structure should retain the additional symmetry elements (if there are any) in addition to the minimal symmetry (see 3.3.). These additional elements should be perpendicular to the stacking direction for a linear structure series, perpendicular to the mosaic plane for a two-dimensional series and in the case of a three-dimensional structure series the segments should contain the symmetry centre or/and symmetry axes.

ii) The segment interfaces cannot hang in space, they necessarily pass through atom centers; in this way the composition of each segment is proportional to the stoichiometry of the parent structure and a simple addition may be applied to obtain the compositions of all possible structures of a selected series.

iii) The segments selected from different parent structures must have at least one topologically equal interface in order for an intergrowth to be possible between them.

iv) If the atom arrangement on the interface is such that it permits more than one possiblity of intergrowing two parent structure segments, it is necessary to introduce the so-called hybride segments where the mutual orientation of the intergrown parent structure segments is determined.

These requirements have been formulated on the basis of a systematical study of linear structure series. The requirements i) and ii) apply to the linear homogeneous structure series. In the description of the inhomogeneous series the requirements iii) and iv) have to be added. It may be possible, that in two- and three-dimensional structure series the segments must conform to additional requirements.

The more parent structures are involved in a selected structure series (excluding the linear homogeneous series; they always have two parent structures), the more basic segments are necessary for its description. However, the number of fragments necessary for a complete and unambigous description of each selected series is always limited.

The identification of the initial parent structure segments in an intergrowth structure may be difficult because of their possible deformation and their occasionally unusual orientation in the intergrowth structure. As a first step in their identification, it is useful to have a list of common basic parent structure segments sliced along different planes having different interface meshes. A list of segments with orthohexagonal, triangular and Kagome as well as rectangular centred meshes has been prepared by Parthé et al. (1985). Segments with square-primitive or square-centred meshes at the interface have been listed by Grin' et al. (1982). The study was later extended to segments found in the superconducting oxide structures (Grin' & Akselrud, 1990). For the interpretation of the so-called "chimney-ladder" structures, a list of basic segments has also been prepared (Grin', 1986). If the identification of a segment by means of the lists mentioned above is unsuccessful, a study of the basic requirements for intergrowth i) - iv) may offer a clue to the nature of the parent structure.

3. The linear intergrowth structure series

A great number of the ternary and binary intermetallic compounds can be considered as members of linear homogeneous and inhomogeneous structure series. The classification of the linear series and their members using numeric symbols and one-dimensional symmetry groups will be made in this section.

3.1. THE COMPOSITIONS OF LINEAR INTERGROWTH STRUCTURES

The composition of a member of a selected structure series can be calculated from the compositions of the segments according to equation (1), where the segments are assumed to consist of three different kinds of components.

$$(A_{x_1}B_{y_1}C_{z_1})_{m_1} + (A_{x_2}B_{y_2}C_{z_2})_{m_2} + .. + (A_{x_n}B_{y_n}C_{z_n})_{m_n} =$$

$$A_{m_1 \cdot x_1 + m_2 \cdot x_2 + .. + m_n \cdot x_n} B_{m_1 \cdot y_1 + m_2 \cdot y_2 + .. + m_n \cdot y_n} C_{m_1 \cdot z_1 + m_2 \cdot z_2 + .. + m_n \cdot z_n} \quad (1)$$

where A, B, C are crystal chemically independent components, x_n, y_n, z_n are the numbers of atoms of each component in segment n, m_n is the number of segments n in the unit cell of the intergrowth structure.

For the structures of the series $BaAl_4$ - AlB_2 equation (1) can be written as follows :

$$(RX_4)_{m_1+m_2} + (RX_2)_{m_3+m_4} = R_{m_1+m_2+m_3+m_4} X_{(m_1+m_2) \cdot 4 + (m_3+m_4) \cdot 2} \quad (2)$$

where m_1 and m_2 are the numbers of the two kinds of $BaAl_4$ segments used, m_3 and m_4 are the numbers of the two kinds of AlB_2 segments used, R and X are the two kinds of atoms with a clearly different coordination (CN for R is usually 18 - 22, for X it is 8 - 12). In some cases it might be advantageous to separate the X atoms into two groups, i.e. X' : CN = 11 - 12, cuboctahedra, occasionally with defects and X" : CN = 9 - 10, tricapped trigonal prisms or capped square antiprisms. In this case equation (3) is applicable.

$$(RX'_2X''_2)_{m_1+m_2} + (RX''_2)_{m_3+m_4} =$$

$$R_{m_1+m_2+m_3+m_4} X'_{(m_1+m_2) \cdot 2} X''_{(m_1+m_2+m_3+m_4) \cdot 2} \quad (3)$$

where the symbols have the same meaning as in equation (2).

3.2. NUMERIC CODE NOTATION FOR THE MEMBERS OF LINEAR STRUCTURE SERIES

The numeric code notation for the intergrowth structures depends on the choice of the initial fragments in the parent structures. As a consequence of this choice, the structures of a linear homogeneous series can be considered in two different ways.

As an example let us consider the different close packed element structures which can be denoted either by the Zhdanov symbols or by the Jagodzinski stacking symbols. In the first case, the parent structure segments are identical except for having one of two possible different orientations. The numeric code in this case (Zhdanov symbol) consists of digits, each of them denoting the number of segments in succession having the same orientation. In the second case, the parent structures are cut into two different kinds of segments, which have the same interfaces, but differ in their symmetry elements perpendicular to the stacking direction. The numeric code for such notation consists of digits, each of them denoting the number in succession of one particular kind of segments.

Both methods of numeric notation can be used for the classification of the linear homogeneous structure series. However, for the derivation of a numeric code notation for the structures of a linear inhomogeneous structure series we prefer the second notation method. The reason is that in order to permit the prediction of the symmetry of a member of the structure series the numeric code must have the following properties :

i) It has to show uniquely the position of the segment in the structure.
ii) Each digit must correspond to only one kind of segment.
iii) It has to contain information on the presence in the intergrowth structure of symmetry elements in addition to the minimal symmetry of the selected series.

The numeric codes for structures of an inhomogeneous structure series are more complicated in comparison to the codes for structures of homogeneous structure series. A numeral (digit) indicates the succeeding number of segments of one kind. A subscripted index shows from what parent structure(s) the segment is taken. A superscripted index specifies the symmetry of the segment.

The following superscripted indexes were used in previous works :

1 : Segment without symmetry elements in addition to the minimal symmetry
2 : Segment with a glide plane (n) perpendicular to the stacking direction
3 : Segment with a mirror plane (m) perpendicular to the stacking direction
4 : Segment with a glide plane (a) in the plane of projection
5 : Segment with a glide plane (c) perpendicular to the plane of projection
6 : Segment with a two-fold axis perpendicular to the stacking direction
7 : Segment with a two-fold screw axis perpendicular to the stacking direction.

This list can be extended by considering segments with other symmetry if it is necessary for the description of a particular series. If the segment contains two kinds of symmetry elements, its superscripted index can consist of two-digit numbers. For example, the segments with the superscripted indices 12, 13, 14, 15 (without symmetry elements normal to the stacking direction) contain an additional translation component according to symmetry planes of type 2, 3, 4 and 5.

To make the numeric code more easily understandable from a crystal chemical point of view the subscripted index can be substituted by the formula(e) of the parent structure(s) (Parthé et al., 1985).

The unit cell of some structures can be described by two or three repetitions of the same digit sequence in the numeric code (see 3.3.). For example, the structure of $NdNiGa_2$ belongs to the series $BaAl_4$ - AlB_2. Its unit cell contains two $BaAl_4$ segments of group 3 (with a mirror plane) and two AlB_2 segments of group 4 (or 5). The numeric code of this structure can be written as follows :

$$(1_1^3 1_2^4)_2 \quad \text{or} \quad (BaAl_4^3 \mid AlB_2^4)_2.$$

Only a numeric code notation is useful for the derivation of the symmetry of an intergrowth structure (see 3.3.). The other notation with the formula of the parent structures offers the possibility to study the crystal chemical relationships of the intergrowth structure with structures of compounds in chemically analogous systems.

3.3. THE SYMMETRY OF STRUCTURES IN LINEAR SERIES

All the segments used for the description of a linear series usually contain some symmetry elements, mostly parallel to the stacking direction, which are retained in any stacking sequence. These symmetry elements form the minimal symmetry of a linear series.

Additional symmetry elements occur if the numeric code is symmetric in relation to one digit. Three general relations have been found between the symmetry of the numeric code and the symmetry of the intergrowth structure :

i) If the numeric code is symmetric relative to an odd digit, then the structure contains the symmetry element located perpendicular to the stacking direction in the segment corresponding to this digit.

ii) If the numeric code is symmetrical relative to an even digit, then the structure usually contains a symmetry element (usually a mirror plane) perpendicular to the stacking direction located between the segments, corresponding to this digit.

iii) If the digit sequence in the numeric code is repeated twice or four times, then the structure contains symmetry elements (translation, glide plane or screw axis) parallel to the stacking direction. If the same digit sequence is repeated three times in the unit cell this indicates a rhombohedral symmetry of the unit cell.

In some linear series (with segments of high intrinsic symmetry) the antisymmetry of the numeric code relative to some digits gives an indication that additional symmetry elements exist in the structure.

All the different space groups possible for the different members of a selected series can be derived by adding in various combinations the intrinsic symmetry elements contained in the segments to the minimal symmetry of the series.

3.4. THE APPLICATION OF THE ONE-DIMENSIONAL OR LINEAR SYMMETRY GROUPS TO THE SYSTEMATIZATION OF LINEAR STRUCTURE SERIES AND THEIR REPRESENTATIVES

Each intergrowth structure and its numeric code can be represented graphically by means of a zigzag line constructed by first drawing a line between topologically similar atoms (or points) on the opposite interfaces of each segment and by further, considering how the segments are intergrown, joining these lines to an uninterrupted zigzag line. Finally, the usually three-dimensional zigzag lines are projected on a plane. Examples of the projections of zigzags lines for different structure series are shown in Figure 1. For the homogeneous zigzag lines no subscripted indices are used. The first zigzag line (left, a) corresponds to the most symmetrical linear group pmg2. The other samples show the evolution of the zigzag line symmetry in different cases. A short review of one-dimensional or linear symmetry groups is given in the Appendix.

Each chain segment of the zigzag line corresponds to one parent structure segment. The number of equal chain segments in succession corresponds to one digit in the numeric code. The different interfaces between the structure segments are indicated by different polygons. If the end point of a chain is above or below the plane of projection, the corresponding polygon is shaded. The number of different shades needed for the graphical representation of selected linear structure series is limited by the number of different levels (Ω) where the end points of the chain segments may be located in respect to the projection plane.

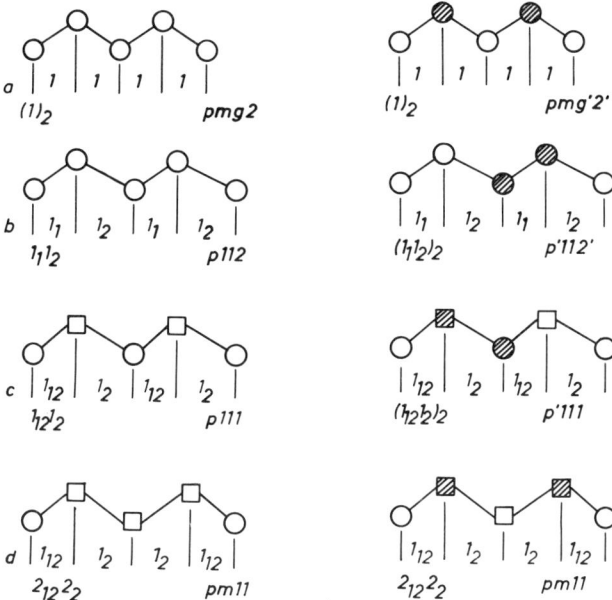

Figure 1. Examples of zigzag lines, numeric codes and linear symmetry groups for structures with $\Omega = 1$ (left) and $\Omega = 2$ (right) : a) Example of a homogeneous structure; b - d) Inhomogeneous structures of different construction.

3.4.1. *The linear structures with $\Omega = 1$.* The zigzag lines have their own intrinsic symmetry of one-dimensional space groups. For the series with $\Omega = 1$ (the zigzag lines are all in the projection plane) only five one-coloured one-dimensional symmetry groups are possible and to each corresponds one three-dimensional space group for the members of a particular linear structure series. Examples of homogeneous series have been discussed by Kripyakevich et al. (1978). Selected inhomogeneous linear series with $\Omega = 1$ are shown in Table 1. The first example in Table 1 concerns the structure series where α-ITl and UPt$_2$ type slabs are intergrown. The real existing intergrowth representatives of this series are the structure types Mo$_2$B$_2$Ni, W$_3$B$_3$Co, Y$_4$Co$_4$Ga and Y$_5$Co$_5$Ga of which projections are shown in Figure 2.

If there is only one connection possibility between the segments with equal intrinsic symmetry the symmetry of the intergrowth structures can be degenerated and there may, for example, exist only one space group for all members of the structure series. One known example is the structure series with composition $R_{m+n}X'_{5m+3n}X''_{2n}$ consisting of intergrown CaCu$_5$ and CeCo$_3$B$_2$ type slabs (Kuzma & Bilonizhko, 1974). Members are the ternary boride structure types Nd$_3$Ni$_{13}$B$_2$ (m = 2, n = 1), CeCo$_4$B (m = 1, n = 1), Lu$_5$Ni$_{19}$B$_6$ (m = 2, n = 3), Ce$_3$Co$_{11}$B$_4$ (m = 1, n = 2) and Ce$_2$Co$_7$B$_3$ (m = 1, n = 3). All of them have one of the possible space groups in this series, i.e. P6/mmm.

TABLE 1. Symmetry, numeric codes and structures of selected inhomogeneous linear structure series with $\Omega = 1$

Linear group	Space group	Numeric code		Structure type
Series α-ITl - UPt$_2$				
p111	Cm	$2_1{}^72_2{}^61_1{}^71_2{}^6$		
p112	C2/m	$1_1{}^71_2{}^6$	ITl^7 | UPt$_2{}^6$	Mo$_2$B$_2$Ni
		$3_1{}^71_2{}^6$	3ITl^7 | UPt$_2{}^6$	Y$_4$Co$_4$Ga
pm11	Amm2	$2_1{}^72_2{}^6$		
p1g1	Cmc2$_1$	$(2_1{}^72_2{}^62_1{}^71_2{}^6)_2$		
pmg2	Cmcm	$(2_1{}^71_2{}^6)_2$	$(2ITl^7 \| UPt_2{}^6)_2$	W$_3$Co$_3$B
		$(4_1{}^71_2{}^6)_2$	$(4ITl^7 \| UPt_2{}^6)_2$	Y$_5$Co$_5$Ga
Series YNiAl$_4$ - R$_2$M$_4$X$_{15}$				
p111	Cm	$2_1{}^62_2{}^61_1{}^61_2{}^6$		
p112	C2/m	$1_1{}^61_2{}^6$	Y$_2$Ni$_2$Al$_8{}^6$ | R$_2$M$_4$X$_{15}{}^6$	Ho$_4$Ni$_6$Al$_{23}$
pm11	Amm2	$1_1{}^31_2{}^63_1{}^31_2{}^6$		
p1g1	Cmc2$_1$	$(2_1{}^62_2{}^62_1{}^61_2{}^6)_2$		
pmg2	Cmcm	$(1_1{}^31_2{}^6)_2$	$(YNiAl_4{}^3 \| R_2M_4X_{15}{}^6)_2$	Y$_3$Ni$_5$Al$_{19}$
Series FeB - UPt$_2$				
p111	Pm	$2_1{}^33_2{}^41_1{}^31_2{}^4$		
p112	P2/b	$1_1{}^33_2{}^41_1{}^31_2{}^4$		
pm11	Pmc2$_1$	$1_1{}^32_2{}^4$		
p1g1	Pbc2$_1$	$(2_1{}^32_2{}^41_1{}^31_2{}^4)_2$		
pmg2	Pbcm	$(1_1{}^31_2{}^4)_2$	$(Fe_2B_2{}^3 \| UPt_2{}^4)_2$	La$_3$Ni$_2$Ga$_2$
Series BaAl$_4$ - CaF$_2$				
p111	P4mm	$1_1{}^22_2{}^33_1{}^21_2{}^3$		
p112	P4/nmm	$2_1{}^22_2{}^2$	2BaAl$_4{}^2$ | 2CaF$_2{}^2$	HfCuSi$_2$
pm11	P4/mmm	$2_{12}{}^{13}$	2BaAl$_4$-CaF$_2{}^{13}$*	CeCu$_{0.6}$Ga$_6$
		$(2_{12}{}^{13})_2$**	4BaAl$_4$-CaF$_2{}^{13}$	LaNi$_{0.6}$Ga$_6$
p1g1	I4mm	$(1_1{}^22_2{}^23_1{}^32_2{}^3)_2$		
pmg2	I4/mmm	$(1_1{}^21_{12}{}^{13})_2$	(BaAl$_4{}^2$ | *BaAl$_4$-CaF$_2{}^{13}$*)$_2$	
				SrZnBi$_2$
		$(2_1{}^21_2{}^2)_2$	$(2BaAl_4{}^2 \| CaF_2{}^2)_2$	Zr$_3$Cu$_4$Si$_6$
		$(1_1{}^22_{12}{}^{13})_2$	(BaAl$_4{}^2$ | *2BaAl$_4$-CaF$_2{}^{13}$*)$_2$	
				Ce$_2$NiGa$_{10}$
		$(3_1{}^22_{12}{}^{13})_2$	(3BaAl$_4{}^2$ | *2BaAl$_4$-CaF$_2{}^{13}$*)$_2$	
				Ce$_4$Ni$_2$Ga$_{17}$

* Hybride segments are written in italics. They have the composition RX$_3$(1/2 RX$_4$ + 1/2 RX$_2$)

** The doubling of the unit cell is caused by the arrangement of the Ni defects

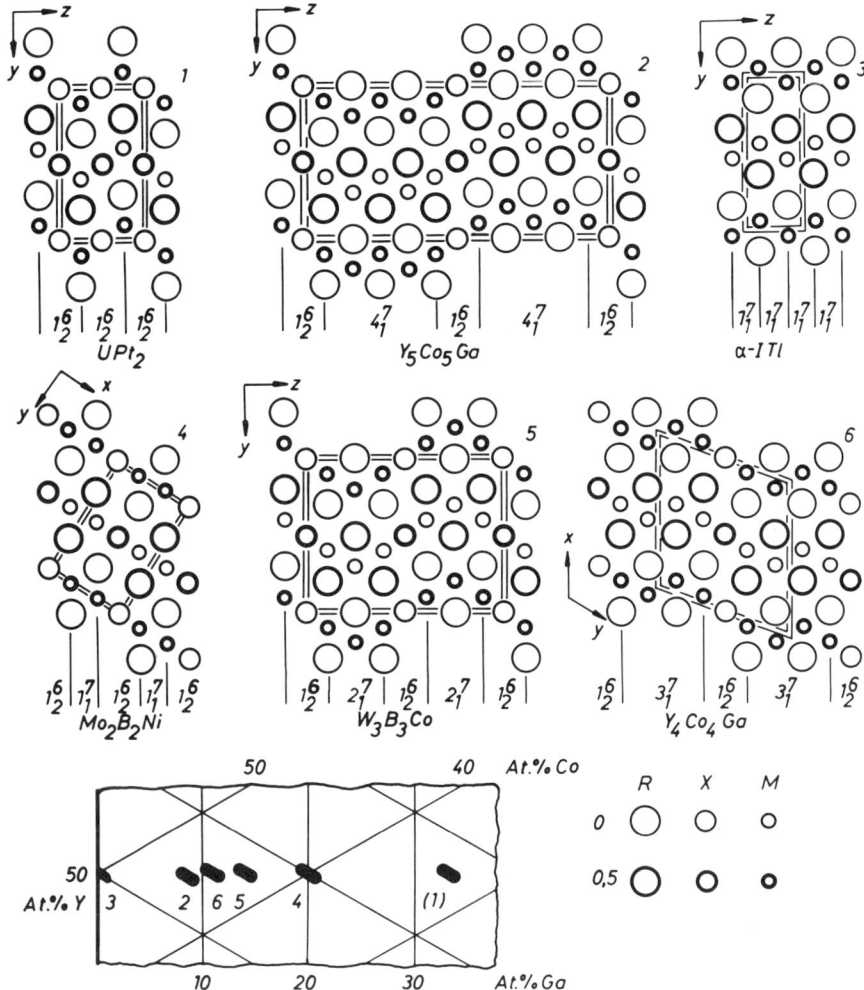

Figure 2. The representatives of the linear inhomogeneous series α-ITl - UPt$_2$ and the homogeneity ranges of the compounds YCo (3), Y$_5$Co$_5$Ga (2), Y$_4$Co$_4$Ga (6), Y$_3$Co$_3$Ga (5), Y$_2$Co$_2$Ga (4) and YCoGa (1) with corresponding structure types in the ternary system Y - Co - Ga.

3.4.2. *The linear series with* $\Omega = 2$. The homogeneous series with $\Omega = 2$ were described mainly by Kripyakevich et al.(1978). Several examples of inhomogeneous series, based on segments having interfaces with square-primitive and square-centred meshes, can be found in the papers mentioned in the introduction.

The inhomogeneous series AuCu-I - AuCu$_3$, based on close packed segments with rectangular primitive interface meshes, is presented in Figure 3. The structure types Pd$_5$Ga$_2$, Rh$_5$Ge$_3$, Pt$_2$Ga and Cl$_2$Pb belong to this series. Their representatives can be found among the noble metals gallides. The real and possible structures of this series may have one of 13 space groups. In addition to the five linear one-coloured groups, mentioned in the Table 1, two-coloured ones have to be considered for the derivation of the space groups (Table 2).

TABLE 2. Symmetry, numeric codes and known representatives in the series AuCu-I - AuCu$_3$.

Linear	Space group	Numeric code		Structure type
p111	Pm	$2_1 6 2_2 4 1_1 6 1_2 4$		
p112	P2/m	$1_1 6 1_2 4 3_1 6 1_2 4$		
p1g1	Pmc2$_1$	$(1_1 6 2_2 4 1_1 6 1_2 4)_2$		
pm11	Pmm2	$1_1 7 2_2 4 1_1 7 4_2 4$		
pmg2	Pmma	$(1_1 6 2_2 4)_2$	(AuCu^6	2AuCu$_3^4$)$_2$ Pt$_2$Ga
p'111	Cm	$(2_1 6 2_2 4 1_1 7 1_2 4)_2$		
p112'	P2$_1$/m	$1_1 7 1_2 4 3_1 7 1_2 4$		
p'112(2')	C2/m	$(1_1 6 1_2 4 1_1 7 1_2 4)_2$		
p1g'1	Pmn2$_1$	$(1_1 6 2_2 4 1_1 6 1_2 4)_2$		
pmg'2'	Pmmn	$(1_1 7 1_2 4)_2$		
p'1g1	Pbam	$(1_1 6 1_2 4)_2$	(AuCu^6 | AuCu$_3^4$)$_2$	Rh$_5$Ge$_3$
p'1g'1	Pnma	$(1_1 7 1_2 4)_2$	(AuCu^7 | AuCu$_3^4$)$_2$	Cl$_2$Pb
		$(1_1 7 3_2 4)_2$	(AuCu^7 | 3AuCu$_3^4$)$_2$	Pd$_5$Ga$_2$
p'm11	Amm2	$(2_1 6 1_2 4)_2$		

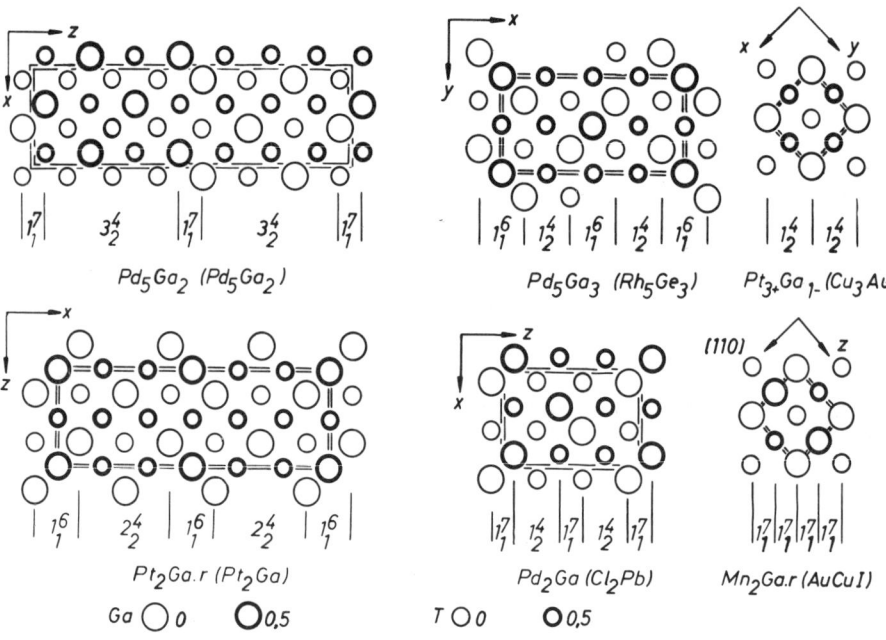

Figure 3. The binary transition metal gallides with structures belonging to the inhomogeneous structure series AuCu-I - AuCu$_3$.

The same space groups are possible for the intergrowth structures of the series BaAl$_4$ - YIrGe$_2$. The only one representative of this series is Y$_3$Pt$_4$Ge$_6$ (numeric code $1_1 7 1_2 4$) which was found by Venturini & Malaman (1990).

Another example of a series with $\Omega = 2$ are the classical "chimney-ladder" structures, which can be described by the overall formula $M_{p+2q+3r}X_{2p+2q+4r}$. For their construction it is necessary to select only three simple segments, one with composition MX_2 from the TiSi$_2$ structures type and two hypothetical slabs of composition M_2X_2 and M_3X_4. All 10 known "chimney-ladder" structures can be constructed by means of these three segments. The possible space groups are : P2, B2 (in two settings), P2/b, B2/b, Pcc2, Pnn2, P222$_1$, C222, I2$_1$2$_1$2$_1$, F222, Pcca, Pnna, Fddd, P-4, I-4, I4$_1$, I4$_1$/a, P-4n2, P-4c2, I-42d and I4$_1$22. Structure proposals for the phases Tc$_4$Si$_7$ and Mo$_9$Ge$_{16}$ have been made on the basis of these considerations.

3.4.3. *The linear series with $\Omega = 3$*. The known series with three levels in the zigzag line representation ($\Omega = 3$) are usually hexagonal or trigonal, based on segments with triangular Kagome mesh interfaces. According to the possible zigzag line symmetries the intergrowth structures can have one of seven possible space groups : P3m1, P-3m1, P-6m2, P6$_3$mc, P6$_3$/mmc, R3m and R-3m.

The series with the biggest number of known binary and ternary members consists of an intergrowth of Laves phase and CaCu$_5$ type slabs (Parthé et al., 1985). Ternary structure types which belong here are : Ce$_3$Co$_8$Si, Dy$_3$Ni$_7$B$_2$, YRh$_2$Si, Ce$_2$Co$_5$B$_2$ and Y$_4$Rh$_9$Si$_5$.

The latest complex series of this kind to be found is the Th$_2$Zn$_{17}$ - HoNi$_{2.4}$Ga$_{2.6}$ series with overall formula $R_{2m+3n}X_{17m+15n}$. The only known structure type of this series is Sm$_{15}$Ni$_{52}$Ga$_{44}$ (m = 3, n = 3).

A new example of a series with $\Omega = 3$ is the AlB$_2$ - RX$_5$ series with overall formula $R_{4m+4n}X_{8m+20n}$. As parent structures serve the AlB$_2$ and the defect CeRe$_4$Si$_2$ or PuGa$_6$ structure types. For the construction of the intergrowth structures in this series one uses three kinds of segments with the same square primitive mesh interface : two slabs from the AlB$_2$ type with the trigonal prisms in two settings and a third slab consisting of filled and empty tetragonal antiprisms and empty cubes which can be taken from the second parent structure. The compositions of the known representatives of the AlB$_2$ - RX$_5$ series are La$_2$(Ge,Al)$_7$ (m = 1, n = 1), Ce$_2$(Ge,Ga)$_7$ (m = 2, n = 2) and La$_4$(Ge,Ga)$_{11}$ (m = 6, n = 2). In Figure 4 is shown, for the known intergrowth structures, the packing of the polyhedra which represent the segments. In spite of the similarity of the selected segments, a larger set of space groups must be used to describe the symmetry of all possible members of this structure series : P1, P2$_1$, C2, Pc, Cm, P2$_1$/m, P2$_1$/c, C2/m, C222$_1$, Cmm2, Cmc2$_1$, Amm2, Abm2, Fmm2, Cmmm, Cmma, Cmcm, Cmca, Fmmm, P-42m, P42nm, P4$_2$22, P4$_2$/mnm and P4$_2$/nnm. The reason for such a diversity of possible space groups is that the segments can be in three different stacking positions along two normal directions. As a consequence nine topologically different positions are available for the positioning of the segments in this series.

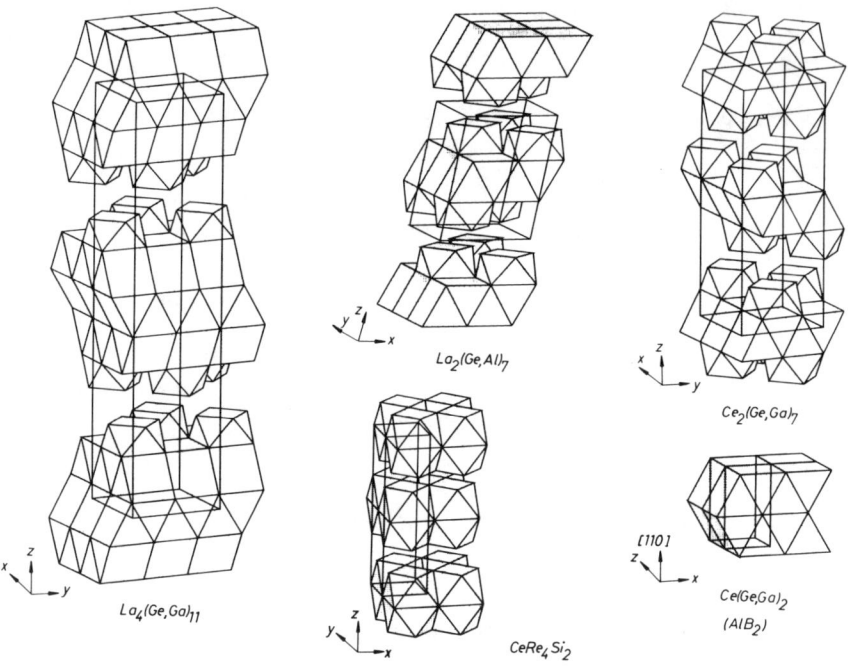

Figure 4. The packing of the different slabs in the structures of the AlB_2 - RX_5 structure series.

3.5. THE CHEMICAL COMPOSITION OF THE SEGMENTS AND THE LOCATION OF THE INTERGROWTH STRUCTURES IN THE PHASE DIAGRAMS

The compounds with linear intergrowth structures are often found in the ternary systems R - T - M, where R is usually a rare earth, alkaline earth, transuranium or transition element of the IV group, T is a late transition element and M is a main group element (B, Al, Ga, Si, Ge etc.). In the case of borides and silicides - B and Si have a relatively high electronegativity value - one often finds an ordered occupation of the different positions in the segments by T and M elements. For example, the centers of the trigonal prisms in the segments cut from the AlB_2 structure type are occupied in these compounds by main group elements and in the centres of the tetragonal antiprisms in the $BaAl_4$ structure segments are located the T atoms. In this case, the ternary compounds with linear intergrowth structures are found in the ternary phase diagrams on the tie lines joining the compounds having the parent structure types (Parthé et al., 1985). For such series the overall formula can be written as a three component formula (see 3.1.).

In the case of the less electronegative gallium the occupation of segment sites is often only partially ordered. The trigonal prismatic sites in the segments are still occupied by the main group elements, but the tetragonal-antiprismatic coordinated sites may be occupied by M as well as T elements. Due to this variation in occupation, the compounds with intergrowth structures can be observed in different regions of the ternary R - T - Ga phase diagrams. If segments of the $BaAl_4$ structure type are not present, the intergrowth structures can be located, like in boron and silicon systems, on the tie lines (for example, see Figure 2 with the compositions of the members of the α-ITl - UPt_2 series structures in the Y - Co - Ga system). If other segments are present then the parent structure of the $BaAl_4$ type has, as mentioned above, a wide homogeneity range. The transition elements can occupy both possible sites (with tetragonal-antiprismatic and cuboctahedral coordination) depending on the T : M ratio (Figure 5, compound 6). In the intergrowth structures of the series $BaAl_4$ - CaF_2, occuring in the Ce - Ni - Ga system (Figure 5, compounds 1, 3, 4, 5), the T atoms seldom occupy the tetragonal-antiprismatic coordinated sites. These sites are mainly occupied by Ga. The T atoms move to the sites with cubic coordination (CaF_2 type segments). As a result the compounds with intergrowth structures are not located on the tie lines only, but may be found on a wide band between the compositions of the parent structures (Figure 5).

4. The two-dimensional inhomogeneous structure series

The formula $R_{n2+3n+2}T_{n2-n+2}M_{n2+n}$ has been assigned to the first described two-dimensional structure series. Its parent structures are the AlB_2, NiAs and α-Fe types and its members the $Ce_6Ni_2Si_3$, $Ce_5Ni_2Si_3$ and $Pr_{15}Ni_7Si_{10}$ types. The simplest representative of this series, Er_3Ru_2, has recently been found by Fornasini & Palenzona (1990).

The next example of a two-dimensional series is the $BaAl_4$ - $AuCu_3$ series which has the overall formula $R_{m+n}X_{4m+3n}$. The only one intergrowth structure now known is La_3Al_{11} (m = 2, n = 1). The representatives of all three structure types can be found in the chemically analogous systems R - T - Ga. Compounds with light rare earth elements form the $BaAl_4$ structure, those with heavy rare earth elements the La_3Al_{11} and $AuCu_3$ structures.

TABLE 3. The structures of the two-dimensional structure series AlB_2 - α-Fe with general formula $R_{m+0.5n}X_n$.

Structure type	m	n
Eu_3Ga_5	2	20
La_3Ge_2Ni	6	12
o-Ni_4B_3	10	12
$Y_{10}Ga_3Co_7$	10	20
$Ce_{14}Ni_8Si_9$	11	34
$Ru_{11}B_8$	14	16
$Ce_7Ni_2Si_5$	14	28

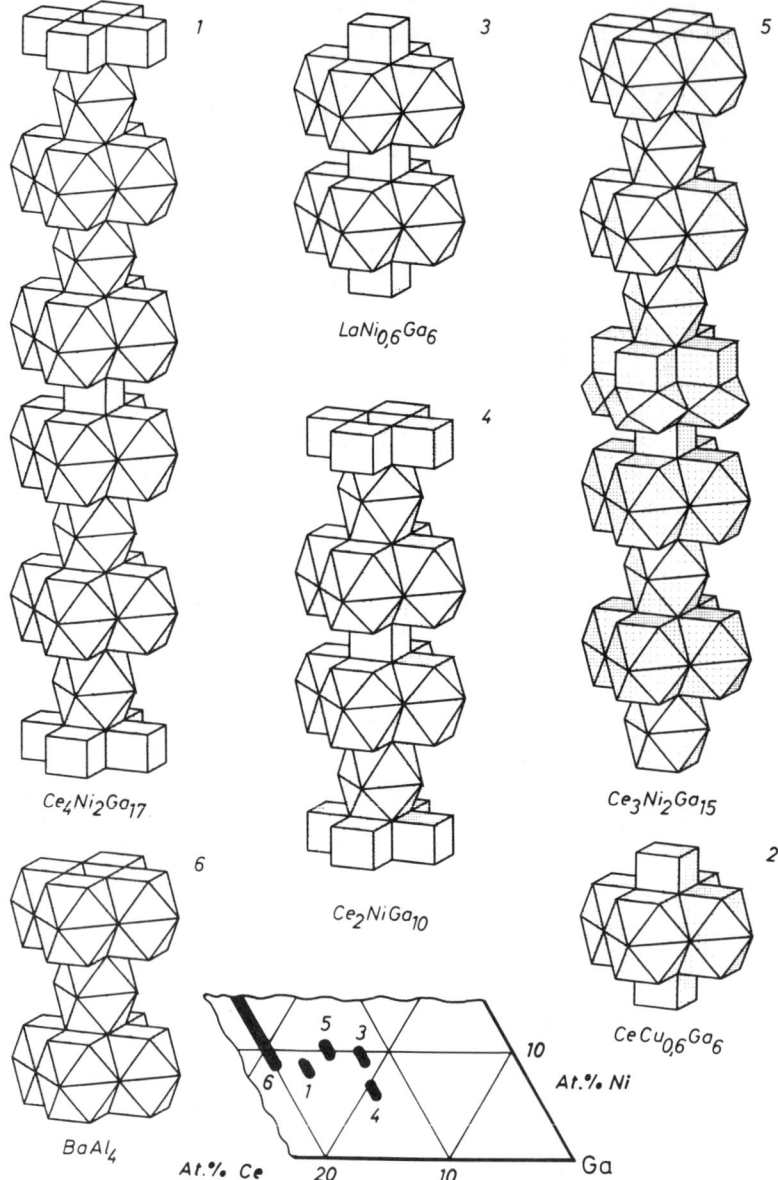

Figure 5. The structures of the linear inhomogeneous series $BaAl_4$ - CaF_2 (1 - 4) and $BaAl_4$ - CaF_2 - Cu (5) together with the parent structure $BaAl_4$ and the location of the corresponding ternary compounds in the ternary system Ce - Ni - Ga.

Several inhomogeneous two-dimensional series occur with the structure types characterized by the trigonal prismatic coordination of the X atoms. The parent structures are AlB_2 and α-Fe. The overall formula of these series is $R_{m+0.5n}X_n$. One example is the series containing two-layered structures with non-coplanar axes of the trigonal prisms. The known intergrowth structures are presented in Table 3.

5. Three-dimensional inhomogeneous structure series

The construction of homogeneous three-dimensional (as well as two-dimensional) intergrowth structures seems to be impossible. At least, there are no structure examples known which can be interpreted in such a way. However, inhomogeneous intergrowth structures exist. The interpretation of the structures as three-dimensional packing of blocks is closely related to the concept of concentric polyhedra packing in cubic structures with large unit cells (Chabot et al., 1981; Hellner & Koch, 1981; Nyman & Hyde, 1981). For example, the Th_6Mn_{23} structure type (or its ternary derivatives) can be interpreted as being constructed from finite structure fragments of the Ca_3Ag_8 and the α-Fe types (Parthé et al., 1985).

Several complex structures of the intermetallic compounds can be interpreted as a three-dimensional packings of finite AlB_2 and α-Fe fragments of different shapes (Y_8Co_5, Er_3Ni_2, $Ce_{24}Co_{11}$, $Ce_{16}Ru_9$).

In this context the intergrowth concept connects to other interpretation models of inorganic structures. For example, if the requirement iii) (see 2.2.) is relaxed (that means that the structural fragments do not necessarily have common interface planes) and if considerations of valence states are included one can see here a connection with the very useful metal cluster model (Simon, 1981).

Acknowledgement
The author is grateful to the Alexander-von-Humboldt-Stiftung for the science fellowship.

6. References

Chabot, B.A., Cenzual, K. & Parthé, E. (1981). 'Nested polyhedra units : a geometrical concept for describing complicated cubic structures.' Acta Cryst. **A37**, 6-11.

Fornasini, M.L. & Palenzona, A. (1990). 'The crystal structure of Er_3Ru_2.' Z. Kristallogr. **192**, 249-254.

Grin', J.N. (1986). 'Ein Aufbaumodell für "Chimney-Ladder"-Strukturen.' Monatsh. Chem. **117**, 921-932.

Grin', Yu.N. & Akselrud, L.G. (1990). 'Use of inhomogeneous linear structure series for the crystal-chemical description of the superconducting oxides.' Acta Cryst. **A46** Suppl., C-338.

Grin', Yu.N., Yarmolyuk, Ya.P. & Gladyshevskii, E.I. (1982). 'The crystal chemistry of series of inhomogeneous linear structures. I. Symmetry and numeric symbols of the structures composed of fragments of the structure types $BaAl_4$, CaF_2, AlB_2, $AuCu_3$, Cu, α-Fe and α-Po.' Sov. Phys. Crystallogr. **27**, 413-417.

Hellner, E. & Koch, E. (1981). 'Cluster or framework considerations for the structures of Tl_7Sb_2, α-Mn, Cu_5Zn_8 and their variants $Li_{22}Si_5$, $Cu_{41}Sn_{11}$, $Sm_{11}Cd_{45}$, Mg_6Pd and Na_6Tl with octuple unit cells.' Acta Cryst. **A37**, 1-6.

Kripyakevich, P.I. (1976a). 'Structures of the AlB_2 - α-$ThSi_2$ series, their symmetry and relationship to the other linear structures.' Sov. Phys. Crystallogr. **21**, 273-276.

Kripyakevich, P.I. (1976b). 'Application of line symmetry groups to the description of homogeneous linear structures.' Sov. Phys. Crystallogr. **21**, 696-704.

Kripyakevich, P.I. & Gladyshevskii, E.I. (1972). 'Homologous series including the new structure types of ternary silicides.' Acta Cryst. **A28** Suppl., S97.

Kripyakevich, P.I. & Grin', Yu.N. (1978). 'On the systematization of series of homogeneous linear structures.' Sov. Phys. Crystallogr. **23**, 45-49.

Kripyakevich, P.I. & Grin', Yu.N. (1979). 'The series of the degenerate homogeneous linear structures.' Sov. Phys. Crystallogr. **24**, 41-44.

Kuzma, Yu.B. & Bilonizhko, N.S. (1974). 'New boride structural types in the homologous series based on the $CaCu_5$ and $CeCo_3B_2$ types.' Sov. Phys. Crystallogr. **18**, 447-449.

Lima-de-Faria, J., Hellner, E., Liebau, F., Makovicki, E. & Parthé, E. (1990). 'Nomenclature of inorganic structure types. Report of the International Union of Crystallography Commission on Crystallographic Nomenclature subcommittee on the nomenclature of inorganic structure types.' Acta Cryst. **A46**, 1-11.

Nyman, H. & Hyde, B.G. (1981). 'The related structures of α-Mn, sodalite, Sb_2Tl_7, etc.' Acta Cryst. *A37*, 11-17.

Pani, M. & Fornasini, M.L. (1990). 'Examples of linear structures of intermetallic compounds described as intergrowth of segments of simple basic structures.' Z. Kristallogr. **190**, 127-133.

Parthé, E., Chabot, B.A. & Cenzual, K. (1985). 'Complex structures of intermetallic compounds interpreted as intergrowth of segments of simple structures.' Chimia **39**, 164-174.

Simon, A. (1981). 'Condensed metal clusters.' Angew. Chemie (Int.Edit.Engl.) **20**, 1-22.

Venturini, G. & Malaman, B. (1990). 'Crystal structure of $Y_3Pt_4Ge_6$: an intergrowth of $BaAl_4$ and $YIrGe_2$ slabs.' J. Less-Common Met. **167**, 45-52.

7. Appendix : Short review of linear symmetry groups

The linear symmetry groups (symmetry groups of bands, frieze groups, border groups) describe the symmetry of two-dimensional objects which are infinite only along one translation direction (the $G_1{}^2$ groups). The known example of such objects are friezes, used by people for decorating buildings. Due to the above mentioned restreints, there are only three symmetry operators possible in these groups (in addition to the translation itself) : the mirror plane m (along and normal to the translation direction), glide plane g (along the translation direction) and two-fold symmetry axis (normal to the frieze plane). There are only seven combinations of these symmetry operators; these are the 7 linear symmetry groups : p111, p112, pm11, p1g1, pmg2, p1m1 and pmm2. As in the space symmetry groups, the notation of linear groups contains four positions. On the first is written p (primitive) as translation lattice symbol, on the second is indicated the symmetry operator normal to the translation direction (x-axis), on the third the symmetry operator parallel to it (y-axis) and on the fourth the symmetry operator normal to the frieze plane (z-axis).

Among the seven linear groups only the first five in the above list are suitable for describing the symmetry of the zigzag lines.

If the symmetry operators change the geometric coordinates and the colour of the frieze parts, the two coloured linear symmetry groups are useful for describing such objects. The colour changing operators are notated in the linear symmetry group symbol by means of an apostrophe '. In the case of there being only two colours there are 31

possible combinations of symmetry operators (black-white or antisymmetry linear groups).
More information about these groups can be found in :
> Shubnikov, A.V. & Koptsik, V.A. (1974). "Symmetry in Science and Art". New York : Plenum Press.
> Shubnikov, A.V. & Belov, N.V. (1964). "Coloured Symmetry". New York : Pergamon Press.

8. Problems

Problem 1 : In Figure 6 are shown 14 decorative drawings ("friezes") obtained by the repetition in one direction of a simple figure. Indicate in each drawing the symmetry elements m (mirror line), g (glide reflection line) and/or 2 (two fold axis). Have each drawing on the right hand side correspond to the one on the left side which has the same symmetry. Write beside the number the letter of the corresponding drawing.

Figure 6 : Fourteen friezes obtained by the repetition in one dimension of a simple figure.

Problem 2 : In Figure 7 are presented the trigonal prisms packings in the 13 representatives of the linear homogeneous structure series α-ITl - FeB :
  1 α-ITl   2 $CaCu_{0.8}Ag_{0.2}$ 3 TbNi HT   4 SrAg
  5 $Gd_{0.7}Y_{0.3}Ni$ 6 $CaCu_{0.975}Ga_{0.025}$ 7 $Gd_{0.75}Y_{0.25}Ni$ 8 $Gd_{0.4}Tb_{0.6}Ni$
  9 $Gd_{0.55}Dy_{0.45}Ni$ 10 TbNi LT   11 β-CaCu   12 α-CaCu
  13 FeB.

All prism centres in the real structures are occupied by the smaller atoms (not shown in Figure 7).

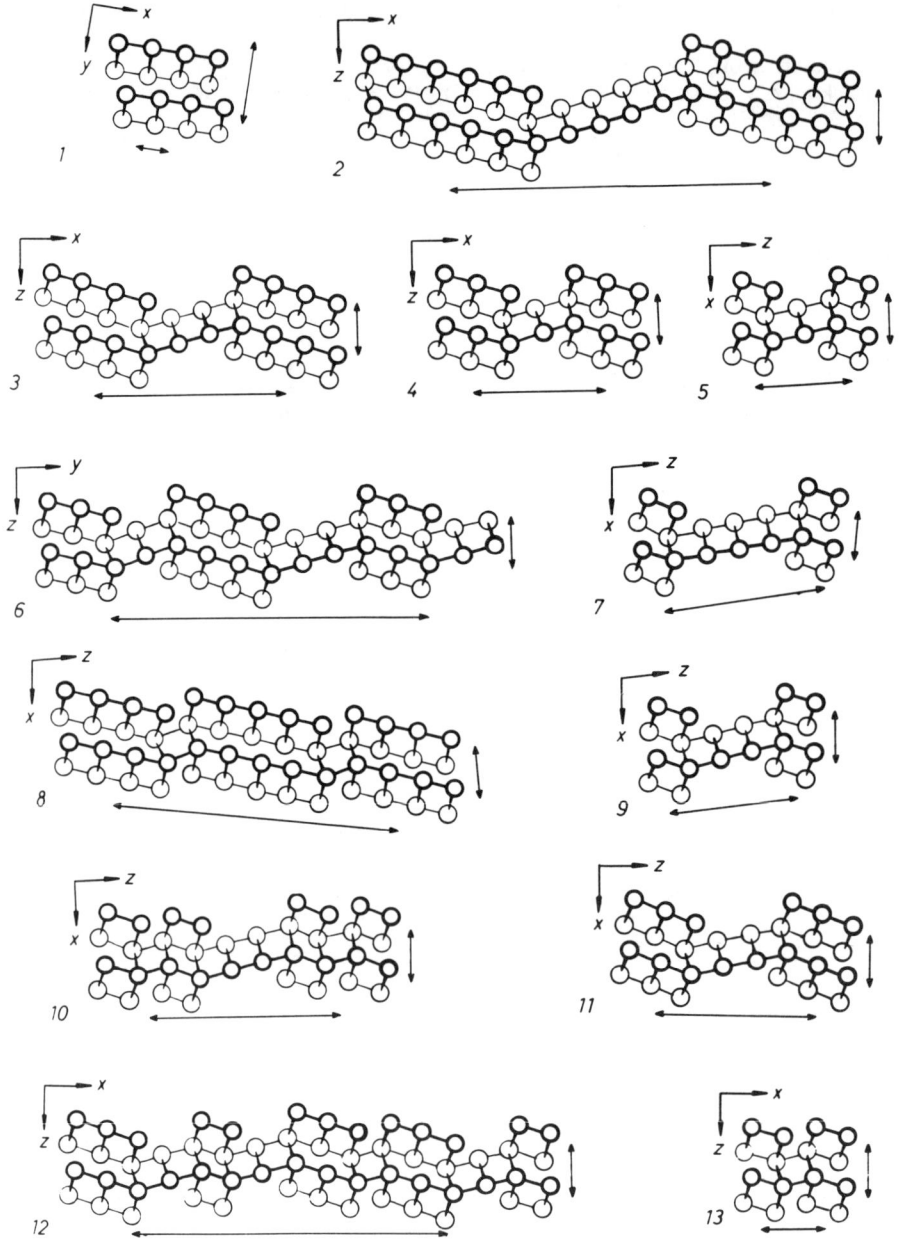

Figure 7 : Thirteen representatives of the linear homogeneous structure series α-ITl - FeB.

Draw the zigzag lines of the two kinds described in sections 3.2. and 3.4., find the numeric codes and determine the linear symmetry group for each structure. Give the space group for each structure considering the following correspondence between the linear symmetry group and the three-dimensional space group : p111 → Pm, p1g1 → Pmc2_1, p112 → P2_1/m, pm11 → Pmn2_1, pmg2 → Pnma.

Problem 3 : The structures BaAl$_4$, AlB$_2$ and RX$_4$ (a binary variant of the CeMg$_2$Si$_2$ structure type), shown in Figure 8, are the parent structures of the inhomogeneous linear structure series BaAl$_4$ - AlB$_2$. Find the initial segments necessary for the description of this series using the intergrowth concept.

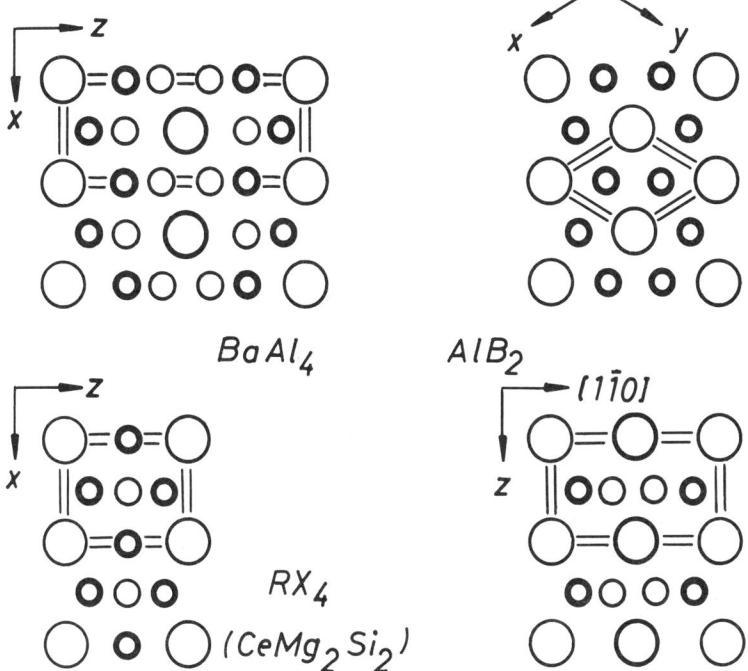

Figure 8 : Projections of the BaAl$_4$, AlB$_2$ and RX$_4$ structures.

8.1. SOLUTION OF THE PROBLEMS

Problem 1 : The symmetry corresponding to linear group p111 is found with friezes 1 and g, p112 with 2 and e, pm11 with 3 and f, p1g1 with 4 and a, pmg2 with 5 and c, p1m1 with 6 and b, pmm2 with 7 and d.

Problem 2 : The one kind of zigzag line, corresponding to the Zhdanov notation, can be found by connecting the atoms at the top of the trigonal prisms at the same height (different heights are distinguished by different line thicknesses). The other kind of zigzag line, which corresponds to the Jagodzinski notation, can be obtained by connecting the mid-points between the atoms at same height. This zigzag line can have a different number of breaks points.

TABLE 4 : Numeric codes, linear symmetry groups and space groups for the known representatives of the linear homogeneous structure series α-ITl - FeB.

	Structure type	Numeric code	Linear group	Space group
1	α-ITl	∞	pmm2	Cmcm
2	$CaCu_{0.8}Ag_{0.2}$	$(5)_2$	pmg2	Pnma
3	TbNi HT	$(3)_2$	pmg2	Pnma
4	SrAg	$(2)_2$	pmg2	Pnma
5	$Gd_{0.7}Y_{0.3}Ni$	13	p112	$P2_1/m$
6	$CaCu_{0.975}Ga_{0.025}$	3223	pm11	$Pmn2_1$
7	$Gd_{0.75}Y_{0.25}Ni$	14	p112	$P2_1/m$
8	$Gd_{0.4}Tb_{0.6}Ni$	1314	p112	$P2_1/m$
9	$Gd_{0.55}Dy_{0.45}Ni$	13	p112	$P2_1/m$
10	TbNi LT	1113	p112	$P2_1/m$
11	β-CaCu	23	p112	$P2_1/m$
12	α-CaCu	$(122)_2$	p1g1	$Pmc2_1$
13	FeB	$(1)_2$	pmg2	Pnma

Problem 3 : The parent structures $BaAl_4$, AlB_2 and RX_4 have only one common interface with a square primitive mesh. According to 2.2. it is necessary to choose four initial segments having such interface mesh : the slabs of trigonal prisms from the AlB_2 structure in two settings (containing the glide planes with different glide directions and belonging to groups 4 and 5) and two kinds of slabs constructed from tetragonal antiprisms in two different connections containing the glide plane n and the mirror plane m (groups 2 and 3).

INTRODUCTION TO THE ELECTRONIC STRUCTURE OF EXTENDED SYSTEMS

ROALD HOFFMANN
Department of Chemistry
Cornell University
Baker Laboratory
Ithaca, NY 14853-1301
USA

ABSTRACT. The concepts of a band structure, Bloch functions, the wave vector, and densities of states are introduced, making as strong connections as possible to quantum chemical ideas. The number of orbitals in a unit cell determines the number of bonds, orbital overlap is responsible for the dispersion of the bands and the topology of orbital interactions for the way the bands run. A measure of bond strength in the solid, the Crystal Orbital Overlap Population, is introduced.

1. Orbitals and Bands in One Dimension

It's usually easier to work with small, simple things, and one-dimensional infinite systems are particularly easy to visualize (Burdett (1984), Whangbo (1982)). Much of the physics of two- and three-dimensional solids is there in one dimension. Let's begin with a chain of equally spaced H atoms, **1**, or the isomorphic π-system of a non-bond-alternating, delocalized polyene **2**, stretched out for the moment. And we will progress to a stack of Pt(II) square planar complexes, **3**, $Pt(CN)_4^{2-}$ or a model PtH_4^{2-}.

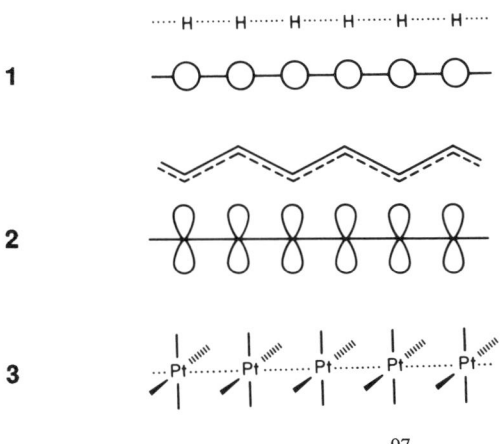

E. Parthé (ed.), *Modern Perspectives in Inorganic Crystal Chemistry*, 97–116.
© 1992 *Kluwer Academic Publishers. Printed in the Netherlands.*

A digression here : every chemist would have an intuitive feeling for what that model chain of hydrogen atoms would do, if we were to release it from the prison of its theoretical construction. At ambient pressure, it would form a chain of hydrogen molecules, **4**.

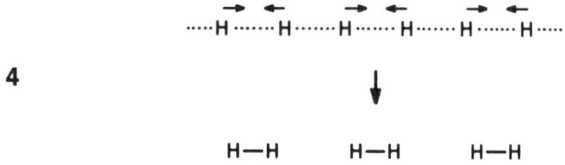

4

This simple bond-forming process would be analyzed by the physicist by calculating a band for the equally spaced polymer, then seeing that it's subject to an instability, called a Peierls distortion. Other words around that characterization would be "strong electron-phonon coupling", "pairing distortion", or "a $2k_F$ instability". And the physicist would come to the conclusion that the initially equally spaced H polymer would form a chain of hydrogen molecules. I mention this thought process here to make the point, which I will do again and again, that the chemist's intuition is really excellent. But we must bring the languages of our sister sciences into correspondence. Incidentally, whether distortion **4** will take place at 3 megabars is not obvious, an open question.

Let's return to our chain of equally spaced H atoms. It turns out to be computationally convenient to think of that chain as an imperceptibly bent segment of a large ring (this is called applying cyclic boundary conditions). The orbitals of medium-sized rings on the way to that very large one are quite well known. They are shown in **5**.

5

For a hydrogen molecule (or ethylene) there is bonding $\sigma_g(\pi)$ below an antibonding $\sigma_u^*(\pi^*)$. For cyclic H_3 or cyclopropenyl we have one orbital below two degenerate ones; for cyclobutadiene the familiar one below two below one, and so on. Except for the lowest (and occasionally the highest) level, the orbitals come in degenerate pairs. The number of nodes increases as one rises in energy. We'd expect the same for an infinite polymer - the lowest level nodeless, the highest with the maximum number of nodes. In between the levels should come in pairs, with a growing number of nodes. The chemist's representation of the band for the polymer is given at right in **5**.

2. Bloch Functions, k, Band Structures

There is a better way to write out all these orbitals, making use of the translational symmetry. If we have a lattice whose points are labelled by an index n = 0, 1, 2, 3, 4 as shown in **6**, and if on each lattice point there is a basis function (a H 1s orbital), χ_0, χ_1, χ_2 etc., then the appropriate symmetry adapted linear combinations (remember translation is just as good a symmetry operation as any other one we know) are given in **6**.

6

$$\psi_k = \sum_n e^{ikna} \chi_n$$

Here a is the lattice spacing, the unit cell in one dimension, and k is an index which labels which irreducible representation of the translation group Ψ transforms as. We will see in a moment that k is much more, but for now, k is just an index for an irreducible representation, just like a, e_1, e_2 in C_5 are labels.

The process of symmetry adaptation is called in the solid state physics trade "forming Bloch functions" (Ashcroft & Mermin (1976); Harrison (1980)). To reassure a chemist that one is getting what one expects from **5**, let's see what combinations are generated for two specific values of k, k = 0 and k = π/a. This is carried out in **7**.

7

$$k = 0 \quad \psi_0 = \sum_n e^0 \chi_n = \sum_n \chi_n$$
$$= \chi_0 + \chi_1 + \chi_2 + \chi_3 + \cdots$$

$$k = \frac{\pi}{a} \quad \psi_{\frac{\pi}{a}} = \sum_n e^{\pi in} \chi_n = \sum_n (-1)^n \chi_n$$
$$= \chi_0 - \chi_1 + \chi_2 - \chi_3 + \cdots$$

Referring back to **5**, we see that the wave function corresponding to k = 0 is the most bonding one, the one for k = π/a the top of the band. For other values of k we get a neat description of the other levels in the band. So k counts nodes as well. The larger the absolute value of k, the more nodes one has in the wave function. But one has to be careful - there is a range of k and if one goes outside of it, one doesn't get a new wave function, but repeats an old one. The unique values of k are in the interval -π/a ≤ k < π/a or |k| ≤ π/a. This is called the first Brillouin zone, the range of unique k.

How many values of k are there ? As many as the number of translations in the crystal, or, alternatively, as many as there are microscopic unit cells in the macroscopic crystal. So let us say Avogadro's number, give or take a few. There is an energy level for each value of k (actually a degenerate pair of levels for each pair of positive and negative k values). There is an easily proved theorem that $E(k) = E(-k)$. Most representations of E(k) do not give the redundant E(-k), but plot $E(|k|)$ and label it as E(k)). Also the allowed values of k are equally spaced in the space of k, which is called reciprocal or momentum space. The relationship between $k = 2\pi/\lambda$ and momentum derives from the de Broglie relationship $\lambda = h/p$. Remarkable k is not only a symmetry label and a node counter, but it is also a wave vector, and so measures momentum.

So what a chemist draws as a band in **5**, repeated at left in **8** (and the chemist tires and draws ~35 lines or just a block instead of Avogadro's number), the physicist will alternatively draw as an E(k) vs k diagram at right.

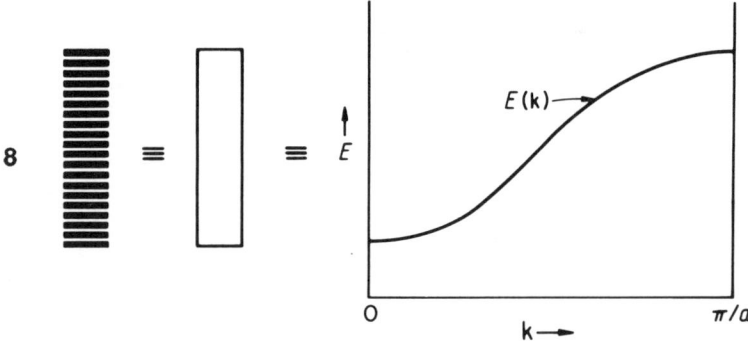

Recall that k is quantized, and there is a finite but large number of levels in the diagram at right. The reason it looks continuous is that this is a fine "dot matrix" printer - there are Avogadro's number of points jammed in there, and so it's no wonder we see a line.

Graphs of E(k) vs k are called band structures. You can be sure that they can be much more complicated than this simple one, but no matter how complicated, they can be understood.

3. Band Width

One very important feature of a band is its **dispersion**, or **band width**, the difference in energy between the highest and lowest levels in the band. What determines the width of bands ? The same thing that determines the splitting of levels in a dimer, ethylene or H_2, namely the overlap between the interacting orbitals (in the polymer the overlap is that between neighboring unit cells). The greater the overlap between neighbors, the greater the band width. Figure 1 illustrates this in detail for a chain of H atoms spaced 3, 2 and 1 Å apart. That the bands extend unsymmetrically around their "origin", the energy of a free H atom at -13.6eV, is a consequence of the inclusion of overlap in the calculations. For two levels, a dimer:

$$E_\pm = \frac{H_{AA} \pm H_{AB}}{1 \pm S_{AB}}$$

The bonding E_+ combination is less stabilized than the antibonding one E_- is destabilized. There are nontrivial consequences in chemistry, for this is the source of 4-electron repulsions and steric effects in one-electron theories. A similar effect is responsible for the bands "spreading up" in Figure 1.

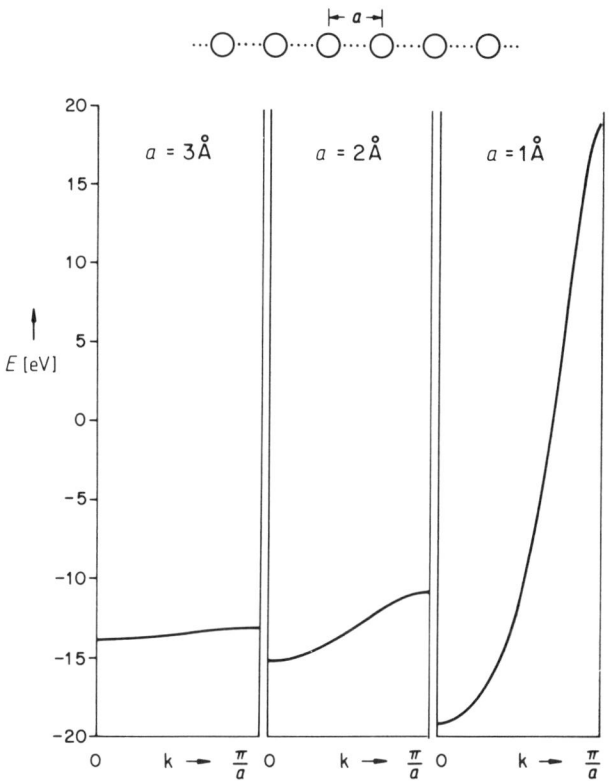

Figure 1. The band structure of a chain of hydrogen atoms spaced 3, 2, and 1 Å apart. The energy of an isolated H atom is -13.6 eV.

4. See How They Run

Another interesting feature of bands is how they "run". The lovely mathematical algorithm 6 applies in general; it does not say anything about the energy of the orbitals at the center of the zone ($k = 0$) relative to those at the edge ($k = \pi/a$). For a chain of H atoms it is clear that $E(k = 0) < E(k = \pi/a)$. But consider a chain of p functions, 9. The same combinations are given to us by the translational symmetry, but now it is clearly $k = 0$ which is high energy, the most antibonding way to put together a chain of p orbitals.

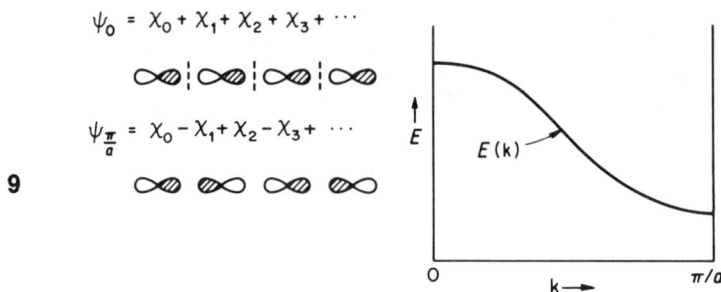

The band of s functions for the hydrogen chain "runs up", the band of p orbitals "runs down" (from zone center to zone edge). In general, it is the topology of orbital interactions which determines which way bands run.

5. An Eclipsed Stack of Pt(II) Square Planar Complexes

Let us test the knowledge we have acquired on an example a little more complicated than a chain of hydrogen atoms. This is an eclipsed stack of square planar d^8 PtL$_4$ complexes, **10**. The normal platinocyanides (e.g. K$_2$Pt(CN)$_4$) indeed show such stacking in the solid state, at the relatively uninteresting Pt⋯Pt separation of 3.3 Å. More exciting are the partially oxidized materials, such as K$_2$Pt(CN)$_4$Cl$_{0.3}$, K$_2$Pt(CN)$_4$(FHF)$_{0.25}$. These are also stacked, but staggered, with a much shorter Pt⋯Pt contact of 2.7 → 3.0 Å. The Pt - Pt distance had been shown to be inversely related to the degree of oxidation of the material (Williams, 1983).

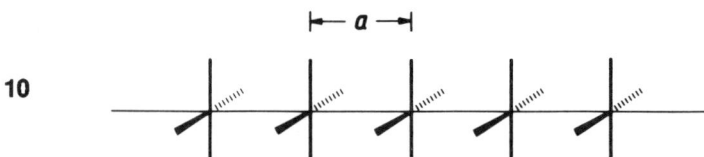

The real test of understanding is prediction. So, let's try to predict the approximate band structure of **10** without a calculation, just using the general principles we have at hand. Let's not worry about the nature of the ligand - it is usually CN$^-$, but since it is only the square planar feature which is likely to be essential, let's imagine a theoretician's generic ligand, H$^-$.

One always begins with the monomer. What are its frontier levels ? The classical crystal field or molecular orbital picture of a square planar complex (Figure 2) leads to a 4 below 1 splitting of the d block. For 16 electrons we have z^2, xz, yz, and xy occupied and $x^2 - y^2$ empty. Competing with the ligand-field-destabilized $x^2 - y^2$ orbital for being the lowest unoccupied molecular orbital (LUMO) of the molecule is the metal z. These two orbitals can be manipulated in understandable ways : π-acceptors push z down, π-donors push it up. Better σ-donors push $x^2 - y^2$ up.

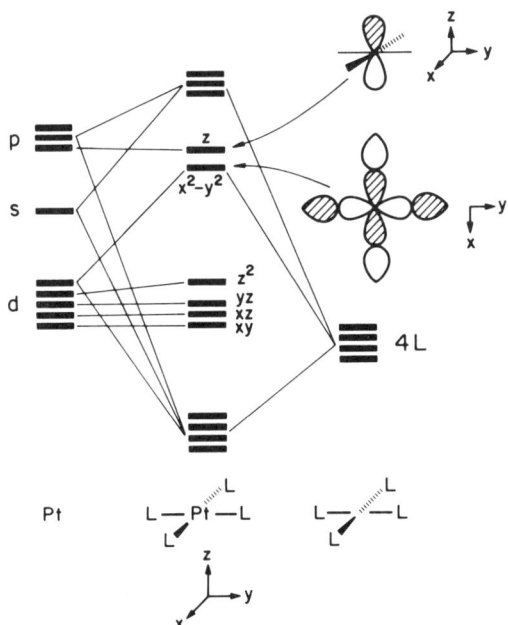

Figure 2. Molecular orbital derivation of the frontier orbitals of a square planar PtL$_4$ complex.

We form the polymer. Each MO of the monomer generates a band. There may (will) be some further symmetry-conditioned mixing between orbitals of the same symmetry in the polymer (e.g. s and z and z^2 are of different symmetry in the monomer, but certain of their polymer MO's are of the same symmetry). But a good start is made by ignoring that secondary mixing, and just developing a band from each monomer level independently.

First a chemist's judgment of the band widths that will develop: the bands that will arise from z^2 and z will be wide, those from xz, yz of medium width, those from x^2-y^2, xy narrow, as shown in **11**.

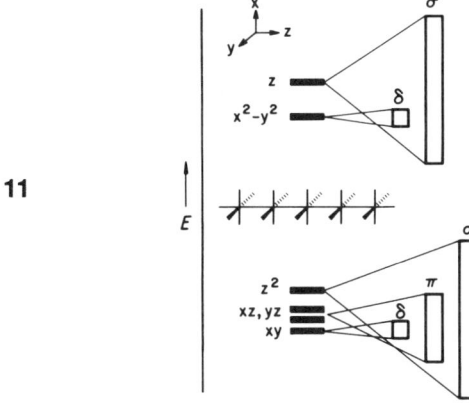

This characterization follows from the realization that the first set of interactions (z, z^2) is σ type, thus has a large overlap between unit cells. The xz, yz set has a medium π overlap, and the xy and $x^2 - y^2$ orbitals (the latter of course has a ligand admixture, but that doesn't change its symmetry) are δ.

It is also easy to see how the bands run. Let's write out the Bloch functions at the zone center (k = 0) and zone edge (k = π/a). Only one of the π and δ functions is represented in **12**.

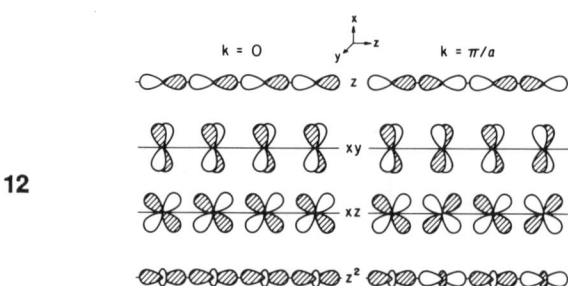

12

The moment one writes these down, one sees that the z^2 and xy bands will run "up" from the zone center (the k = 0 combination is the most bonding) while the z and xz bands will run "down" (the k = 0 combination is the most antibonding).

The predicted band structure, merging considerations of band width and orbital topology, is that of **13**. To make a real estimate one would need an actual calculation of the various overlaps, and these in turn would depend on the Pt⋯Pt separation.

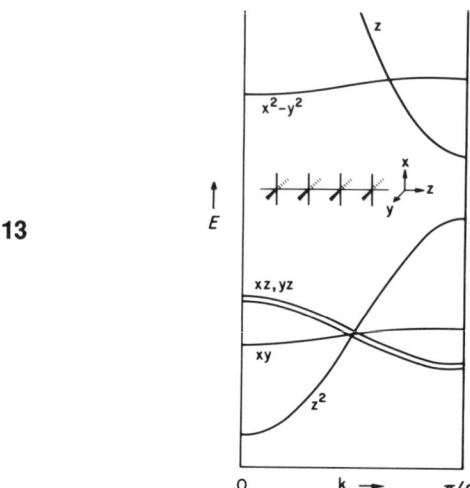

13

The actual band structure, as it emerges from an extended Hückel calculation, matches our expectations very precisely.

6. The Fermi Level

It's important to know how many electrons one has in one's molecule. Fe(II) has a different chemistry from Fe(III), and CR_3^+ carbocations are different from CR_3 radicals and CR_3^- anions. What about the $[PtH_4^{2-}]_\infty$ polymer **11** ? Each monomer is d^8. If there be Avogadro's number of unit cells, there will be Avogadro's number of levels in each bond. And each level has a place for two electrons. So the first four bands are filled, the xy, xz, yz, z^2 bands. The Fermi level, the highest occupied molecular orbital (HOMO), is at the very top of the z^2 band.

Is there a bond between platinums in this $[PtH_4^{2-}]_\infty$ polymer ? We haven't introduced, yet, a formal description of the bonding properties of an orbital or a band, but a glance at **11** will show that the bottom of **each** band, be it made up of z^2, xz, yz or xy, is bonding, and the top antibonding. Filling a band completely, just like filling bonding and antibonding orbitals in a dimer (think of He_2, think of the sequence N_2, O_2, F_2, Ne_2) provides no net bonding. In fact, it gives net antibonding. So why does the unoxidized PtL_4 chain stack ? It could be van der Waals attractions, not in our quantum chemistry at this primitive level. I think there is also a contribution of orbital interaction, i.e. real bonding, involving the mixing of the z^2 and z bands. We will return to this soon.

The band structure gives a ready explanation for why the Pt···Pt separation decreases on oxidation. A typical degree of oxidation is 0.3 electrons per Pt. These electrons must come from the top of the z^2 band. The degree of oxidation specifies that 15% of that band is empty. The states vacated are not innocent of bonding. They are strongly Pt - Pt σ antibonding. So it's no wonder that removing these electrons results in the formation of a partial Pt - Pt bond.

The oxidized material also has its Fermi level in a band, i.e. there is a zero band gap between filled and empty levels. The unoxidized platinocyanides have a substantial gap - they are semiconductors or insulators. The oxidized materials are good low-dimensional conductors, which is a substantial part of what makes them interesting to physicists.

7. Density of States

In the solid, or on a surface, both of which are just very large molecules, one has to deal with a very large number of levels or states. If there are n atomic orbitals (basis functions) in the unit cell, generating n molecular orbitals, and if in our macroscopic crystal there are N unit cells (N is a number that approaches Avogadro's number), then we will have N·n crystal levels. Many of these are occupied and, roughly speaking, they are jammed into the same energy interval in which we find the molecular or unit cell levels. In a discrete molecule we are able to single out one orbital or a small subgroup of orbitals (HOMO, LUMO) as being the frontier, or valence orbitals of the molecules, responsible for its geometry, reactivity, etc. There is no way in the world that a single level among the myriad N·n orbitals of the crystal will have the power to direct a geometry or reactivity.

There is, however, a way to retrieve a frontier orbital language in the solid state. We cannot think about a single level, but perhaps we can talk about bunches of levels. There are many ways to group levels, but one pretty obvious one is to look at all the levels in a given energy interval. The density of states (DOS) is defined as follows :

DOS(E)dE = number of levels between E and E + dE

For a simple band of a chain of hydrogen atoms, the DOS curve takes on the shape of **14**. Note that because the levels are equally spaced along the k axis, and because the E(k) curve, the band structure, has a simple cosine curve shape, there are more states in a given energy interval at the top and bottom of this band. In general, DOS(E) is proportional to the inverse of the slope of E(k) vs k, or to put it into plain English, the flatter the band, the greater the density of states at that energy.

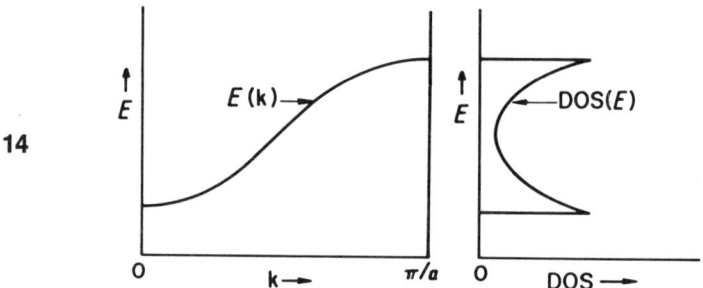

14

The shapes of DOS curves are predictable from the band structures. Figure 3 shows the DOS curve for the PtH_4^{2-} chain. The Figure could have been sketched by hand from the band structure.

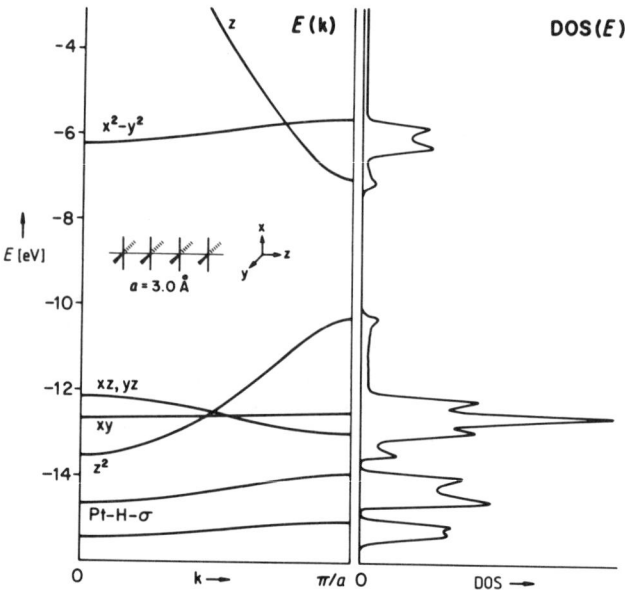

Figure 3. Band structure and density of states for an eclipsed PtH_4^{2-} stack. The DOS curves are broadened so that the two-peaked shape of the xy peak in the DOS is not resolved.

The density of states curve counts levels. The integral of DOS up to the Fermi level is the total number of occupied MO's. Multiplied by two, it's the total number of electrons. So, the DOS curves plot the distribution of electrons in energy.

One important aspect of the DOS curves is that they represent a return from reciprocal space, the space of k, to real space. The DOS is an average over the Brillouin zone, over all k that might give molecular orbitals at the specified energy. The advantage here is largely psychological. If I may be permitted to generalize, I think chemists (with the exception of crystallographers) by and large feel themselves uncomfortable in reciprocal space. They'd rather return to, and think in, real space.

There is another aspect of the return to real space that is significant : *chemists can sketch the DOS of any material, approximately, intuitively.* All that's involved is a knowledge of the atoms, their approximate ionization potentials and electronegativities, and some judgment as to the extent of inter-unit-cell overlap (usually apparent from the structure).

Let's take the PtH_4^{2-} polymer as an example. The monomer units are clearly intact in the polymer. At intermediate monomer-monomer separations (e.g. 3 Å) the major inter-unit-cell overlap is between z^2 and z orbitals. Next is the xz, yz π-type overlap; all other interactions are likely to be small. **15** is a sketch of what we would expect. In **15** I haven't been careful in drawing the integrated areas commensurate to the actual total number of states, nor have I put in the two-peaked nature of the DOS each level generates -- all I want to do is to convey the rough spread of each band. Compare **15** to Figure 3.

15

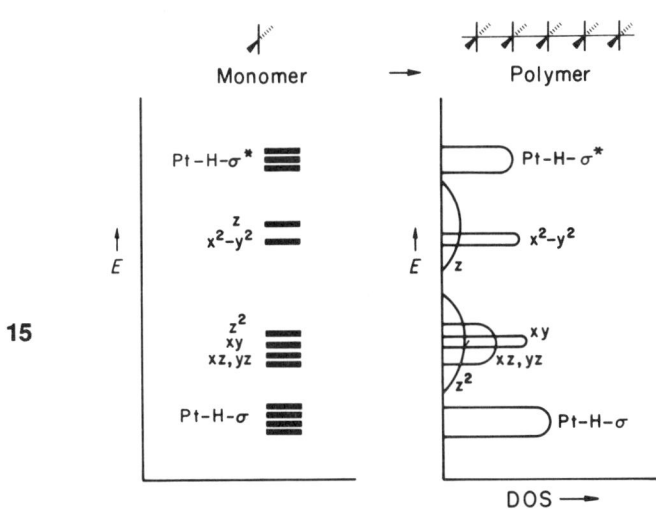

The chemists' ability to write down approximate density of states curves should not be slighted. It gives us tremendous power, and qualitative understanding, an obvious connection to local, chemical viewpoints such as the crystal or ligand field model.

In **15** the qualitative DOS diagrams for PtH_4^{2-}, there is, however, much more than a guess at a DOS. There is a chemical characterization of the localization in real space of the states (are they on Pt, on H ?) and a specification of their bonding properties (are they Pt - H bonding, antibonding, nonbonding ?). The chemist sees right away, or asks -- where in space are the electrons ? Where are the bonds ? There must be a way that these inherently chemical, local questions can be answered, even if the crystal molecular orbitals, the Bloch functions, delocalize the electrons over the entire crystal.

8. Where Are the Electrons ?

One of the interesting tensions in chemistry is between the desire to assign electrons to specific centers, deriving from an atomic, electrostatic view of atoms in a molecule, and the knowledge that electrons are not as localized as we would like them to be. Let's take a two-center molecular orbital

$$\Psi = c_1\chi_1 + c_2\chi_2$$

where χ_1 is on center 1, χ_2 on center 2 and let's assume centers 1 and 2 are not identical, and that χ_1 and χ_2 are normalized, but not orthogonal.

The distribution of an electron in this MO is given by Ψ^2. Ψ should be normalized, so

$$1 = \int |\Psi|^2 d\tau = \int |c_1\chi_1 + c_2\chi_2|^2 d\tau = c_1^2 + c_2^2 + 2c_1 S_{12}$$

where S_{12} is the overlap integral between χ_1 and χ_2. This is how one electron in Ψ is distributed. Now it's obvious that c_1^2 of it is to be assigned to center 1, c_2^2 to center 2. $2c_1c_2S_{12}$ is clearly a quantity that is associated with interaction. It's called the overlap population, and we will soon relate it to the bond order. But what are we to do if we persist in wanting to divide up the electron density between centers 1 and 2 ? We want all the parts to add up to 1 and $c_1^2 + c_2^2$ won't do. We must assign, somehow, the "overlap density" $2c_1c_2S_{12}$ to the two centers. Mulliken suggested (and that's why we call this a Mulliken population analysis, Mulliken (1955)) a democratic solution, splitting $2c_1c_2S_{12}$ equally between centers 1 and 2. Thus center 1 is assigned $c_1^2 + c_1c_2S_{12}$, center 2 $c_2^2 + c_1c_2S_{12}$ and the sum is guaranteed to add up to 1. It should be realized that the Mulliken prescription for partitioning the overlap density, while uniquely defined, is quite arbitrary.

What a computer does is just a little more involved, for it sums these contributions for each atomic orbital on a given center (there are several), over each occupied MO (there may be many). And in the crystal, it does that sum for several k points in the Brillouin zone, and then returns to real space by averaging over these. The net result is a partitioning of the total DOS into contributions to it by either atoms or orbitals. We have also found very useful a decompostion of the DOS into contributions of fragment molecular orbitals (FMO's) of MO's of specified molecular fragments of the composite molecule. In the solid state trade these are often called "projections of the DOS" or "local DOS". Whatever they're called, they divide up the DOS among the atoms. The integral of these projections up to the Fermi level then gives the total electron density on a given atom or in a specific orbital. Then, by reference to some standard density, a charge can be assigned.

Figure 4 give the partitioning of the electron density between Pt and H in the PtH_4^{2-} stack. Everything is as **15** predicts, as the chemist knows it should be - the lower orbitals are localized in the more electronegative ligands (H), the higher ones on the metal.

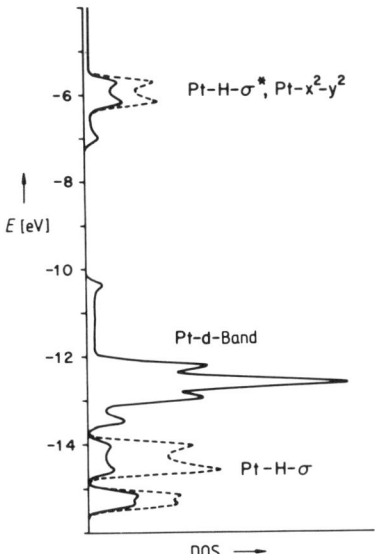

Figure 4. The solid line is the Pt contribution to the total DOS (dashed line) of an eclipsed PtH_4^{2-} stack. What is not on Pt is on the four H's.

9. Where Are the Bonds ?

Local bonding considerations (see **15**) trivially lead us to assign bonding characteristics to certain orbitals and, therefore, bands. There must be a way to find these bonds in the bands that a fully delocalized calculation gives.

It's possible to extend the idea of an overlap population to a crystal. Recall that in the integration of Ψ^2 for a two-center orbital, $2c_1c_2S_{12}$ was a characteristic of bonding. If the overlap integral is taken as positive (and it can always be arranged so) then this quantity scales as we expects of a bond order : it is positive (bonding) if c_1 and c_2 are of the same sign, and negative if c_1 and c_2 are of opposite sign. And the magnitude of the "Mulliken overlap population", for that is what $2c_1c_2S_{12}$ (summed over all orbitals on the two atoms, over all occupied MO's) is called, depends on c_i, c_j, S_{ij}.

A bond indicator is thus easily constructed for the solid. An obvious procedure is to take all the states in a certain energy interval and interrogate then as to their bonding proclivities, measured by the Mulliken overlap population, $2c_ic_jS_{ij}$. What we are defining is an overlap-population-weighted density of states. The beginning of the obvious acronym (OPWDOS) unfortunately has been preempted by another common usage in solid state physics. For that reason, we have called this quantity COOP, for **crystal orbital overlap population** (COOP was introduced for extended systems by Hughbanks & Hoffmann (1983); Wijeyesekera & Hoffmann (1984); Kertesz & Hoffmann (1984)). It's also nice to think of the suggestion of orbitals working together to make bonds in the crystal, so the word is pronounced "co-op".

To get a feeling for this quantity, let's think what a COOP curve for a hydrogen chain looks like. The simple band structure and DOS were given earlier, **14**; they are repeated with the COOP curve in **16**.

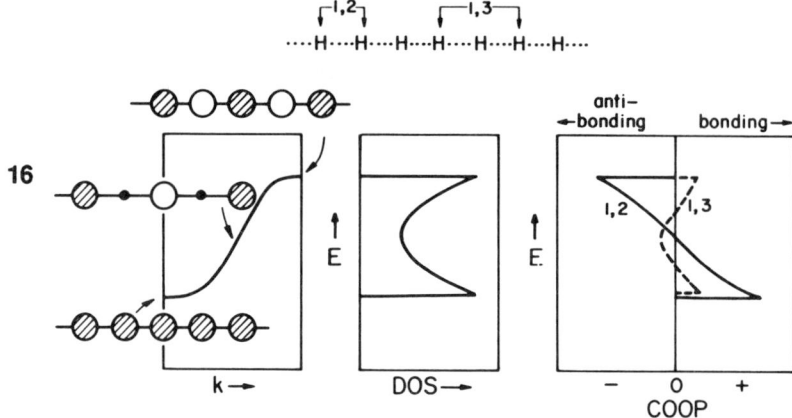

To calculate a COOP curve, one has to specify a bond. Let's take the nearest neighbor 1,2 interaction. The bottom of the band is 1,2 bonding, the middle nonbonding, the top antibonding. The COOP curve obviously has the shape shown at right in **16**. But not all COOP curves look that way. If we specify the 1,3 next-nearest-neighbor bond (silly for a linear chain, not so silly if the chain is kinked), then the bottom **and** the top of the band are 1,3 bonding, the middle anti-bonding. That curve, the dashed line in the drawing **16**, is different in shape. And, of course, its bonding and antibonding amplitude is much smaller because of the rapid decrease of S_{ij} with distance.

Note the general characteristics of COOP curves - positive regions which are bonding, negative regions which are antibonding. The amplitudes of these curves depend on the number of states in that energy interval, the magnitude of the coupling overlap, and the size of the coefficients in the MO's.

The integral of the COOP curve up to the Fermi level is the total overlap population of the specified bond. This points us to another way of thinking of the DOS and COOP curves. These are the differential versions of electronic occupation and bond order indices in the crystal. The integral of the DOS to the Fermi level gives the total number of electrons; the integral of the COOP curve gives the total overlap population, which is not identical to the bond order, but which scales like it. It is the closest a theoretician can get to that ill-defined but fantastically useful simple concept of a bond order.

To move to something a little more complicated than the hydrogen or polyene chain, let's examine the COOP curves for the PtH_4^{2-} chain, Figure 5 shows both the Pt - H and Pt - Pt COOP curves.

The DOS curve for the polymer is also drawn. The characterization of certain bands as bonding or antibonding is obvious, and matches fully the expectations of the approximate sketch **15**. The bands at -14, -15eV are Pt - H σ bonding, the band at -6eV Pt - H antibonding (this is the crystal-field destabilized $x^2 - y^2$ orbital). It is no surprise that the mass of d-block levels between -10 and -13eV doesn't contribute anything to Pt - H bonding. But of course it is these orbitals which are involved in Pt - Pt bonding. The rather complex structure of the -10 to -13eV region is easily understood by thinking of it as a superposition of σ (z^2-z^2), π (xz, yz)-(xz, yz) and δ (xy-xy) bonding and antibonding, as shown in **17**. Each type of bonding generates a band, the bottom of which is bonding and the top antibonding. The δ contribution to the

COOP is small, because of the poor overlap involved. The large Pt - Pt bonding region at -7eV is due to the bottom of the Pt z band.

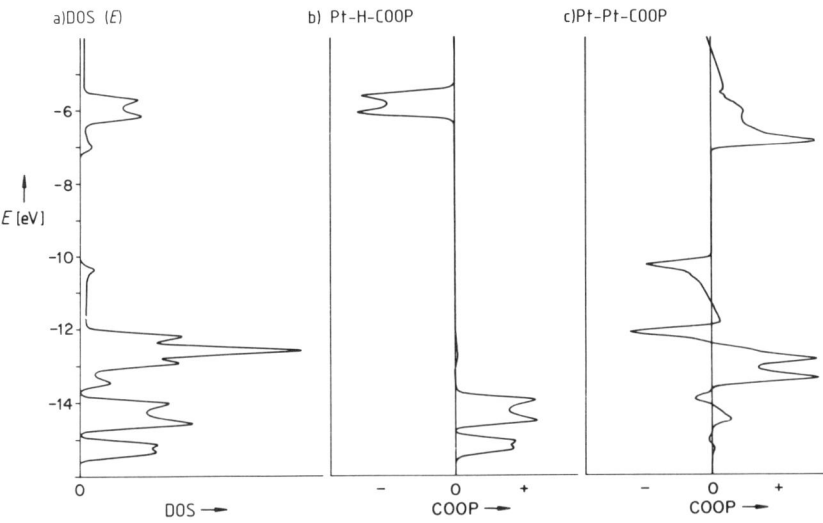

Figure 5. Total density of states (left), and Pt - H (middle) and Pt - Pt (right) crystal orbital overlap population curves for the eclipsed PtH_4^{2-} stack.

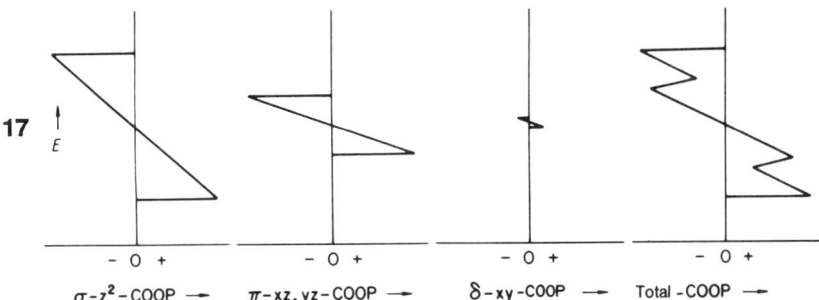

We now have a clear representation of the Pt - H and Pt - Pt bonding properties as a function of energy. If we are presented with an oxidized material, then the consequences of the oxidation on the bonding are crystal clear from Figure 5. Removing electrons from the top of the z^2 band at ~-10eV takes them from orbitals that are Pt - Pt antibonding, Pt - H nonbonding. So we expect the Pt - Pt separation, the stacking distance, to decrease, as it does.

The tuning of electron counts is one of the strategies of the solid state chemists. Elements can be substituted, atoms intercalated, non-stoichiometries enhanced.

Oxidation and reduction, in solid state chemistry as in ordinary molecular solution chemistry, are about as characteristic (but experimentally not always trivial) chemical activities as one can conceive. The conclusions we reached for the Pt - Pt chain were simple, easily anticipated. Other cases are guaranteed to be more complicated. The COOP curves allow one, at a glance, to reach conclusions about the local effects on bond length (will bonds be weaker, stronger) upon oxidation or reduction.

There is much more than this, of course, to bonding in the solid state. A recent book continues the story at this simplistic level (Hoffmann (1988)).

10. References

Ashcroft, N.W. & Mermin, N.D. (1976). "Solid State Physics.", New York : Holt, Rinehart and Winston.

Burdett, J.K. (1984). 'From Bonds to Bands and Molecules to Solids.' Progr. Sol. State Chem. **15**, 173-255.

Harrison, W.A. (1980). "Solid State Theory." New York : Dover

Hoffmann, R. (1988). "Solids and Surfaces." New York : VCH

Hughbanks, T. & Hoffmann, R. (1983). 'Chains of Trans-Edge-Sharing Molybdenum Octahedra : Metal-Metal Bonding in Extended Systems.' J. Amer. Chem. Soc. **105**, 3528-3537.

Kertesz, M. & Hoffmann, R. (1984). 'Octahedral vs. Trigonal-Prismatic Coordination, and Clustering in Transition-Metal Dichalcogenides.' J. Amer. Chem. Soc., **106**, 3453-3460.

Mulliken, R.S. (1955). 'Electronic Population Analysis on LCAO-MO Molecular Wave Functions. II. Overlap Populations, Bond Orders, and Covalent Bond Energies.' J. Chem. Phys. **23**, 1841-1846.

Whangbo, M.-H. (1982). 'Band Structures of One-Dimensional Inorganic, Organic, and Polymeric Conductors.' In J.S. Miller (ed.), "Extended Linear Chain Compounds", Volume 2, pp. 127-158. New York : Plenum Press.

Wijeyesekera, S.D. & Hoffmann R. (1984). 'Transition-Metal Carbides : A Comparision of Bonding in Extended and Molecular Interstitials.' Organometallics **3**, 949-961.

Williams, J.R. (1983). 'One-Dimensional Inorganic Platinum-Chain Electrical Conductors.' Adv. Inorg. Chem. Radiochem. **26**, 235-268.

11. Problems

Problem 1 : Sketch the band structure of a one-dimensional linear chain of main group element atoms, e.g. carbons. Discuss the expected band widths and the way the bands run.

Problem 2 : Rutile, TiO_2, has the structure shown below. There is an approximate octahedral environment around each Ti and infinite chains of edge-sharing MO_6 octahedra running in one direction. The metal - metal separation is relatively long. Sketch the density of states and the M-O COOP curve for rutile.

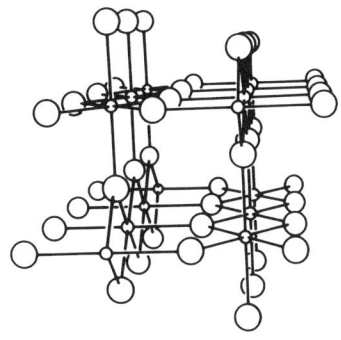

Figure 6. The structure of rutile TiO_2.

11.1. PROBLEM SOLUTIONS

Problem 1 : The band structure of a one-dimensional chain of main group atoms.

Let's take a carbon chain as a starting point, **18**.

Taking a single atom as the repeat unit, we can easily draw the most bonding and most antibonding combinations of the carbon's one s and three p orbitals. These special crystal orbitals for a linear carbon chain are shown in **19** and **20**.

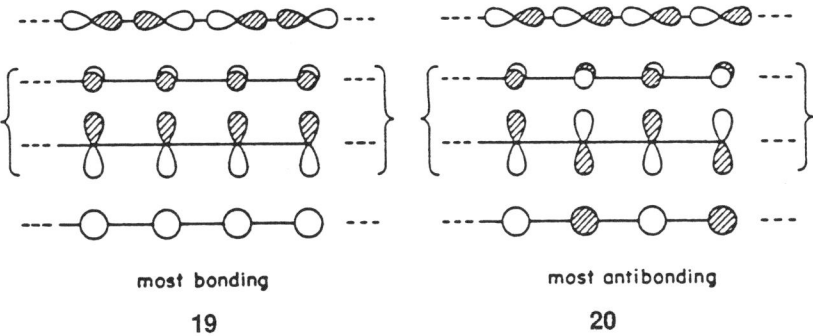

The orbital combinations for p_x and p_y (hereafter abbreviated as x and y, respectively), the p orbitals perpendicular to the chain propagation axis, are of course of the same energy (or degenerate). We see from **19** that s, x, and y give the most bonding crystal orbital when the unit cell (here a single carbon atom) orbitals are all in-phase, i.e., have the same sign. This is the k = 0 (termed Γ) Bloch function.

Conversely, the most antibonding crystal orbital, **20**, has the unit cell orbitals all out-of-phase, or, in other words, each second unit cell has a different sign. This is the wave function at the Brillouin zone edge, traditionally called Z. For the p_z (hereafter z) orbitals on the other hand, the reverse is true in both cases : Its out-of-phase combination in **19** (k = π/a, Z) is most bonding while the in-phase one in **20** (k = 0, Γ) is most antibonding.

We can now construct the band structure for a linear carbon chain. The s and the degenerate x/y band run up from the all-in-phase combination at k = 0 to the all-out-of-phase combination at k = π/a. The z band, of course, shows the opposite behavior -- running down from k = 0 to k = π/a. This is indicated in Figure 7, which is an actual calculation for a long C - C distance of 220 pm. At a shorter separation we'd get much s-z mixing.

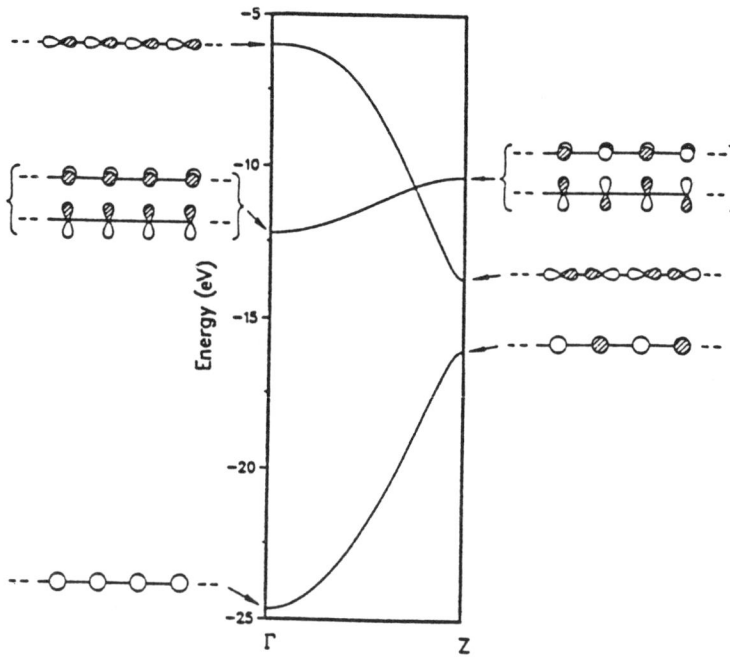

Figure 7. Band structure diagram for a linear carbon chain with C - C of 220 pm and one carbon atom per unit cell. The crystal orbitals are sketched schematically at Γ and Z.

Like the bonding/antibonding splitting in molecular systems, the band width is determined by the interaction (overlap) between orbitals on neighboring atoms (actually unit cells), which in turn depends on the type of orbitals involved and the inter-unit-cell distance. Other things being equal, σ bands generally have a larger dispersion than π, and these in turn are wider than δ bands. Figure 7 clearly shows that the degenerate π band has a smaller width than the two σ (s and z) bands.

Problem 2 : The density of states and M-O crystal orbital overlap population for rutile.

The structure has a nice octahedral environment of each metal center, each ligand bound to three metals. There are infinite chains of edge-sharing MO_6 octahedra running in one direction in the crystal, but the metal - metal separation is relatively long. There are no monomer units here, just an infinite assembly. Yet there are quite identifiable octahedral sites. At each, the metal d block must split into t_{2g} and e_g combinations, the classic three-below-two crystal field splitting. The only other thing we need is to realize that O has quite distinct 2s and 2p levels, and that there is no effective O···O or Ti···Ti interaction in this crystal. We expect something like **21**.

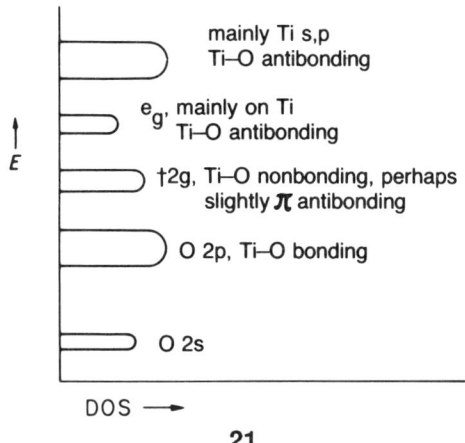

21

The band structure computed in Figure 8 has 12 O 2p bands, 6 t_{2g} bands, 4 e_g bands simply because the unit cell contains two formula units, $(TiO_2)_2$. There is not one reciprocal space variable, but several lines ($\Gamma \to X$, $X \to M$, etc.) that refer to directions in the three-dimensional Brillouin zone. If we glance at the DOS, we see that it does resemble the expectations of **21**. There are well-separated O 2s, O 2p, Ti t_{2g} and e_g bands.

The corresponding COOP is pretty simple. It is shown in Figure 9. Note the bonding in the lower oxygen bands and antibonding in the e_g crystal field destabilized orbitals. The t_{2g} band is, as expected, Ti - O antibonding, for the O 2p's mix into the metal d's in a π - antibonding fashion.

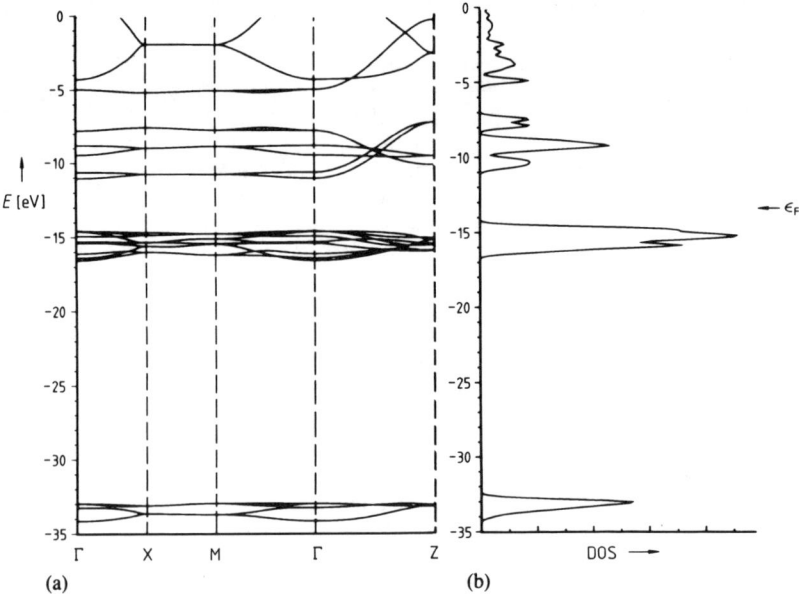

Figure 8. Band structure and density of states for rutile, TiO_2.

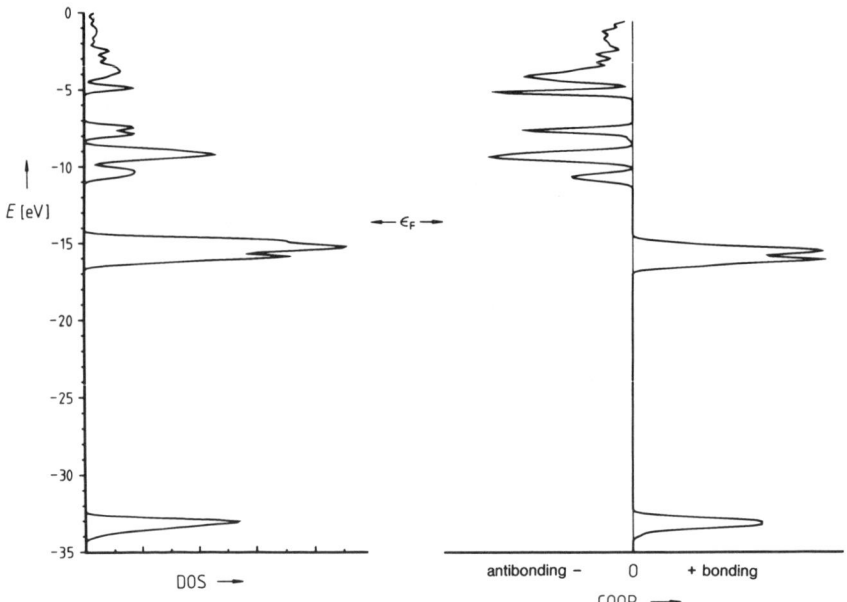

Figure 9. DOS and Ti-O COOP for rutile.

CRYSTAL CHEMICAL FORMULAE FOR INORGANIC STRUCTURE TYPES

JOSÉ LIMA-DE-FARIA
Centro de Cristalografia e Mineralogia
Instituto de Investigaçao Científica Tropical
Alameda D. Afonso Henriques, 41 - 4° Esq.
P - 1000, Lisbon
Portugal

ABSTRACT. Based on the nomenclature report of the IUCr (1990) the concepts of isopointal, isoconfigurational, isotypic and homeotypic structures are described and illustrated by several examples. The notation proposed in this report for the crystal - chemical formulae is also developed, including the symbologies for close packings which are not closest, and for the condensation process of the structural units.

1. Relations among crystal structures and the concept of structure type

The structure type concept refers to crystal structures which are isotypic, that is, have very similar structures. Two crystal structures are similar when they correspond to similar structural arrangements of the atoms, however there are several criteria for establishing relationships among crystal structures, to which correspond different degrees of similarity. For instance, thorianite, ThO_2, has a structure similar to that of fluorite, CaF_2, these two structures having the same symmetry (space group), the same corresponding atoms occupying the same equivalent positions, only differing by the parameters of the unit cell. In this case the equivalent positions are invariant, and the atomic arrangement is exactly the same. However, when the equivalent positions are not invariant, the radius ratio between corresponding atoms are different, or the bonds change, then the relationship becomes more complicated.

1.1. DEFINITIONS OF PRIMARY DEGREES OF STRUCTURAL SIMILARITY

In a report presented by Lima-de-Faria, Hellner, Liebau, Makovicky, & Parthé (1990) (henceforth denoted 1990 Report) a detailed analysis of this problem is made, and the following hierarchy of terms is proposed, based on the degree of structural similarity: *isopointal, isoconfigurational, crystal chemically isotypic,* and *homeotypic structures*:

1) Two structures are *isopointal*, if they have the same space group, or belong to a pair of enantiomorphous space groups, and corresponding atoms occupy the same equivalent positions, either fully or partially at random. Examples are FeS_2 and CO_2, Mg_2SiO_4 (olivine) and $AlMgBO_4$ (sinhalite), where means randomly distributed, $LiNiO_2$ and $NaHF_2$.

2) Two structures are *isoconfigurational* if they are isopointal, and their geometrical properties are similar, such as axial ratios and interaxial angles. Examples are CaF_2 and Li_2O, $Ca_3Al_2Si_3O_{12}$ and $Y_3Fe_5O_{12}$.

3) Two structures are *crystal chemically isotypic*, or *belong to the same structure type* if they are isoconfigurational and the corresponding atoms and bonds (interactions) have similar physical/chemical characteristics. Examples are CaF_2 and ThO_2, NaCl and MgO, Mg_2SiO_4 (olivine) and Al_2BeO_4 (chrysoberyl).

In this last sentence we have to explain what we mean by similar physical/chemical characteristics. The word "similar" arises from the inherent difficulty in defining a priori limits to the similarity of geometrical configurations or physical/chemical characteristics. The physical/chemical characteristics may also vary according to the factors taken into consideration, such as bond strength distribution, bond character, electronegativities assigned to atoms, radius ratios assigned to pairs of atoms and electronic states. Consequently, crystal chemical isotypism may be defined in different ways.

1.2. OTHER IMPORTANT PARENTAL STRUCTURES

Apart from these three main categories of related structures, there are other parental structures which are interesting to consider, the so-called homeotypic structures. They correspond to the relaxation of some of the isotypism conditions, such as :
   (i)   identical or enantiomorphic space-groups, allowing for group-subgroup and group-supergroup relationships;
   (ii)   limitations imposed on the similarity of geometric properties, i.e. axial ratios, interaxial angles, values of adjustable positional parameters and the coordination of corresponding atoms;
   (iii)   site occupancy limits, allowing given sites to be occupied by different atomic species.

Examples of homeotypic structures are :
   Distortion variants (or distortion derivatives) : $CaTiO_3$ (ideal perovskite), $KCuF_3$, $BaTiO_3$, $GdFeO_3$, etc., with subgroup symmetries.
   Substitution derivatives : ZnS (sphalerite) → Cu_3SbS_4 (famatinite) → Cu_2FeSnS_4 (stannite).

Other parental structures that may also be considered are the
   Polytypic structures, examples are SiC polytypes, micas, etc.,
   Interstitial (or stuffed) derivatives, examples are ReO_3 → $CaTiO_3$, SiO_2 (tridymite) → $KNa_3(AlSiO_4)_4$ (nepheline), and
   "*Recombination structures*" which are formed when topologically simple parent structures are periodically divided into blocks, rods or slabs which in turn are recombined into derivative structures by means of one or more structure building operations (Makovicky & Hyde, 1981). Some of these operations are twinning on the unit cell scale, shear on crystallographic planes and intergrowth of blocks, rods or slabs of different structure types. Examples are for shear derivatives : TiO_2 (rutile) → Nb_2O_5; for an intergrowth on the unit cell scale : the olivine - norbergite homologous series, $n(Mg,Fe)_2SiO_4 \cdot Mg(OH,F)_2$.

2. Crystal chemical formulae for inorganic structure types

2.1. THE SYMBOLOGY FOR STRUCTURE TYPES

There is a difference between the crystal chemical formula of a structure and the symbol of a structure type. The crystal chemical formula of a structure (or structural formula of a compound) corresponds to the chemical formula of the compound with the addition of superscripts, subscripts, parentheses, etc., expressing its various structural characteristics. The symbol of a structure type represents a certain kind of structural arrangement and may be based on various notations such as alphabetic letters, lattice complexes or crystal chemical formulae.

The first nomenclature for inorganic structure types has been proposed by Ewald & Hermann in 1931. They used letters to designate the kind of chemical compound and numerals to distinguish among compounds with the same general formula. The chemical elements are designated by the letter A, the binary compounds by the letter B, the AB_2 compounds by C and so on. Examples are A1 for Cu, A10 for Hg, B1 for NaCl, B3 for ZnS (sphalerite), H12 for Mg_2SiO_4 (olivine), G1 for $CaCO_3$ (calcite). This notation has not received much acceptance, possibly due to its lack of self explanatory structural information.

The nomenclature committee of the American Society for Testing Materials (1957) proposed a nomenclature which expresses the chemical formula followed by the number of atoms per unit cell and the Bravais lattice symbol. This notation was later developed by Schubert (1964) and by Pearson (1967). This symbology may be used for classification purposes and might help to find out if a particular structure type had been determined before, but it does not contain information on the geometry of the structure.

Hellner (1965) has proposed a nomenclature for inorganic structure types based on the lattice complex notation, which is useful for compounds with higher symmetry, but which becomes very complicated for compounds of lower symmetry. Examples are Fd-3m, $F'''_{222}+D,T'$ for Al_2MgO_4 (spinel), and Pmcn, (h)n $C_{22}+00_{1/2}12_{xy}A_{21}1$ 1/4 1/4F for $(Mg,Fe)_2SiO_4$ (olivine).

In 1965 Lima-de-Faria proposed a symbology for close-packed structure types based on general structural formulae. For instance, the NaCl structure type was represented by the symbol A^oX^c, where A meant an interstitial atom in octahedral coordination (o), and X a packing atom forming a cubic closest packing (c). This notation was later extended to all the inorganic structures by Lima-de-Faria & Figueiredo in 1976. No general formulae were used, but only the structural formula of the compound, most representative of the assemblage of compounds having the same structure, the so-called aristotype according to Megaw (1973). The NaCl structure type was then represented by $Na^o[Cl]^c$, and not by A^oX^c.

2.2 STRUCTURAL UNITS AND THE FIVE MAIN CATEGORIES OF STRUCTURES

As already pointed out the crystal chemical isotypy may be defined in different ways according to the choice of physical/chemical characteristics considered as more important. The concept of structure type adopted here is based on the strength distribution and directional character of the bonds among the atoms. It is in agreement with that used in the 1990 Report and with the work of Lima-de-Faria & Figueiredo (1976). This nomenclature is a kind of extension of the Machatschki notation (1947).

In inorganic structures there is a tendency for close packing (first Laves principle) and the bonds are predominantly non directional. The bond strength distribution being, in most cases, homogeneous. However, if the character of certain bonds is directional then there is a deviation from the close packing. That is, the structures tend to form less dense atomic arrays and certain atoms are more tightly linked together than others. This results in finite groups or in assemblages that are infinite in one, two or three dimensions. These assemblages are called "structural units", and the remaining atoms are the "interstitial atoms".

There are listed in Table 1 the five main categories of structural units.

TABLE 1. The five main categories of structural units.

Dimensionality of structural unit		Category of structural unit
0-dimensional	↗	individual atoms
	↘	groups (i.e. rings, chain fragments, cages)
1-dimensional	→	chains
2-dimensional	→	sheets
3-dimensional	→	frameworks

The corresponding kinds of structures are designated by : atomic (or close-packed), group, chain, sheet and framework structures (Lima-de-Faria & Figueiredo, 1976) (Figure 1).

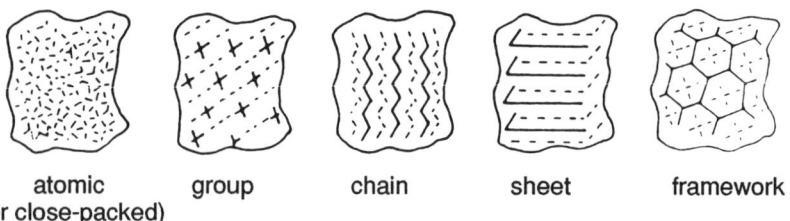

atomic (or close-packed) group chain sheet framework

Figure 1 : Very schematic representation of the bond strength distribution among the five categories of inorganic structures.

If the spatial bond strength distribution is homogeneous two limiting situations may be discerned : either the structure is based on a three-dimensional framework (examples are diamond and cristobalite with directional bonds) or it is simply a packing of individual atoms (examples are helium, copper or sodium chloride with non-directional bonds). The corresponding structural units are thus either a framework or the individual atoms, respectively.

In special cases the bond strength distribution may be heterogeneous but the character of the bonds non directional and the structure is then based on the close packing of the atoms. This heterogeneity may only be a consequence of a particular stable distribution of the interstitial atoms and in such cases the structure should be considered as a close-packed structure. An example is $Cd^o[Cl_2]^c$ which is formed by the closest packing of Cl⁻ ions, with Cd⁺⁺ ions occupying one half of the octahedral voids in two alternate layers, one completely filled and another empty. This is one of the possible stable distributions for the case in which half of the octahedral voids are occupied. $Cd^o[Cl_2]^c$ should therefore be considered as a close-packed structure and not as a sheet structure.

A structural unit may be imagined as subdivided into subunits such as single atoms, polyhedra, single rings, single chains or single layers.

A structure can be considered to consist of structural units, packed together with interstitial atoms between them. Since the strengths of the bonds cannot always be accurately quantified, some ambiguity may exist in assigning a structure to a given category.

2.3. THE CRYSTAL CHEMICAL FORMULAE OF STRUCTURES

The crystal chemical formulae should give detailed structural information on the structural units, their constitution and way of packing, the interstitial atoms and the coordination of the atoms, both interstitial and those contained in the structural units. The nomenclature should be as simple and self-explanatory as possible.

Symbols for atoms belonging to the structural unit(s) are placed between square brackets, [ ] and the packing information between angle brackets, < >. The information on constitution which relates to the structural unit as a whole is placed within curly brackets, { }. However, the constitutional information which relates to subunits of the structural units(s) may be expressed either within curly brackets or as trailing superscripts to the chemical elements or subunits inside the structural unit.

Curly brackets with constitutional information precede and angle brackets for packing information immediately follow the structural unit to which they refer.

Information concerning interstitial atoms and/or groups of atoms should generally be placed before or after that on the structural unit(s) in the sequence that chemical formulae are usually written.

The coordination of each atom, either interstitial or in the structural unit, is expressed within square brackets as a trailing superscript to the chemical symbol.

In accordance with IUPAC (1990) rules, the valency state of each atom is expressed immediately after its chemical symbol by a Roman numeral in parentheses (e.g. Fe(III)), a superscripted Roman numeral (e.g. Fe^{III}), or by a superscripted Arabic numeral followed by the + or − sign (e.g. Fe^{3+}).

The coordination number [N] of an atom is given by the number of "coordinating atoms". The coordination polyhedron of an atom is the polyhedron that has vertices coincident with the centers of the coordinating atoms. The complete symbol for coordination gives the total number of atoms coordinated to a central atom A, and the type of coordination polyhedron (indicated by lower-case letters). The simplified symbol requires only the coordination number [N] without specifying the polyhedron type. The notation must, however, be able to describe coordination by different sets of atoms, or coordination at different distances, or self-coordination, and coordination polyhedra composed of several distinct atomic species. In the general case the coordination of atom A in a compound $A_aB_bC_c$ is written as $A^{[m,n;p]}$ where m and n denote the number of atoms B and C (i.e., always in the sequence given in the formula) which are coordinated to atom A. The coordination numbers are separated by commas; the self-coordination number p of A by atoms A follows the semicolon. The coordination of B is written as $B^{[m',n';p']}$, where m', n', and p' denote the number of atoms A, C and B around atoms B, respectively, etc. As an example we may consider $CaTiO_3$:

$$Ca^{[8cb,12co;6o]}Ti^{[8cb,6o;]}O_3^{[4l,2l;8p]}$$

which can be simplified as follows

$$Ca^{[8,12;6]}Ti^{[8,6;]}O_3^{[4,2;8]}$$

or still

$$Ca^{[12]}Ti^{[6]}O_3^{[;8]}$$

Additional constitutional information related to subunits may be given within the square brackets for the structural unit. It should be placed between Japanese quotation marks, 「 」, as an additional trailing superscript : $A^{[「」]}$. Examples of the more common coordinations and their corresponding symbols are given in Table 2.

TABLE 2. Symbols for some more common coordination polyhedra

Coordination polyhedron around atom A	Complete symbol	Simplified symbols
Triangle coplanar with atom A	[3l]	[3]
Triangle non-coplanar with atom A	[3n]	[3]
Tetrahedron	[4t]	[t] or t
Square coplanar with atom A	[4l] or [4s]	[s] or s
Octahedron	[6o]	[o] or o
Trigonal prism	[6p]	[p] or p
Trigonal antiprism	[6ap]	[ap] or ap
Cube	[8cb]	[cb] or cb
Anticube	[8acb]	[acb] or acb
Dodecahedron with triangular faces	[8do]	[do] or do
Cuboctahedron	[12co]	[co] or co
Anticuboctahedron	[12aco]	[aco] or aco
Icosahedron	[12i]	[i] or i

The general notation for a compound $A_a B_b C_c D_d E_e F_f G_g$ could thus be :

$$A_a^{[\alpha]} B_b^{[\beta]} \{\ \} [C_c^{[\gamma「\varphi」]} D_d^{[\delta「\chi」]} E_e^{[\epsilon]}] <\ > F_f^{[\zeta]} G_g^{[\eta]}$$

intersti- ⇓ structural unit ⇓ interstitial atoms ⇓ ⇓ tial atoms
  constitution of packing of
  structural unit structural unit

If several distinct structural units are present, each is considered separately with its information in curly brackets followed by that in square brackets, for example :

$$A_a^{[\alpha]} \{\ \} [B_b^{[\beta]} C_c^{[\gamma]}] \{\ \} [D_d^{[\delta]} E_e^{[\epsilon]}] <\ > F_f^{[\zeta]} G_g^{[\eta]}$$

The packing information within angle brackets describes the way the two different structural units pack together.

The hierarchy of bonds leads to a hierarchy of structural units when several degrees of bond strengths may be discerned in a structure. This often leads to weaker bond-strength units incorporating previous more strongly bonded units, and can be expressed by multiple brackets, with the central brackets referring to the structural unit having the strongest bonds :

$$A_a{}^{[\alpha]}\{\quad\}[B_b{}^{[\beta]}\{\quad\}[C_c{}^{[\gamma]}D_d{}^{[\delta]}E_e{}^{[\epsilon]}]<\quad>]<\quad>F_f{}^{[\zeta]}G_g{}^{[\eta]}.$$

The proposed formula can be used with any amount and any selection of structural information depending on the purpose of the study. Examples are given in Table 3.

TABLE 3. Examples of crystal chemical formulae of inorganic structures

Cu	$_\infty^0[Cu]<c>$	$[Cu]^c$	$_\infty^0[Cu^{[12co]}]$
NaCl	$Na^{[6]}{}_\infty^0[Cl^{[6]}]$	$Na^o[Cl]^c$	$_\infty^3[Na^{[6o]}Cl^{[6o]}]$
SiO_2 (cristobalite)	$_\infty^3[Si^{[4t]}O_2]$	$_\infty^3[Si^{[4t]\lceil1;4\rfloor}O_2]$	$_\infty^3[Si^tO_2]$
FeS_2 (pyrite)	$Fe^{[6o]}\{_\infty^0\}[S_2{}^{[3;(1+2)]}]$	$Fe^{[6o]}\wedge[S_2{}^{[(3;1)t]}]$	$Fe^o\{g\}[S_2]^c$
$(Mg,Fe)_2SiO_4$ (olivine)	$(Mg,Fe)_2{}^{[6o]}\{_\infty^0\}[Si^{[4t]}O_4]$		$(Mg,Fe)_2{}^o\,Si^t\,[O]_4{}^h$
$MgAl_2O_4$ (spinel)	$_\infty^3[Mg^{[4t]}Al_2{}^{[6o]}O_4{}^{[1,3;12co]}]$		$Mg^tAl_2{}^o[O]_4{}^c$
$KAl_3Si_3O_{10}(OH)_2$ (muscovite)	$K^{[6+6]}\{2\}[Al_2{}^{[6o]}\{_\infty^2\}[(Al_{0.5}Si_{1.5})^{[4t]\lceil1;3\rfloor}O_5]_2(OH)_2]$		
		$K^{[6]}Al_2{}^o\,_\infty^2[Al^tSi_3{}^t\,O_{10}](OH)_2$	
LaP_2 (HT form)	$La_4\{_\infty^0\}[P_2{}^{[;1]}P^{[;2]}]\{_\infty^0\}[P_2{}^{[;1]}P_3{}^{[;2]}]$		$La_4\,\wedge[P_3]\wedge[P_5]$
$Ca_3Si_2O_7$ (rankinite)	$Ca_3\{_\infty^0\}[Si_2{}^{[4t]\lceil1;1\rfloor}O_7]$		

2.4. THE NOTATION EXPRESSING THE CONSTITUTION OF THE STRUCTURAL UNITS

The constitution of a structural unit expresses its extensional and geometrical 'structure', i.e. the way the structural unit is built from its subunits, which may be polygons, polyhedra or any other clusters. Some of the constitutional aspects are concerned with the structural unit as a whole, whereas other aspects are only concerned with the way each subunit is linked to other subunits. The former include dimensionality, multiplicity, branchedness and periodicity.

The *dimensionality* is the number of dimensions in which a structural unit has infinite extension. It is zero for individual atoms and finite groups, and one, two or three for infinite chains, sheets and frameworks, respectively. The corresponding symbols to be used in a crystal chemical formula are

$^0_\infty$, $^1_\infty$, $^2_\infty$, and $^3_\infty$, the well known Machatschki symbols.

The following specific symbols may be used for 0-dimensional structural units:

individual atoms: {a}

group: {g}	↗	ring:	{r} or	○
	→	chain fragment:	{f} or	∧
	↘	cage:	{k} or	⊘

Examples are: $Cs_2{}^\wedge[S_6]$, $Na_4\oslash[Si_4]$, $Cu_6\{r\}[Si_6O_{18}]\cdot 6\,H_2O$.

If dimensionality is the only information expressed, the $^n_\infty$ and the pictorial symbols may be used without curly brackets. Otherwise, curly brackets are compulsory in order to avoid ambiguity. For details of this notation see Parthé (1980).

The symbol {a} is not needed when several individual atoms, A, B, C, ..., considered as structural units, are written [A] [B] [C]... When only one atom symbol is placed within square brackets, it means that the structural unit is reduced to an individual atom. However, if the same symbols are written [ABC], then it is necessary to add {a} in front of the square brackets.

In the case of a group of structures, e.g. ring, chain fragments and cage structures, the number of atoms of each chemical element within square brackets must be equal to the number of atoms of each chemical element in the finite group.

The *multiplicity* of a structural unit is the number of single subunits, e.g. polyhedra, single rings, single chains or single layers which are linked to form a complex structural unit of the same dimensionality.

With regard to *branchedness*, finite structural units and single chains are called unbranched if they contain no subunits that are linked to more than two other units. They are called branched if they do.

The *periodicity* of a structural unit of infinite extension is the number of subunits, excluding branches, within one repeat unit of the chain from which the structural unit can be generated by successive linking.

For details of these concepts, see Liebau (1982, 1985).

The main constitutional aspects concerned only with the way each subunit is linked to the other subunits are linkedness and connectedness.

The *linkedness* is the number L of peripheral atoms shared between two subunits.

The *connectedness* of a subunit is the total number s of adjacent subunits with which it shares common atoms, irrespective of its linkedness with a particular adjacent subunit.

The specific values L_1, L_2 etc. of linkedness and/or s of connectedness of a subunit are written within 'Japanese brackets' as trailing superscripts to its central atom, by analogy with the coordination symbols. The first entries in the Japanese brackets are the different values of L_n, separated from the value of s by a semicolon. The general formula for a structural unit with only one kind of subunit then reads

$$[A_m{}^\ulcorner L_1, L_2, ...; s \lrcorner B_n]$$

For example, SiO_2 exists in a number of polymorphs having different values of linkedness and connectedness of the SiO_4 tetrahedra:

Fibrous silica:
$$^1_\infty[Si^{[4t]\,\ulcorner 2;2 \lrcorner}O_2]$$

and quartz, cristobalite, or coesite :
$$^{3}_{\infty}[Si^{[4t]\lceil1;4\rfloor}O_2]$$

A structural unit can often be generated from a part of either lower or the same dimensionality by a simple geometrical process that usually represents an infinitely repeated translation. This imaginary geometrical process is called "*condensation*" because it emphasizes the way a chain can be generated from a group, a sheet from a chain and a framework from a sheet. It also reveals certain similarities between different structural units and a specific composite notation for the structural units has been developed which emphasizes this interrelationship (Lima-de-Faria & Figueiredo, 1976).

2.5. THE PACKING OF THE STRUCTURAL UNITS

The *packing* of structural units expresses the three-dimensional arrangement in space. When the structural units are individual atoms, the known nomenclature for describing the packing of atoms (three-dimensional and layer-stacking descriptions) may be used. When the structural units are groups, their centres of gravity may be used with the same nomenclature as for the packing of atoms. However, this will be an incomplete description because of the lack of information on the orientation of the groups.

The layer description can also be applied to structures based on infinite chains or sheets, or on frameworks. Such a layer description consists of slicing the structure into atomic layers which, in general, coincide with the plane direction of highest density of atoms. These layers, called "constructive layers", are formed by the packing atoms and the interstitial atoms which are immediately above them, so that the interstitial atoms which are below will pertain to the constructive layers located underneath. The stacking of constructive layers completely generates the original crystal structure.

With respect to the nomenclature of the packing of structural units, only the symbols for cubic closest packing c, and hexagonal closest packing h, and their sequential combination have been adopted in the 1990 Report mentioned above. When no other packing information is provided these symbols may be given as trailing superscripts to the square brackets which contain the structural unit. In this case, angle brackets are not compulsory. Any other packing information, particulary the packing (or stacking) symbolism used by individual authors should be given in angle brackets on the line.

$$[ABC]^c \text{ or } [ABC] < >$$

If packing information is to be given for a set of atoms which does not constitute a structural unit, the symbol should be placed within vertical bars followed by the packing information :

$$|ABC|^c \text{ or } |ABC| < >$$

3. Further developments of the nomenclature of inorganic structure types.

3.1. THE GENERALIZATION OF THE CLOSE PACKING SYMBOLOGY

We shall now consider some developements of the notation for structure types which have not been included in the 1990 Report mentioned above. These developements regard especially the symbology for several *close packings* which are not closest, but are quite common, and for the *condensation* or *polymerization process* of the structural units (Lima-de-Faria & Figueiredo, 1976).

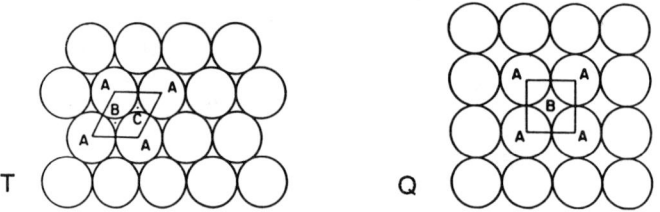

Figure 2. Closest-packed layers (T) and square layers (Q)

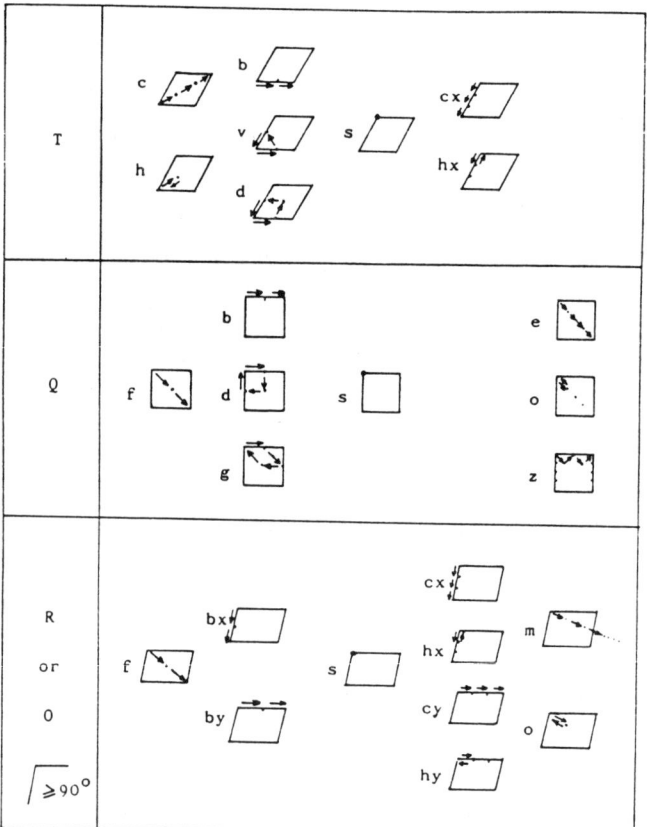

Figure 3. Generalized stacking symbols (adapted from Lima-de-Faria & Figueiredo, 1976).

Many structures are based on less dense stackings of closest layers (T) or of square layers (Q) (Fig. 2), giving rise to several kinds of close packings with specific interstices within the packings. First we shall consider the simple and hexagonal close packings which correspond to a stacking of T and Q layers, respectively and where the packing atoms of one layer superimpose the atoms of the layer below, that is, touch only one atom of the layer below. This stacking mode is symbolized by the letter *s* (superposition) and the simple cubic and the simple hexagonal packing are described by Ts and Qs, respectively. An example of a structure with Ts close packing is Al_2Th, whose crystal chemical formula is written $Al_2^p[Th]^{Ts}$, where Al is located in triangular prismatic voids (p) of the simple hexagonal packing of Th atoms. An example of a structure based on simple cubic packing is CaF_2, whose structural formula is $Ca^{cb}[F_2]^{Qs}$, where Ca atoms occupy interstices with cubic coordination (cb) of the simple cubic packing of F atoms.

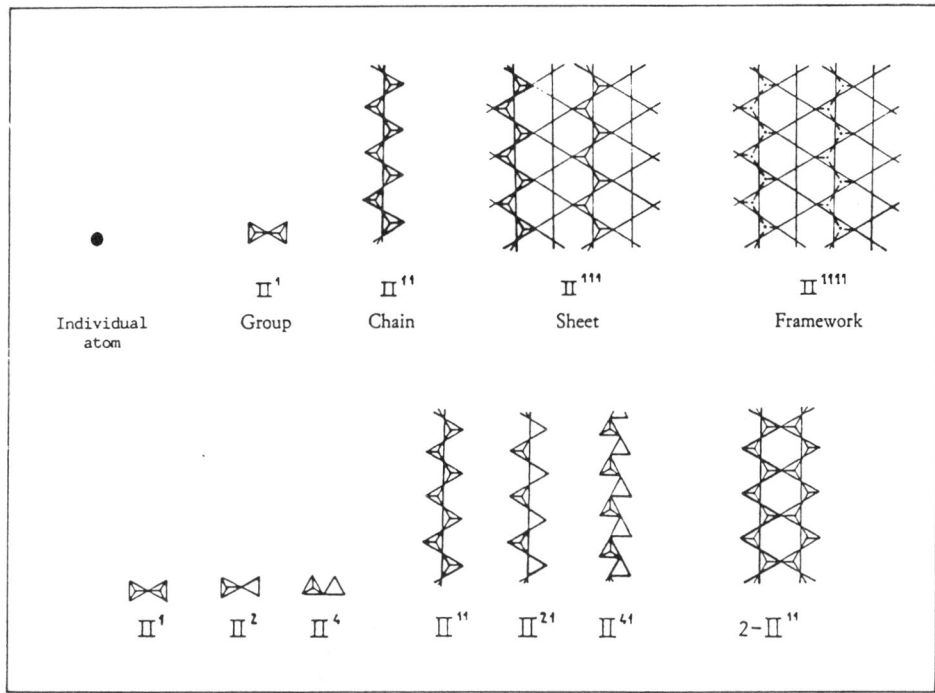

Figure 4. Nomenclature of structural units expressing the condensation process (adapted from Lima-de-Faria & Figueiredo, 1976).

Other stacking of T and Q layers may be considered, where the atoms of a layer touch only two atoms of the layer below. They give rise to pure sequences of type *b*, *v* or *d*. Examples are $[Pa]^{Tb}$ (protactinium), $Zn^t[BaO_2]^{Tv}$ and $[TiSi_2]^{Td}$. For a Q stacking of this sort an example is $Si^o[Th]^{Qd}$. In Fig. 3 are represented the more common pure stacking sequences with their corresponding symbols.

These packings may be combined, giving rise to mixed sequences. For instance, the structure of the mineral molybdenite, MoS_2, has the structural formula $Mo\ P[S_2]^{(2Ts)h}$. The S atoms form double layers with s stacking, which pack together in a h sequence.

3.2. THE NOTATION EXPRESSING THE CONDENSATION PROCESS OF THE STRUCTURAL UNITS

A notation was also proposed to emphasize the condensation process of the structural units.

The number of silica tetrahedra in a group silicate is designated by a Roman numeral, with a superscript to differenciate it from other groups with the same number of silica tetrahedra.

The symbol of a chain derived from a certain group by polymerization is formed by the symbol of the group plus another superscript, to distinguish different chains derived from the same group and so on. In this way the symbol of the structural unit indicates immediately what is the substructural unit from which it derives (Fig. 4).

A framework is in itself an infinite structural unit, but it can be imagined subdivided in parts which can be infinite sheets, infinite chains, or finite groups. These subunits are called "connected units", and the framework may be considered as a condensation of such "connected units".

4. References

Ewald, P.P. & Hermann, C. (1931). "Strukturbericht", Vol.1. Leipzig : Akademische Verlagsgesellschaft m.b.H.

Hellner, E. (1965). 'Descriptive Symbols for Crystal-Structure Types and Homeotypes Based on Lattice Complexes.' Acta Cryst. **19**, 703 - 712.

IUPAC (1990). "Nomenclature of Inorganic Chemistry". Oxford : Blackwell.

Liebau, F. (1982). 'Principles of Silicate Classification Based on Crystal Chemistry.' Sov.Phys.Crystallogr. **27**, 66 - 72.

Liebau, F. (1985). "Structural Chemistry of Silicates : Structure, Bonding and Classification". Berlin : Springer-Verlag.

Lima-de-Faria, J. (1965). 'Systematic Derivation of Inorganic Close-packed Structures : AX and AX_2 Compounds, Sequence of Equal Layers.' Z.Kristall. **122**, 359 - 374.

Lima-de-Faria, J. & Figueiredo, M.O. (1976). 'Classification, Notation and Ordering on a Table of Inorganic Structure Types.' J. Solid State Chem. **16**, 7 - 20.

Lima-de-Faria, J., Hellner, E., Liebau, F., Makovicky, E. & Parthé, E. (1990). 'Nomenclature of Inorganic Structures Types'. Acta Cryst. **A46**, 1 - 11.

Machatschki, F. (1947). 'Konstitutionsformeln für den festen Zustand.' Monatsh. Chem. **77**, 333 - 342.

Makovicky, E. & Hyde, B.G. (1981). 'Non-commensurate (Misfit) Layer Structures.' In "Structure and Bonding", 46, 101 - 175. Berlin : Springer.

Megaw, H.D. (1973). "Crystal Structures. A Working Approach". London : W.B.Saunders Company.

Nomenclature Subcommittee of Committee E-4 on Metallography. (1957). American Society for Testing Materials. Bulletin No. **226**, 27 - 30.

Parthé, E. (1980). 'Crystal Chemical Formulae for Simple Inorganic Crystal Structures.' Acta Cryst. **B36**, 1 - 7.

Pearson, W.B. (1967). "A Handbook of Lattice Spacings and Structures of Metals and Alloys." Oxford : Pergamon Press.

Schubert, K. (1964). "Kristallstrukturen zweikomponentiger Phasen." Berlin : Springer Verlag.

5. Problems

Problem 1 : Considering that in the ZnS wurtzite structure the S atoms form an hexagonal closest packing and the Zn atoms occupy tetrahedral interstices write the corresponding crystal chemical formula.

Problem 2 : Considering that in the AuCu the Au and Cu atoms have similar radius and both form together a cubic closest packing, write its crystal chemical formula.

Problem 3 : Considering that in the structure of the high pressure form of Mg_2SiO_4 the oxygen atoms form a cubic closest packing, and the Mg and the Si atoms occupy the octahedral and tetrahedral voids, respectively, write the structural formula.

Problem 4 : The pyrite structure, FeS_2, is a group structure because the S atoms are linked by pairs. These S_2 pairs pack together as close as possible. The centers of gravity of these S_2 pairs are in the same positions as individual atoms in cubic closest packing, and the Fe atoms occupy the interstices of this packing of groups with an octahedral coordination in relation to the S atoms. Write the structural formula.

Problem 5 : The structure of diopside, $CaMgSi_2O_6$, is a chain structure where the structural units are silicate chains Si_2O_6. These chains pack together in such a way that in a certain plane direction they form layers which in turn pack together in a $\overline{cx}$ stacking sequence. All the other atoms occupy interstices in this packing, Ca with coordination 8 and Mg with octahedral coordination. What is its structural formula ?

Problem 6 : The hornblende structure, $NaCa_2Mg_5Si_7AlO_{22}(OH,F)_2$ is a chain structure where the structural units correspond to the condensation of two silicate chains of the diopside type. However, Al atoms replace some Si atoms in these structural units. These chains form layers which pack together in a $\overline{cx}$ stacking sequence. All the other atoms occupy the interstices of this packing : Na with 10 coordination, Ca with 8 coordination, Mg, with octahedral coordination and (OH) or F occupy the large interstices inside the silicate chains. Write the structural formula.

5.1. SOLUTIONS OF THE PROBLEMS

Problem 1 : $Zn^t[S]^h$

Problem 2 : $[AuCu]^c$

Problem 3 : $Mg_2^oSi^t[O_4]^c$

Problem 4 : $Fe^o[S_2]^c\{g\}$

Problem 5 : $Ca^{[8]}Mg^o\{^1_\infty\}[Si_2{}^tO_6]<\overline{cx}>$

Problem 6 : $Na^{[10]}Ca_2{}^{[8]}Mg_5{}^o(OH,F)_2\{^1_\infty\}[Si_7{}^tAl^tO_{22}]<\overline{cx}>$

CRYSTAL CHEMISTRY OF COMPLEX SULFIDES (SULFOSALTS) AND ITS CHEMICAL APPLICATION

EMIL MAKOVICKY
Geological Institute
University of Copenhagen
Øster Voldgade 10
DK-1350 Copenhagen K
Denmark

ABSTRACT. Crystal structures of many complex sulfides can be obtained by recombination of fragments of archetypal structures by the action of various structure-building operators. Chemical categories, coordination polyhedra, archetypes and recombination operators for these structures are described in the paper. These structures often occur as members of accretional or variable-fit homologous series or as homologous pairs of various kinds. Three case studies supplemented by exercises are described in some detail: the lillianite homologous series, the misfit layer structures and those structures that represent combination of the accretional and variable-fit principles.

1. Crystal Structures of Complex Sulfides and Sulfosalts

Crystal structures of a large number of complex sulfides can be described as composed of rods, blocks or layers of simple, archetypal structures that are joined (recombined) in various ways by the action of one or more structure-building operators, such as the various kinds of unit-cell twinning. The resulting structures were denoted as the recombination structures by the Subcommittee on the Nomenclature of Inorganic Structure Types (Lima-de Faria et al., 1990). It is this kind of complex sulfides we deal with in the present review; those with complex chemistry but simple structural principles (e.g., the substitution-, interstitial- or omission derivatives of archetypal structures) are not included.

1.1. SULFIDE FAMILIES WITH RECOMBINATION STRUCTURES

The principal sulfide family with recombination structures are the so-called "sulfosalts". They are defined as complex sulfides (rarely also selenides and tellurides) $A_xB_yS_z$ with A = Pb, Cu, Ag, Hg, Tl, Fe, Mn, Sn^{2+}, Sn^{4+} and other metals; these may include alkali metals, alkaline earths and lanthanides in some synthetic sulfosalts; B = formally trivalent cations As, Sb and Bi (rarely also Te^{4+}) in non-planar, (fundamentally) three-fold coordination BS_3; and S = S, Se and Te^{2-}. Structurally very close to sulfosalts are on the one hand the complex sulfides of (primarily) octahedral In with Pb, Sn, Bi and other metals and on the other, the layer misfit sulfides, based on a combination of layers that contain large cations (La to Sn, Y, Sn^{2+}, Pb and Bi^{3+}) with layers that contain smaller cations with octahedral (Cr, Ti, V) or trigonal-prismatic (Nb and Ta) coordinations. Yet another group of complex sulfides with recombination structures are the lanthanide sulfides in which large coordination polyhedra of light lanthanides combine with smaller polyhedra of heavier lanthanides.

1.2. TYPICAL COORDINATION POLYHEDRA

At the small-scale level, the crystal chemistry of sulfide recombination structures is determined by the easy and variable combinations of fundamental coordination polyhedra, yielding both the blocks of archetypal structures and the configurations on

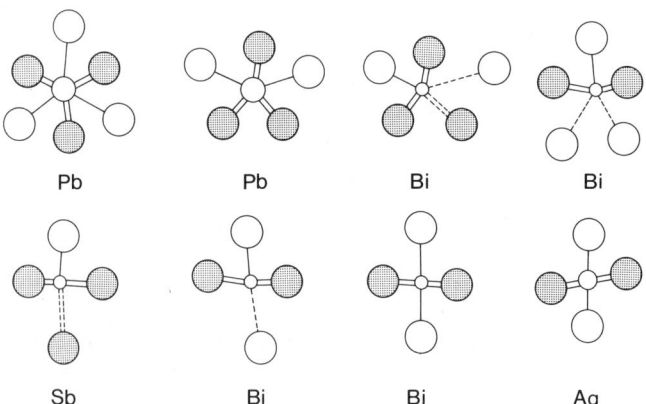

Figure 1. Selected coordination polyhedra in complex sulfides: tri, bi- and 'lying' monocapped trigonal prisms of Pb and metalloids; a quasi-octahedral and octahedral coordination of Bi, [2+4] coordination of Ag. Small circles: cations, large circles: sulfur; void and filled circles denote atoms at 2 distinct height levels, ~2 Å apart.

their boundaries. With the exception of largest cations such as Tl and K that have high coordination numbers and mostly irregular coordination polyhedra, the large cations typically assume trigonal prismatic coordination with additional anions capping all three, two, or one of the prismatic faces (Fig. 1). Such prisms may be "standing", i.e. parallel to, or "lying", i.e. perpendicular to the short (~4Å) repetition period that is characteristic for the majority of these structures. Such trigonal prisms are typical for the most characteristic chemical element of recombination sulfosalt structures, lead. Moreover, they are also the typical coordination polyhedron for As^{3+}, Sb^{3+}, Sn^{2+} and a large part of Bi^{3+}. In these cases it is the lone electron pair of the cations that occupies the volume of the prism whereas the cation resides in (or just inside of) one of the approximately square-shaped prismatic faces. All short bonds are concentrated in the square pyramid that caps this face. The length differences between the three shortest ($\underline{p}^3$) bonds in the pyramid, two additional, longer bonds ($\underline{d}^2$) in the pyramidal base and the remaining distances, to the S atoms shielded away by the lone electron pair $\underline{s}^2$ of the cation, decrease from As to Bi. For bismuth, cases of irregular, quasi-octahedral coordinations $BiS_{(3+2)+1}$ grading into fairly regular octahedral coordinations BiS_6 are frequent; in fact, all intermediate stages between the limiting coordinations BiS_3 and BiS_6 are observed.

The prism caps of Pb coordination prisms are fairly well commensurable with the dimensions of square pyramids of As, Sb and Bi, the latter (or the groups BS_3) are in turn commensurable with the coordination octahedra around Pb, Mn, In, Fe, Ag, (Hg), etc., or with tetrahedral and trigonal coordinations of primarily Ag and Cu. The latter may fill interspaces in the above arrays or form polyhedral arrangements of their own, with only subdued amounts of prismatic, pyramidal or octahedral cations present.

The interplay of variously capped (and oriented) trigonal coordination prisms with coordination octahedra is essential also for the complex lanthanide sulfides and the layer misfit compounds. Through the entire range of structures and compositions the relative plasticity (ability for minor adjustments) of large coordination polyhedra created around large or lone-electron-pair cations is of primary importance in the interlayer and interblock adjustments.

1.3. LARGE-SCALE CRYSTAL CHEMISTRY: ARCHETYPES AND STRUCTURE BUILDING PRINCIPLES

For the recombination structures aggregation is typical of a number of coordination polyhedra into rods, blocks or layers with internal structure similar to a simple 'archetypal' structure.

The structure archetype occurs as an independent structure and encompasses all the geometric as well as bonding properties of a given atomic array. In the case of weakly expressed steric lone electron pair activity of cations (e.g. in the case of Bi^{3+} or Pb^{2+}) the internal structure of blocks (etc.) represents little deformed PbS (i.e., B1) archetype (Fig. 2a). In the structures with stereochemically active lone electron pairs (As-, Sb-, Sn^{2+}-, and some Bi-containing structures) this archetype is often replaced by the SnS archetype (Fig. 2b) with the coordinations $BS_{(3+2)+1+1}$ or the TlI archetype (Fig. 2c) with the coordinations BS_{5+2}. The relationship between the PbS and TlI (approx. also SnS) archetypes can be expressed as 2Å-shear between double layers M_2S_2 along the ~4Å direction. This shear alters coordination octahedra into monocapped trigonal coordination prisms, offering thus better accommodation for the lone electron pairs of cations.

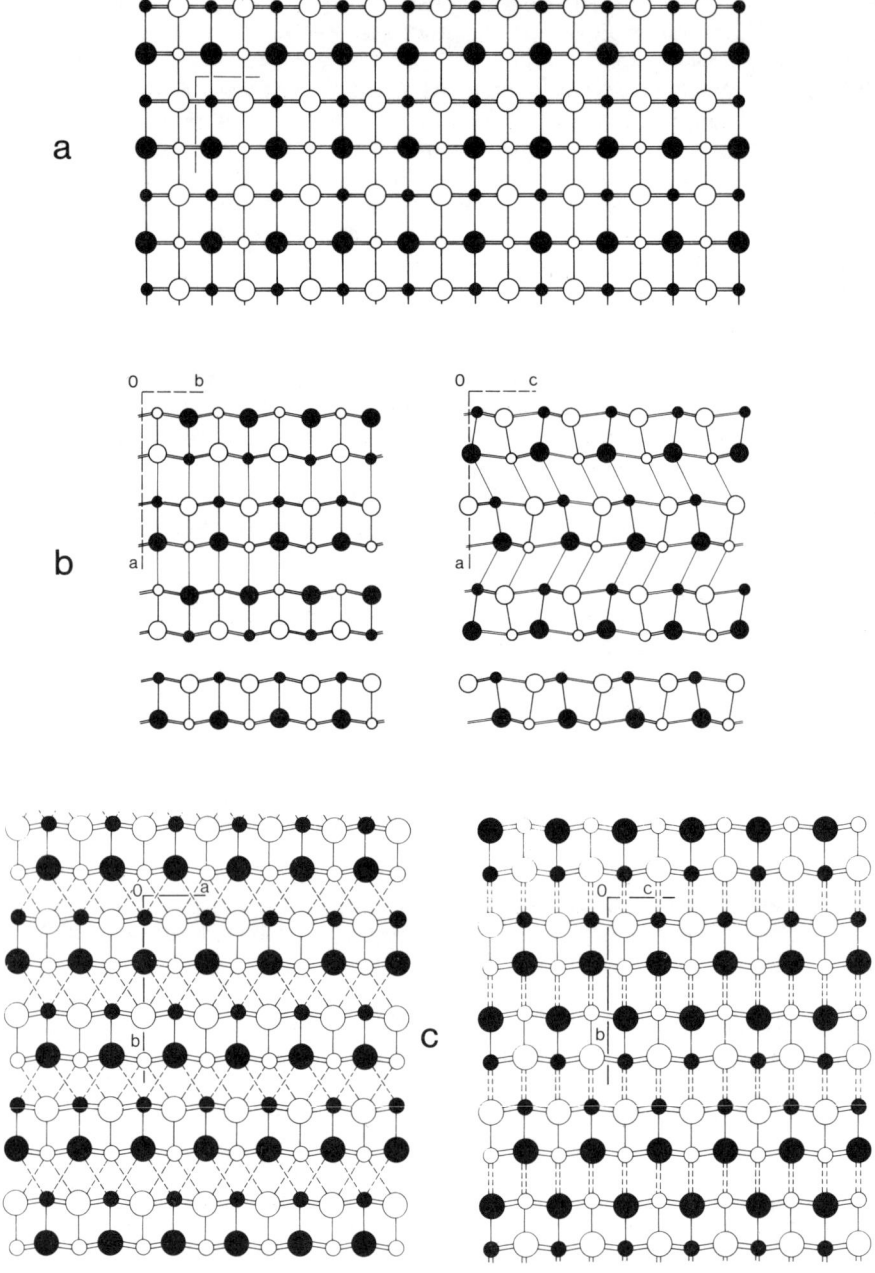

Figure 2. Three principal archetype structures for complex sulfides: (a) PbS, (b) SnS, and (c) TlI. Drafting conventions as in Fig. 1.

The rods, blocks or layers of archetypal structures are limited by surfaces that can be indexed in terms of the archetypal structure. For the majority of cases these surfaces are $(100)_{PbS}$, $(100)_{SnS}$ (with pseudotetragonal motifs of metal and S atoms; the so-called Q surfaces), as well as $(111)_{PbS}$ (with pseudohexagonal, H, motif of S atoms) and the analogous planes $(210)_{SnS}$ and $(301)_{SnS}$ (sheared H surfaces).

The more complicated surfaces (e.g. $(101)_{SnS}$, $(501)_{SnS}$ or $(311)_{PbS}$) can be interpreted as combinations of stripes of two fundamental, simplest surfaces. The two atomic planes thick $(100)_{PbS}$ layers can be interpreted as belonging to either PbS or TlI (SnS) archetype.

For the large part, the structure-building principles (operations) according to which the rods, blocks or layers of archetypal structure are recombined into a complex sulfide structure are various types of unit-cell twinning: reflection-(i.e. mirror-)twinning that acts either on a full set of atoms (Andersson & Hyde, 1974) or on a "contracted" (i.e. partial) set of atoms out of the larger one of the two mirror-related portions (Takéuchi, 1978), glide-reflection twinning (l.c.) that can consecutively be applied on different (hkl) planes of the archetypal structure (swinging twinning of Bovin & Andersson, 1977), and cyclic twinning (Hyde et al. 1974). For complex sulfides non-commensurability between adjacent building blocks, etc. (the vernier principle) comes next in the order of importance (Makovicky & Hyde, 1981), followed by antiphase and out-of-phase boundaries, crystallographic shear (Hyde et al., 1974), the 2Å-shear (or slip) in the structures with 4Å periodicity, and the intergrowth of two different structure types on a unit cell scale (ibid.). Detailed definitions of these operations were made by other contributors to this volume.

Coordination states and, in the case of cations, often also the chemical species of atoms on the block surfaces, rarely also in the interfaces, differ from those inside the building blocks.

2. The homologous series

A large number of sulfosalts and other complex sulfides represent members of homologous series. There are two principal types of homologous series which the present author defined as the accretional and the variable-fit series, respectively (Makovicky, 1989).

2.1. THE ACCRETIONAL SERIES

The accretional series is a series of structures in which the type(s) of building blocks (rods, layers) and the principles that define their mutual relationships remain preserved but the size of these blocks varies incrementally by varying the number of fundamental coordination polyhedra in them. The order N of a homologue in this type of series can be defined by the number of coordination polyhedra (polyhedral layers) across a suitably defined diameter of the building block (rod, layer). The ratio of atoms on the block surfaces (interfaces) to those inside the blocks varies with N as also does the overall M/S ratio.

Every member of the accretional series has its own chemical formula, unit cell parameters and symmetry; a general chemical formula for the entire series can be devised as well. A given homologue can represent a single compound or a (dis)continuous solid solution which can exsolve into a series of structurally ordered phases with well defined compositions (or compositional ranges) but with the same $\underline{N}$. The ideal space group can then be reduced to subgroups in the process of (cation) ordering.

Thompson (1978) described accretional homologous series using a different name (polysomatic series) and different approach: the coordination polyhedra on the surfaces and in the interfaces of the blocks (in his case, layers) are treated as one type of (layer) module whereas the incrementally accreted polyhedral layers in the layer (block) interior as another type of layer modules. Thus, all accretional homologues are treated as ordered intergrowths of two structure types which occur in different proportions in different homologues (polysomes). The two descriptions are equally valid; the polysomatic approach presents difficulties for either rudimentary or very elaborate interfaces (Makovicky, 1989) as well as for rod structures with two intersecting systems of interfaces (Veblen, 1991).

Some accretional series are <u>extensive</u> series with $\underline{N}$ varying over a range of values. Besides the members with equal widths ($\underline{N}_1 = \underline{N}_2$) across the unit-cell twinning plane or interface of another kind, those with unequal widths ($N_1 \neq N_2$) can occur. A number of accretional series are limited to only <u>pairs of homologues</u> (N_1, and $N_2 = N_1 + 1$) because of various local or global crystal chemical reasons. These pairs can be extended into <u>combinatorial series</u>, the members of which represent regular intergrowths of the above two accretional homologues: $N_1N_2N_1N_2..., N_1N_1N_2N_1N_1N_2..., N_1N_2N_2N_2N_1N_2N_2N_2...$, etc.

2.2. THE VARIABLE-FIT SERIES

This type of homologous series occurs for crystal structures that are composed of two kinds of alternating, mutually non-commensurate layers (rarely of such columns or of a matrix/infilling combination). Each kind of layer has its own short-range (sub)periodicity and it takes $\underline{m}$ periods of one layer and $\underline{n}$ periods of the other layer before they meet in the same configuration as at the origin. These are so-called <u>semi-commensurate</u> cases for which the two layer (sub)periodicities comprise a ratio of two not very large integers. Besides them, <u>incommensurate</u> cases exist for which these short-range periodicities comprise irrational fractions or ratio of very large integers. Non-commensurability of layers may occur in one or two interplanar directions. With minor compositional changes in the cations, the m/n ratio and the M/S ratio vary within certain, rather narrow limits, leading to a series of closely related compounds.

Those cases for which the pile of the two alternating non-commensurate layers is periodically sheared, kinked or modified by antiphase boundaries, represent <u>combination of the accretional</u> (separately for each layer) <u>and variable-fit</u> (for the interlayer match) <u>principles</u>. The geometrical constraints in these structures are much more severe than in the pure variable-fit structures and result in a very reduced number of homologues (usually homologous pairs).

3. Case study I: The lillianite homologous series

The lillianite homologous series (Makovicky & Karup-Møller, 1977) is an easily understandable extensive accretional series, eminently suitable for illustrating all the above defined problems and categories. Members of this series are the Pb-Bi-Ag sulfosalts with the structures consisting of alternating layers of PbS-like structure, cut parallel to $(311)_{PbS}$. These planes also represent the reflection- and contact planes of unit-cell twinning. The overlapping octahedra of the adjacent, mirror-related layers are replaced by bicapped trigonal coordination prisms PbS_{6+2} with the Pb atoms positioned on the mirror planes (Otto & Strunz, 1968; Takéuchi et al., 1974, etc.).

Distinct homologues differ in the thickness of the PbS-like layers. This is conveniently expressed as the number $\underline{N}$ of octahedra in the chain of octahedra that runs diagonally across an individual PbS-like layer and is parallel to $[011]_{PbS}$ (Fig. 3). Each lillianite homologue can be denoted as $^{N1,N2}L$ where $\underline{N}_1$ and $\underline{N}_2$ are the (not necessarily equal) values of $\underline{N}$ for the two alternating sets of layers (Fig. 4). Its chemical formula is $Pb_{N-1-2x}Bi_{2+x}Ag_xS_{N+2}$ (Z = 4) where N = $(N_1+N_2)/2$ and $\underline{x}$ is the coefficient of the Ag+Bi = 2Pb substitution. If the trigonal coordination prisms of Pb are not substitutable (which is very close to the real situation), $x_{max} = (N-2)/2$. This structure type is quite frequent also outside the Pb-Bi-Ag compositional space, e.g. for a number of complex lanthanide sulfides; the general formula then becomes $M^{2+}_{N-1}M^{3+}_2S_{N+2}$. For a somewhat different generalized approach the reader should consult Hyde et al. (1979).

The existence of lillianite homologues (sensu lato) depends on the suitable sizes of coordination polyhedra (trigonal prisms vs. octahedra), satisfactory valence balance and feasibility of close-to-regular octahedral (i.e., ccp or PbS-like) arrays. The case with N = 1 is typified by, for example, $NdYbS_3$ (Fig. 5) or $UFeS_3$, N = 2 by MnY_2S_4 or $CrEr_2S_4$, whereas $N_1;N_2 = 1;2$ by $FeHo_4S_7$, MnY_4S_7 and $Er_3Sc_2S_7$ (lanthanide atoms occupy the trigonal coordination prisms as well as they mix statistically with the other component in the octahedral positions). For the $MnS-Y_2S_3$ system, Bakker & Hyde (1978) found that the homologous pair MnY_2S_4 and "$MnYS_3$" (that occurs only as a layer in MnY_4S_7) form a combinatorial series that comprises MnY_4S_7 ($^{1,2}L$), $Mn_2Y_6S_{11}$ ($^{1,2,2}L$), $Mn_4Y_{10}S_{19}$($^{1,2,2,2,2}L$), etc. Both the cases with ideal symmetry and those with subgroup symmetry (Problem 1) are present; the latter caused either by deformations of coordination polyhedra or by the asymetric position of cations in the trigonal coordination prisms. The tetrahedral voids on the mirror planes of unit cell twinning are occupied only in exceptional cases (Eu_2CuS_3, $^{1,1}L$) [for references on individual compounds the reader should consult Structure Reports].

Higher homologues start at N = 4; they allow more pronounced departures from the galena-like array, especially in the form of locally "inflated" interspaces that accommodate lone electron pairs of quasi-octahedral Bi or Sb (so-called lone electron pair micelles, Makovicky & Mumme, 1983). Only $^{4,4}L$ and $^{7,7}L$ are known for the Ag-free subsystem Pb-Bi-S. Reduction of symmetry in $^{7,7}L$ from the usual orthorhombic to monoclinic, caused by Pb/Bi ordering that proceeded differently for the two (originally) mirror-related slabs was exceptionally observed in one natural occurrence of $^{7,7}L$ (Mumme, unpubl.). Allowing for the Ag+Bi = 2Pb substitution in octahedral layers, also the cases $^{4,7}L$, $^{4,8}L$, $^{5,9}L$ and

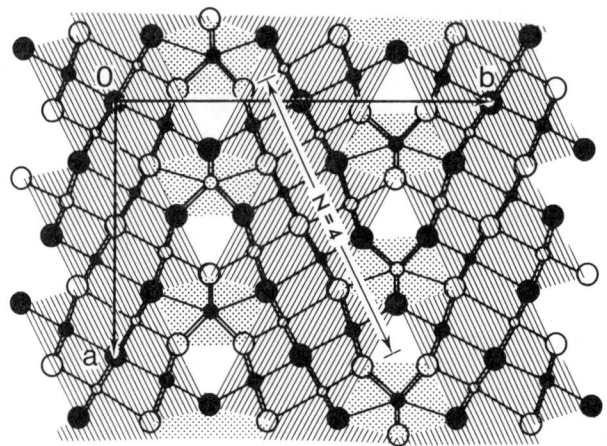

Figure 3. The crystal structure of lillianite, $Pb_3Bi_2S_6$, $N_1;N_2 = 4;4$ (Takagi & Takéuchi, 1972). In order of decreasing size circles denote S, Pb, (Pb,Bi). "Galena-like" layers are ruled, bicapped trigonal coordination prisms of Pb are stippled. Void and filled circles denote atoms at z = 1/2 and 0, respectively, along the vertical, 4.1 Å axis.

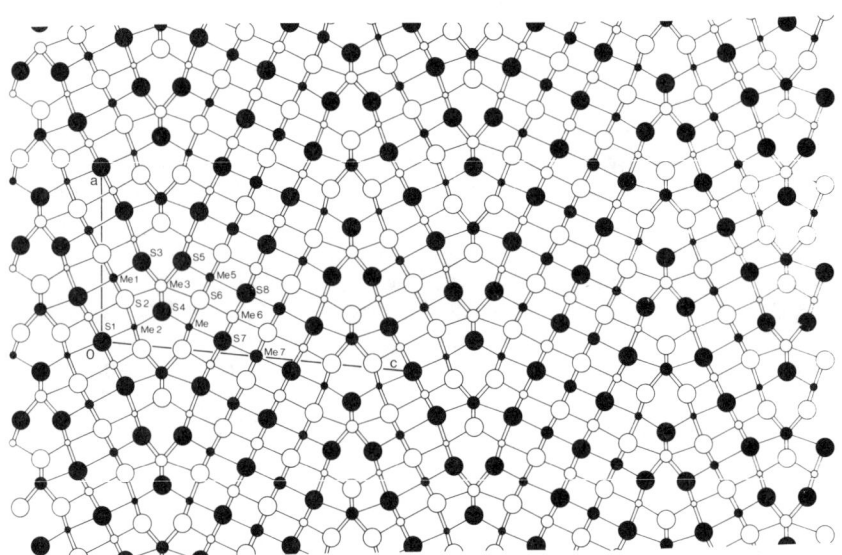

Figure 4. The crystal structure of vikingite, $Pb_{10}Ag_4Bi_{12}S_{30}$ (Makovicky, Mumme & Madsen, in prep.), a lillianite homologue $N_1;N_2 = 4;7$. The same conventions apply as for Fig. 3; the smallest circles represent (Ag,Bi).

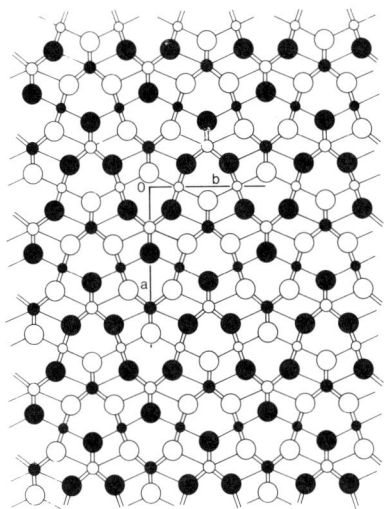

Figure 5. The crystal structure of NdYbS$_3$(Carré & Laruelle, 1974), a lillianite homologue $N_1;N_2$ = 1;1. Nd occupies trigonal prisms, Yb the octahedral positions of the (rudimentary) PbS-like layers. For other conventions see Fig. 3.

Figure 6. The crystal structure of gustavite, PbAgBi$_3$S$_6$ (Kupčík & Steins 1990). In order of decreasing size circles denote S, Pb, Bi and (Ag,Bi); two atomic levels of the orthorhombic subcell, ~2 Å apart, are indicated by void and filled circles. Stippling indicates cores of large lone electron pair micelles in which the $\underline{s}^2$ pairs of Bi are concentrated.

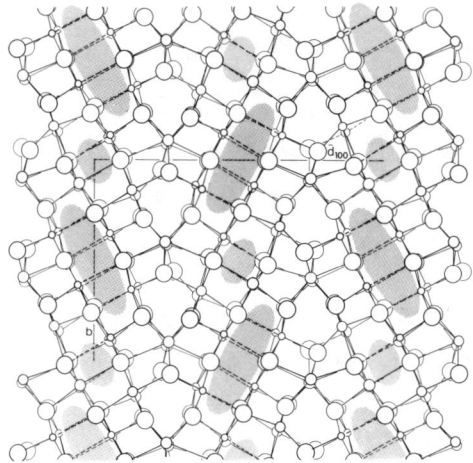

Figure 7. The crystal structure of ramdohrite, $Ag_6Pb_{12}Sb_{22}S_{48}$ (Makovicky & Mumme, 1983). In the order of decreasing size the circles indicate S, Pb, Ag (and mixed positions) and Sb. All atoms are situated about four levels along the 8.7 Å (vertical) axis, ~2.2 Å apart. Stippling indicates cores of large and small lone electron pair micelles in which the $\underline{s}^2$ pairs of Sb are concentrated.

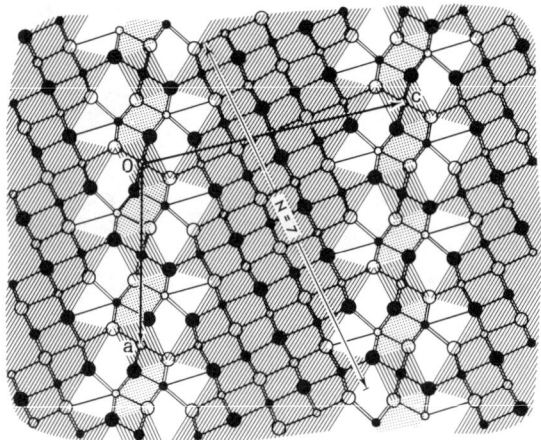

Figure 8. The crystal structure of benjaminite, $Cu_{0.5}Pb_{0.4}Ag_{2.3}Bi_{6.8}S_{12}$ (i.e. $\sim Ag_3Bi_7S_{12}$) (Makovicky & Mumme, 1979), the 7th homologue of the pavonite homologous series. Conventions are analogous to Fig. 6 but the PbS-like portions are ruled and the BiS_{3+2} caps of distorted trigonal prisms on the boundaries of galena-like layers with $N_1 = 1$ and $N_2 = 7$ are stippled.

^{11,11}L (Makovicky & Karup-Møller, 1977) are found, as well as the disordered combinations based on $N_1 = 4$ and $N_2 = 7$ in different proportions (ibid., Skowron & Tilley, 1990). The structures with close-to-ideal PbS-like arrays (Fig. 3) and those with extensive lone electron pair micelles (Fig. 6) do not form continuous solid solutions inspite of the same $\underline{N}$ or/and they separate by exsolution at low temperatures (e.g. $Pb_3Bi_2S_6$-$PbAgBi_3S_6$, Figs. 3 and 6). The match of coordination polyhedra is better in the Pb-Ag-Bi sulfosalts and yields an extensive accretional series. In the parallel system Pb-Ag-Sb-S the mismatch of Pb, Ag and Sb coordination polyhedra appears serious; only members with $N = 4$ were found and they are known only for $\geq 50\%$ of substitution (Ag+Sb) for 2Pb; still, they form a string of intermediate (exsolved) phases with interesting distribution of lone electron pair micelles and superperiods of the 4Å dimension (Fig. 7). The existence of phases close to 50% substitution often depends on incorporation of smaller M^{2+} instead of octahedral Pb; e.g. $AgPb_3MnSb_5S_{12}$ (Moëlo et al., 1988).

The symmetry or asymmetry of the trigonal prismatic cation site varies widely for the lanthanide-based lillianite homologues with lower N values. The difference in the symmetry or asymmetry of this position becomes substantial, and to the large extent unsurmountable by cation substitutions, between the lillianite and pavonite homologous series in the Pb-Bi-Ag-Cu system (Makovicky et al., 1977; Makovicky, 1981, 1989). In the pavonite homologues (Fig. 8) all members have $N_1 : N_2 = 1 : N_{pav}$ where $N_{pav} = 2$ to 8, possibly even 11. The trigonal prisms on the planes of contracted-set unit-cell twinning are occupied by square-pyramidal Bi and its lone electron pair. The extensive PbS-like portions (N_2) contain quasi-octahedral to octahedral Bi combined with Ag,Cu and some Pb. The sole, skewed octahedra in the narrow portions ($N_1 = 1$) represent AgS_{2+4} or contain [3] and [4] coordinated Cu. Also two other lillianite analogues, $BaCu_4S_3$ ($N = 1$) and $Pb_{1-x}Bi_{2+x}Cu_{4-x}S_5I_2$ ($x = 0.88$; 2-Å sheared, with $N \approx 5$) contain Cu atoms in similar coordinations.

Two more homologous series, the extensive meneghinite homologous series and the combinatorial sartorite homologous series, both being based on glide-reflection twinned arrays, are examined in Problems 2 and 3.

4. Case study II: Sulfides of heavy and transition metals with misfit layer structures

These sulfides consist of two types of alternating, mutually non-commensurate layers. One type (Q layers) has pseudotetragonal symmetry and belongs to the PbS or SnS archetype. The other type (H) are octahedral or trigonal prismatic layers.

These compounds can display one of the three following types of interlayer match. If the Q layers are described as (100) and the match is expressed in terms of centered pseudotetragonal and orthohexagonal ($\underline{c}_H = \sqrt{3}\underline{b}_H$) subcells, the first, "cylindrite-like" match (Fig. 15) implies (exactly or approximately) $\underline{b}_Q \parallel \underline{b}_H$ and $\underline{c}_Q \parallel \underline{c}_H$ with $\underline{b}_Q > \underline{b}_H$ and $\underline{c}_Q < \underline{c}_H$ (e.g. in cylindrite ~$FePb_3Sn_4Sb_2S_{14}$, $19\underline{b}_Q \approx 30\underline{b}_H$ and $13\underline{c}_Q = 12\underline{c}_H$, common modulation of both layers $\parallel \underline{c}$); the second, "$LaCrS_3$-like" match implies $\underline{c}_Q = \underline{c}_H$ and $\underline{b}_Q > \underline{b}_H$ ($\underline{b}$ is the semi- or incommensurate modulation direction, e.g. $3\underline{b}_Q = 5\underline{b}_H$ for ~$LaCrS_3$). The third match involves Q layers turned by 45° so that $\underline{b}_H \parallel$ (and equal to) $\underline{b}_{Qprimitive}$ (=

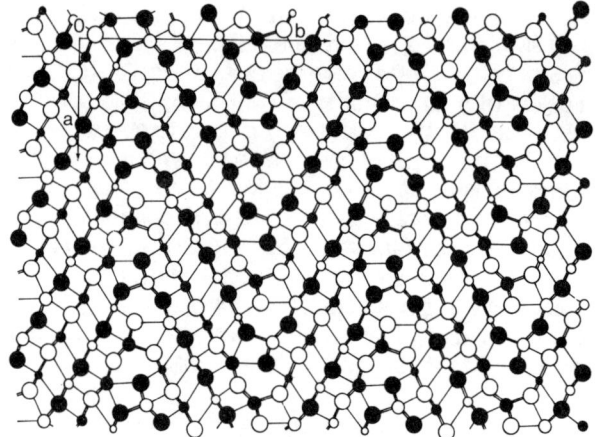

Figure 9. The crystal structure of meneghinite, $CuPb_{13}Sb_7S_{24}$ (Euler & Hellner, 1960), homologue N = 5 of the meneghinite homologous series. Circles in order of decreasing size indicate S, Pb, (Pb,Sb), Sb. The statistically occupied, unrefined positions of Cu in the interfaces of SnS-like blocks were omitted. Filled and empty circles indicate atoms at two z levels, 2.1 Å apart.

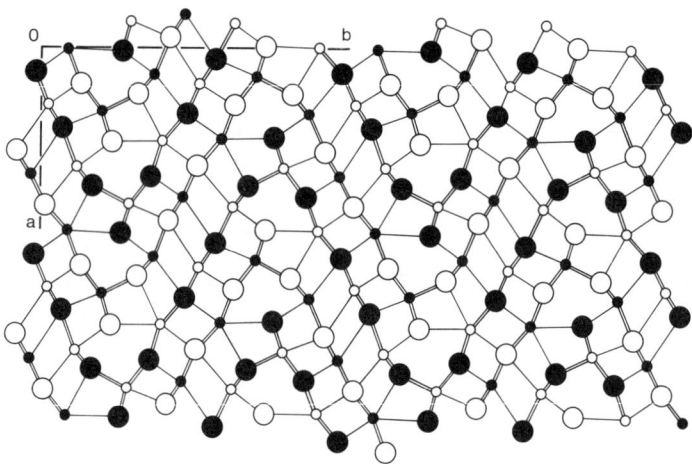

Figure 10. The crystal structure of $Bi_xSb_{2-x}Sn_2S_5$ (0.4 > x > 0.2) (Kupčík & Wendshuh, 1982), N = 4. Metal atoms are not differentiated, remaining conventions as in Fig. 8.

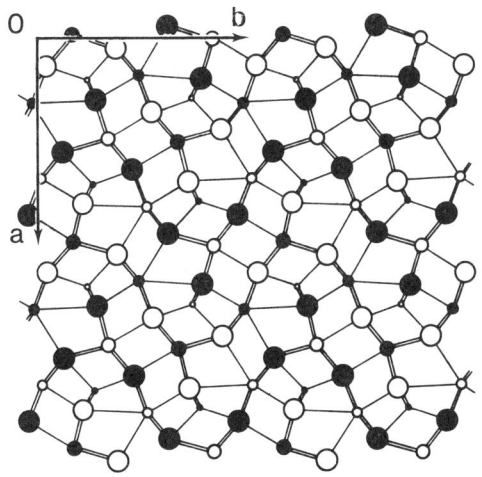

Figure 11. The crystal structure of aikinite, $CuPbBiS_3$ (Ohmasa & Nowacki,1970; Kohatsu & Wuensch,1971). In the order of decreasing size circles indicate S, Pb, Bi and Cu; atoms are at two $\underline{z}$ levels, 2 Å apart.

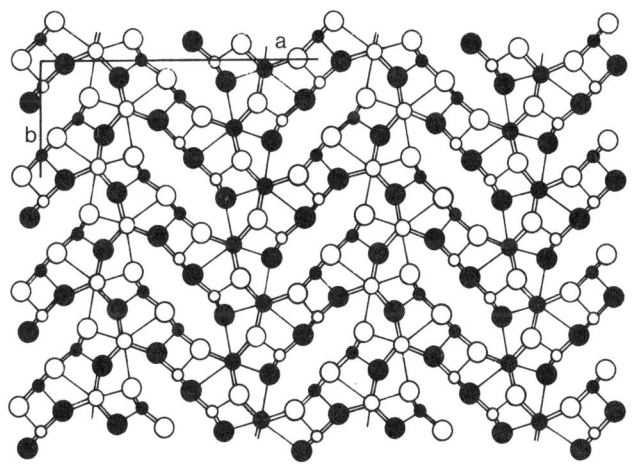

Figure 12. The crystal structure of sartorite, $PbAs_2S_4$ (Iitaka & Nowacki,1961), homologue N = 3 of the sartorite homologous series. Only the 4.2 Å substructure, with atoms at two levels, has been determined. Circles in order of decreasing size indicate S, Pb and As.

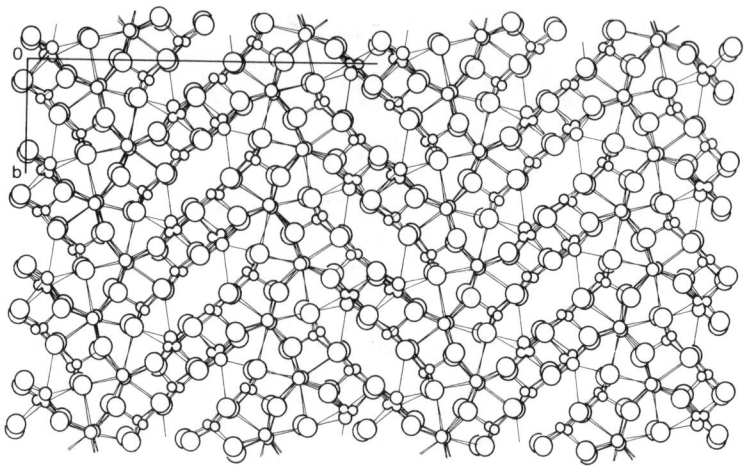

Figure 13. The crystal structure of rathite I $Pb_{11.1}Tl_{0.9}As_{16}(As_{1.6}Ag_{2.3})S_{40}$ (Marumo & Nowacki;1965), homologue $N = 4$ of the sartorite homologous series. Projection of the 8.4 Å structure onto (001). In the order of decreasing size circles indicate S, Pb (and Tl), As (and Ag). Overlapping atoms are 4.2 Å apart in the $\underline{c}$ direction; statistically occupied Ag and As positions are on the same level.

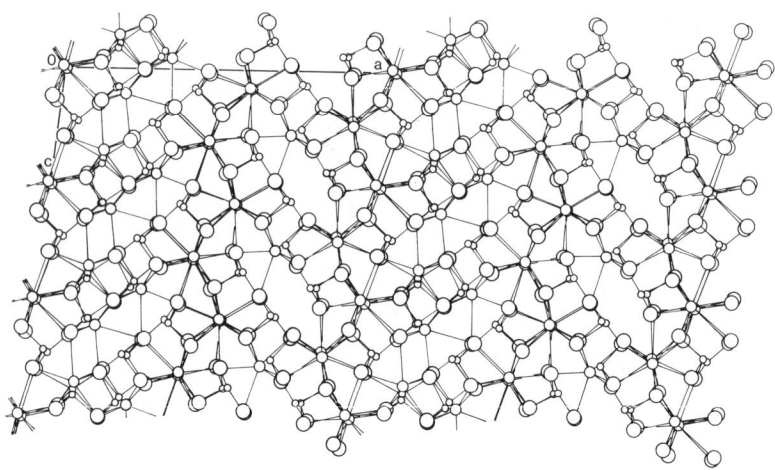

Figure 14. The crystal structure of baumhauerite, $Pb_{11.6}As_{15.7}Ag_{0.6}S_{36}$ (Engel & Nowacki, 1969), a member of the combinatorial sartorite series. The same conventions apply as for Figs. 12 and 25.

1/2 diagonal of Q centred) and $\underline{c}_H > \underline{c}_{Qprimitive}$ (e.g., cannizzarite $\sim Pb_{24}Bi_{28}S_{66}$ with $12\underline{c}_Q \approx 7\underline{c}_H$). The LaCrS$_3$-like match occurs for the largest Q/H cation radius ratio whereas, as the other extreme, the cannizzarite match is typical for about equally large average cations in the Q and H layers (Fig. 16).

The Q layers, MeS, can be 2 to 4 atomic layers thick; the H layers, MeS$_2$, represent single, double- or seldom even triple-octahedral layers; the trigonal prismatic layers occur between two consecutive Q layers as single layers or in pairs. Fine changes in the chemistry of the two layer types lead to variable-fit homologous series for all three match types (Problem 4). Geometric modulation of layers along the non-commensurate direction is ascribed solely to layer interactions for the "ABX$_3$" cases whereas it might be connected with the constant presence of minor cations, Fe and Sb, in the structures of cylindrite family (complex Pb-Sn^{2+}-Sn^{4+}-Sb-Fe sulfosalts). Recent detailed reviews were written by Makovicky & Hyde (1981, 1992).

5. Case study III: Complex sulfides with combination of the accretional and variable-fit principle

Stresses and strains as well as the charge balance problems in the non-commensurate layer structures can be relieved by breaking them up by means of glide planes, antiphase or out-of-phase boundaries, or by exposing alternatively the Q and the H surfaces of the same layer (Fig. 17). With respect to the rest of the structure, these planes of break-up may be composition-and-structure-conservative or non-conservative (Fig. 17). Problem 5 deals with several basic types of this category.

For the majority of these phases, cannizzarite (Fig. 18) acts as a parent structure. Similar to cannizzarite, they represent piles of Q- and H-type layers of different thicknesses (not always the thicknesses observed in cannizzarite itself); they have similar match modes as cannizzarite and similar average cation radii.

6. Epilogue

The rationale behind this contribution is to outline the basic structural features of complex sulfides with recombination structures. Its volume prevents exhaustive discussion of details for the structural families for which detailed reviews have already been written (referenced in Makovicky 1989 and Makovicky & Hyde 1992). Detailed exhaustive descriptions of individual structure-building principles are given by other contributors to this crystal chemistry volume.

7. Acknowledgments

Kind assistance of the editor of this volume, Prof. E. Parthé, (University of Genève), the typing assistance of Mrs. M.L. Johansen and the help of generations of draftsmen and -women are gratefully acknowledged.

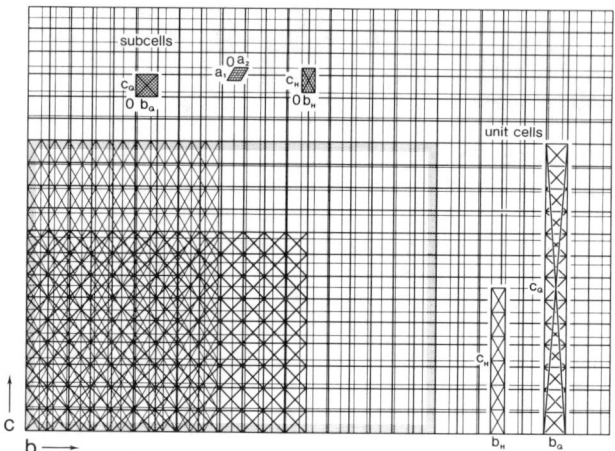

Figure 15. Interlayer match in cylindrite, $\sim FePb_3Sn_4Sb_2S_{14}$ (Makovicky, 1974). The two component layers, Q and H, are indicated by their submesh (pseudotetragonal and orthohexagonal; denoted by the two grids with their centrations partly omitted) and by the two unit cells. The $\underline{b} \times \underline{c}$ (incommensurate x semicommensurate) coincidence mesh is stippled; modulation vector is [001].

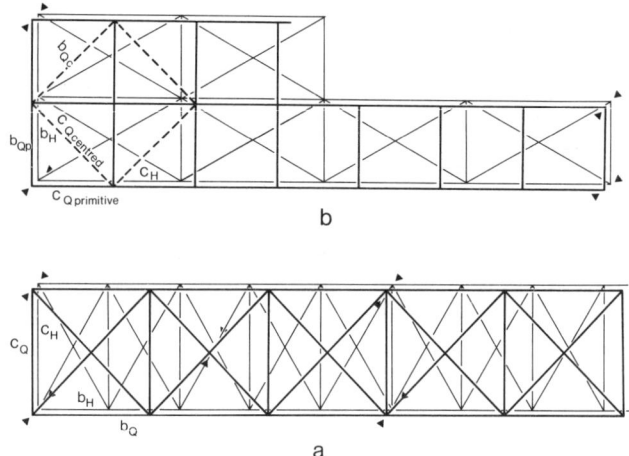

Figure 16. Interlayer match of pseudotetragonal (Q) and orthohexagonal (H) submesh in (a) layer misfit compound $(LaS)_{1.20}CrS_2$ (i.e., "$\sim LaCrS_3$") and (b) in the lanthanide sulfide $Er_9La_{10}S_{27}$ with a match analogous to that in cannizzarite.

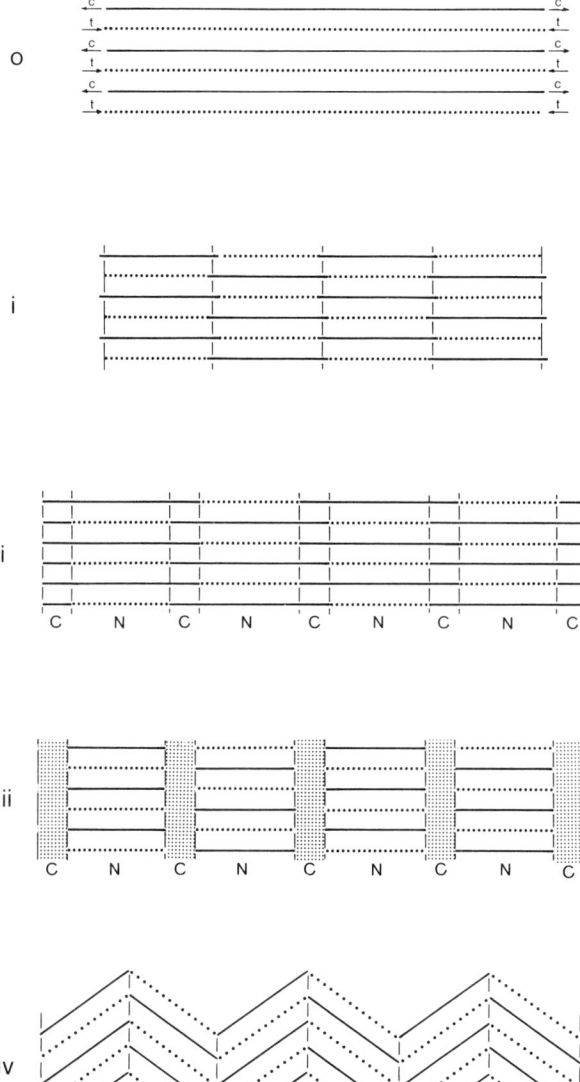

Figure 17. A non-commensurate, layer misfit structure (o), composed of alternating layers of two types, and the recombination structures (i-iii) derived from (o) by means of (i) composition-conservative and of (ii-iii) two kinds of composition non-conservative antiphase boundaries (or glide planes). (iv) A layered non-commensurate structure with corrugated layers that are composed of alternating strips of two kinds that exhibit layer misfit on interfaces.

8. Problems

Problem 1
(a) Determine ideal space groups for lillianite homologues with $N_1 = N_2$ (Fig. 3) and $N_1 \neq N_2$ (Fig. 4). Derive one or more subgroups by displacing various atoms (groups) from their 'ideal' positions or by ordered cation substitutions in the 4Å or 8Å cell.

(b) Calculate chemical formulae of Ag-free and (Ag+Bi)-substituted homologues $^{4,4}L$, $^{7,7}L$, $^{5,9}L$, $^{11,11}L$, $^{4,7}L$ as well as of the hypothetical homologue $^{2,2}L$.

(c) Unit cell parameters of $Pb_3Bi_2S_6$ ($^{4,4}L$) are $\underline{a}$ 13.54 Å, $\underline{b}$ 20.45 Å and $\underline{c}$ 4.10 Å,; those of $Pb_6Bi_2S_9$ ($^{7,7}L$) are $\underline{a}$ 13.70 Å, $\underline{b}$ 31.21 Å and $\underline{c}$ 4.13 Å. Calculate the increment $\underline{\Delta b}$ for $\Delta N = 1$ and estimate the $\underline{b}$ parameters for the hypothetical homologues $^{2,2}L$, $^{5,5}L$, $^{6,6}L$ and $^{11,11}L$. What parameter do we obtain for $^{4,7}L$?

Problem 2
Figures 9-11 show three distinct members of the meneghinite homologous series: meneghinite $CuPb_{13}Sb_7S_{24}$, synthetic $Bi_xSb_{2-x}Sn_2S_5$ and $CuPbBiS_3$.

(a) Outline the slabs $(50\pm1)_{SnS}$ in the SnS-like portions of each structure.

(b) Determine the value of $\underline{N}$ for each of these structures using the number of square coordination pyramids in the SnS-like regions.

(c) Show the statistically occupied Cu site in meneghinite using the structure of $CuPbBi_3S_6$ as a model.

(d) Calculate chemical formulae of meneghinite homologues $N = 2$ to 5 using the general formula you have derived from Figs. 9-11. Calculate these formulae separately for Cu-free and Cu-saturated cases. What is the site occupancy for Cu in real meneghinite, $CuPb_{13}Sb_7S_{24}$?

(e) Unit-cell parameters of $CuPb_{13}Sb_7S_{24}$ are $\underline{a}$ 11.36 Å, $\underline{b}$ 24.06 Å and $\underline{c}$ 4.13 Å. Those of $CuPbBiS_3$ are $\underline{a}$ 11.32 Å, $\underline{b}$ 11.64 Å and $\underline{c}$ 4.04 Å. Determine $\underline{b}$ parameters of the hypothetical phases with $N = 3$ and 4. What is the (ideal) space group for all these structures? Compare the calculated data with the incremental parameter $\underline{b}$ (19.81 Å) of $Cu_{0.2}Pb_{2.2}(Sb,Bi)_{1.8}S_5$ and with that of the closely related compound $FeSb_2S_4$ (berthierite, 14.12 Å).

Problem 3
Figures 12 and 13 show the two known simple members of the sartorite homologous series, $N = 3$ and 4 (i.e., more exactly, 3; 3 and 4; 4). Their structures are based on the SnS archetype, cut along $(301)_{SnS}$. In this family, the SnS archetype is pretwinned on $(011)_{SnS}$ what does not influence our interpretation. This series is of combinatorial character.

(a) Outline slabs (301) of SnS archetype in each structure.

(b) Demonstrate the derivation of N for each of the illustrated structures.

(c) The general formula of sartorite homologues is $Me^{2+}_{4N-8-2x}Me^{3+}_{8+x}Me^{+}_{x}S_{4N+4}$ (Makovicky, 1985) where Me^{+} enters the composition as $2Pb^{2+} = Ag^{+}+As^{3+}$. Derive ideal formulae for the combinations (3;3), (4;4), (3;4), (3;4;4) and the yet hypothetical case (3;3;4); all formulae should be Ag-free.

(d) The incremental direction $\underline{a}$ is 19.62 Å for sartorite and 25.74 Å for $Pb_2As_2S_5$ (dufrenoysite, N = 4 without Ag). Derive $\underline{d}_{100}$ for the above (Ag-free) slab combinations.

(e) Analyse the crystal structure of baumhauerite in Fig. 14 and find its $(N_1;N_2)$ combination.

Problem 4
(a) Derive chemical formulae of the following series of $\sim ABX_3$ compounds with 2 atomic planes thick Q layers MeS and single trigonal prismatic H layers MS_2 (see text for details of the interlayer match). These compounds are described as truly incommensurate. Normalize therefore to 1 orthohexagonal cation.

Cations		$\underline{a}$ (Å)	$\underline{b}$ (Å)	$\underline{c}$ (Å)
Y-Nb	Q	22.28	5.393	5.660
	H	11.13	3.322	5.660
Sm-Ta	Q	22.56	5.562	5.648
	H	22.56	3.292	5.648
Ce-Ta	Q	11.44	5.737	5.750
	H	22.89	3.293	5.750
La-Ta	Q	11.53	5.813	5.778
	H	23.06	3.295	5.778

(b) Misfit layer structures of the cylindrite family have a semicommensurate Q/H match (and common modulation) in the interplanar $\underline{c}$-direction that changes from 12 centred Q subcells: 11 ortho-hexagonal H subcells (valid for Sn^{2+}-rich Q layers in this variable-fit series) to about 16 Q subcells: 15 H subcells (for the Pb-rich Q layers; the Pb-Sn^{2+} substitution in the Q layers is the principal chemical variation in this series; it does not concern the H layers that are based on Sn^{4+}). If we assume (in agreement with available data) that the Q/H match in the - presumably incommensurate - $\underline{b}$-direction is reasonably constant and can be simplified to $7b_Q \approx 11b_H$, try to calculate the generalized formulae $M_QM_HS_{total}$ for the cylindrite homologues 12Q/11H, 13Q/12H, 14Q/13H, 15Q/14H, and 16Q/15H and examine the development of different atomic ratios. Normalize the formulae to 1H subcell. N.B. Cylindrite has 2 atomic planes thick Q layers and single-octahedral H layers.

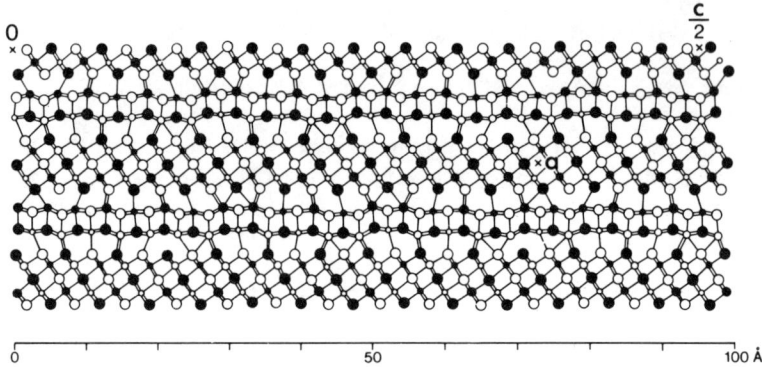

Figure 18. The crystal structure of cannizzarite, $Pb_{46}Bi_{54}S_{127}$ (Matzat, 1979). In order of decreasing size the circles indicate S, Pb, (Pb,Bi) and Bi; filled and open circles indicate atoms at two discrete levels along the 4 Å axis, 2 Å apart. Double-octahedral H layers are intercalated by double Q layers.

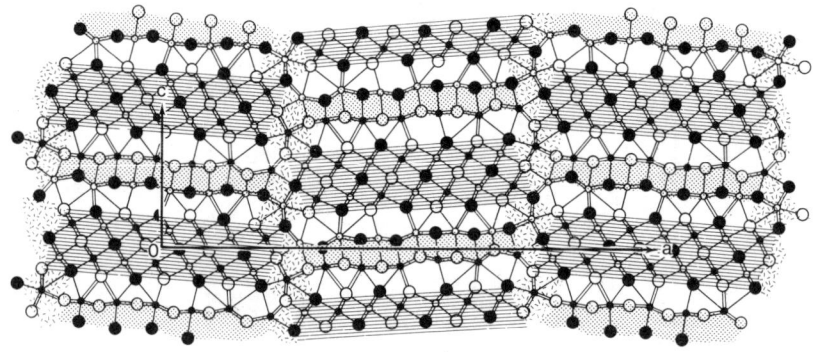

Figure 19. The crystal structure of weibullite, $Ag_{0.33}Pb_{5.33}Bi_{8.33}(S,Se)_{18}$ (Mumme, 1980). Small circles: undifferentiated Me atoms, large circles: undifferentiated S and Se atoms. Two height levels, 2 Å apart, are populated. Remaining features (signatures) are to be determined in Problem 5.

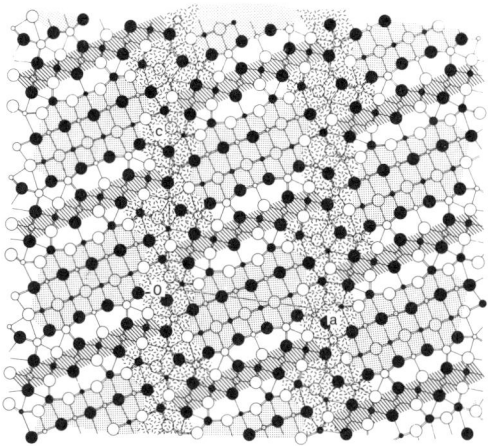

Figure 20. The crystal structure of $Pb_4In_3Bi_7S_{18}$ (Krämer & Reis, 1986). In order of decreasing size circles represent S, Pb, In and Bi. Remaining features (signatures) are to be determined in Problem 5.

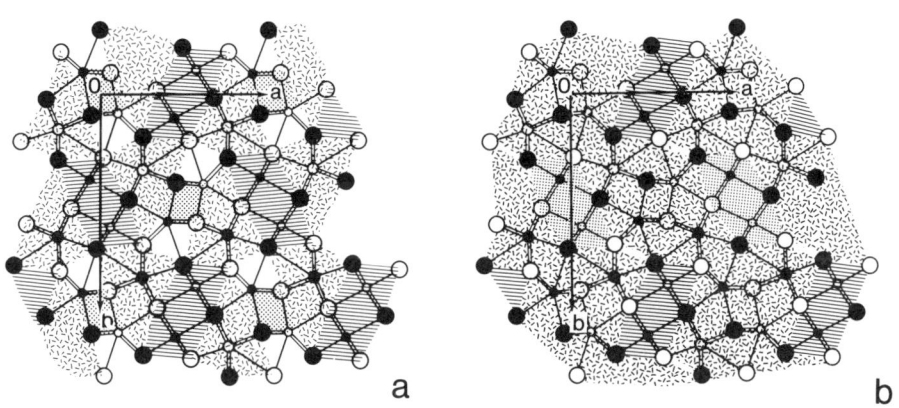

Figure 21. The crystal structure of galenobismutite $PbBi_2S_4$ (Iitaka & Nowacki, 1962). Circles in order of decreasing size signify S, Pb and Bi. The two types of shading (a) and (b) relate to the structures in Figs. 19 and 20.

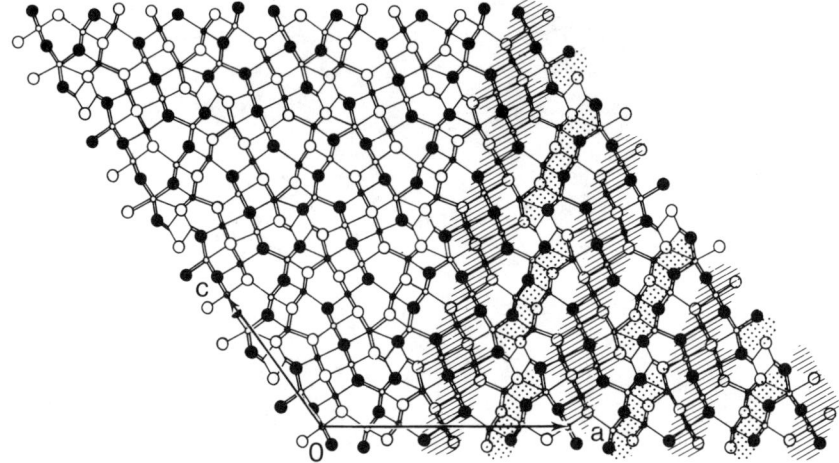

Figure 22. The crystal structure of junoite, $Cu_2Pb_3Bi_8(S,Se)_{16}$ (Mumme,1975), a sheared layer structure. Circles in order of decreasing size indicate (S,Se), Pb and Bi, and Cu. Atoms are at y = 0 and 1/2 of the 4 Å axis. Remaining features and signatures are subject of Problem 5.

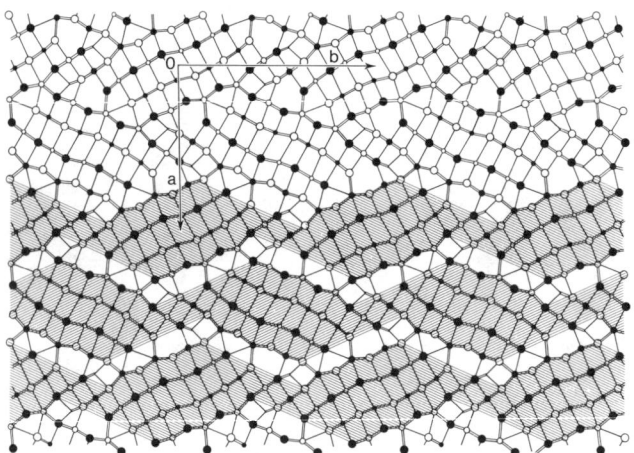

Figure 23. The crystal structure of cosalite, ~$Pb_2Bi_2S_5$ (Srikrishnan & Nowacki, 1979). Circles in order of decreasing size: S, Pb, Bi and Cu, empty and full circles are at two levels, 2 Å apart.

Problem 5

(a) Figures 19 and 20 show the crystal structures of weibullite, $Ag_{0.3}Pb_{5.3}Bi_{8.3}(S,Se)_{18}$, and of $Pb_4In_3Bi_7S_{18}$. Describe their structural relationship to cannizzarite (Fig. 18), define the sizes and types of building blocks, interlayer match and recombination operators. Find and characterize their relationship to the crystal structure of galenobismutite (Fig. 21).

(b) Define the building blocks, interlayer match and recombination operator for the crystal structure of junoite $Cu_2Pb_3Bi_8(S,Se)_{16}$ in Fig. 22.

(c) Characterize the building blocks and their sizes, as well as the extent of the two types (H and Q) of their surfaces (in terms of H and Q subcells/subperiodicites) for an individual, periodically constricted layer in the crystal structure of cosalite $\sim Pb_2Bi_2S_5$ in Fig. 23. Find the interlayer match between two consecutive breakpoints on the surfaces of two adjacent layers. (Hint: resolve ambiguities by preserving the constant thickness of the interface).

(d) Try layer selection and descriptions as sub(c) for the crystal structure of jamesonite $FePb_4Sb_6S_{14}$ (Fig. 24).

8.1. SOLUTIONS OF THE PROBLEMS

Problem 1

(a) For the choice of axes in Fig. 3 the space groups are Bbmm and B2/m, respectively. For gustavite, $PbAgBi_3S_6$ (Fig. 6) with regular sequences Ag-Bi-Ag-Bi in the mixed octahedral positions, Ag (and Bi) in the consecutive Ag-Bi strings that follow each other in the orthorhombic [100] direction are staggered in a monoclinic fashion; the resulting space group is $P2_1/c$ for 8 Å cell (Harris & Chen, 1975). The 8 Å cell of ramdohrite $Ag_6Pb_{12}Sb_{22}S_{48}$ is $P\ 1\ 2_1/n\ 1$ (Fig. 7). Space group for $NdYbS_3$ is $B22_12$ due to minor shifts in atomic positions.

(b)

Homologue	Ag-free	Ag+Bi substituted
$^{2,2}L$	$PbBi_2S_4$	-
$^{4,4}L$	$Pb_3Bi_2S_6$	$PbAgBi_3S_6$
$^{7,7}L$	$Pb_6Bi_2S_9$	$PbBi_{4.5}Ag_{2.5}S_9$
$^{5,9}L = {}^7L$	$Pb_6Bi_2S_9$	$PbBi_{4.5}Ag_{2.5}S_9$
$^{4,7}L = {}^{5.5}L$	$Pb_{4.5}Bi_2S_{7.5}$	$PbBi_{3.75}Ag_{1.75}S_{7.5}$
$^{11,11}L$	$Pb_{10}Bi_2S_{13}$	$PbBi_{5.5}Ag_{4.5}S_{13}$

(c) The average increment of $\underline{b}$ for $\Delta N = 1$ in the case of Ag-free lillianite homologues is

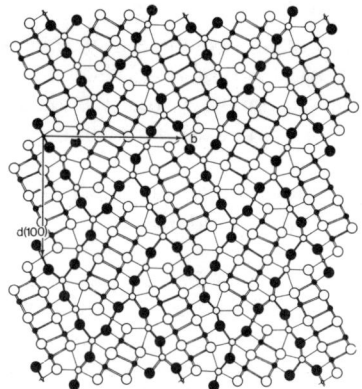

Figure 24. The crystal structure of jamesonite, $FePb_4Sb_6S_{14}$ (Niizeki & Buerger, 1957). In order of decreasing size circles denote S, Pb, Sb and Fe. All atoms are at $\underline{z}$ ~0 and $\underline{z}$ ~1/2, respectively. Remaining features are the subject of Problem 5.

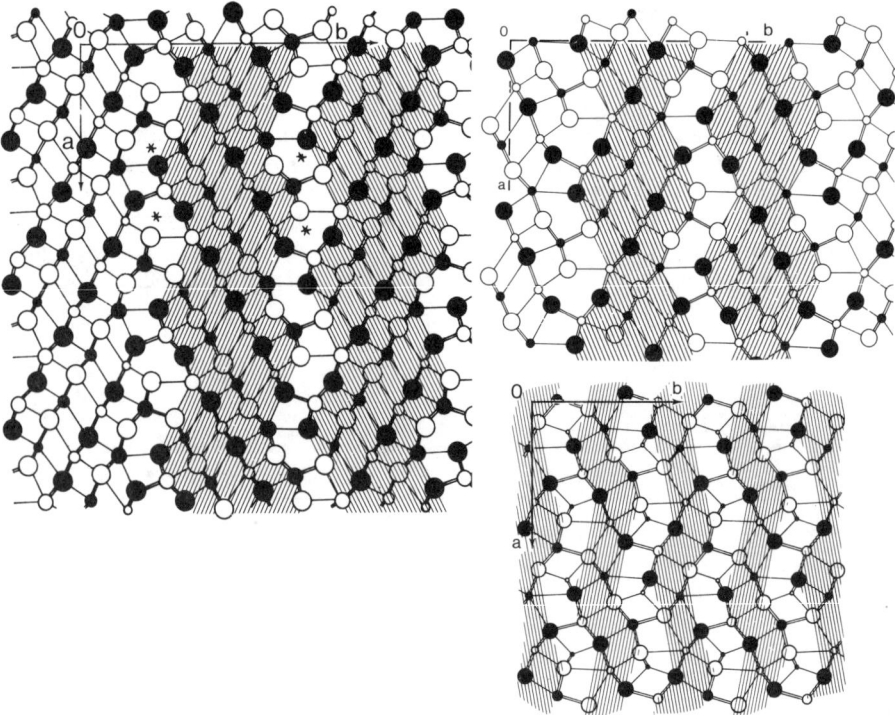

Figure 25. Slabs $(501)_{SnS}$ of SnS archetype in the crystal structures of meneghinite homologues (ruled). (a) $CuPb_{13}Sb_7S_{24}$ (N = 5); (b) $Bi_xSb_{2-x}Sb_2S_5$ (N = 4), and (c) $CuPbBiS_3$ (N = 2). Asterisks denote some of the statistically occupied Cu tetrahedral sites. Compare with Figs 9-11.

3.587 Å. The estimated $\underline{b}$ parameters are 24.04 Å for ^{5,5}L, 27.62 Å for ^{6,6}L and 45.56 Å for ^{11,11}L. $\underline{d_{010}}$ for ^{4,7}L is 25.83 Å. For comparison, $\underline{b}$ of Ag-substituted ^{11,11}L (ourayite) is 44.04 Å; that of Ag-substituted ^{4,7}L (vikingite) is 25.25 Å.

Problem 2

(a) Slabs of SnS archetype and hatched in Fig. 25.

(b) N is determined by the number of square pyramids on either side of a tightly bonded double layer running diagonally across the slab $(501)_{SnS}$.

$CuPb_{13}Sb_7S_{24}$ N = 5

$Bi_xSb_{2-x}Sb_2S_5$ N = 4

$CuPbBiS_3$ N = 2

(c) The statistically occupied Cu site (Euler & Hellner, 1960) is indicated by an asterisk in Fig. 25.

(d) <u>Chemical formulae of meneghinite homologues</u>, $Cu^+_xMe^{2+}_{2N-4+x}Me^{3+}_{4-x}S_{2N+2}$

N	Cu-free	Cu-saturated
2	$Bi_2S_3(Sb_2S_3)$	$CuPbBiS_3$
3	$Pb(Sb,Bi)_2S_4$	$CuPb_2(Sb,Bi)S_4$
4	$Pb_2Sb_2S_5$	$CuPb_3SbS_5$
5	$Pb_3Sb_2S_6$	$CuPb_4SbS_6$

Natural meneghinite $CuPb_{13}Sb_7S_{24}$ displays 25% saturation by Cu.

(e) Increment of $\underline{b}$ for $\Delta N = 1$ is 4.14 Å. This value includes the changes caused by the Bi/Sb exchange which correlates in nature with the decrease in the value of N. Therefore $\underline{b}$ = 15.78 Å for N = 3 (compare with only 14.12 Å for related berthierite $FeSb_2S_4$ with N = 3) and 19.92 Å for N = 4 (compare with 19.81 for jaskolskiite $Cu_{0.2}Pb_{2.2}(Sb,Bi)_{1.8}S_5$). The ideal space group of meneghinite homologues is Pbnm.

Problem 3

(a) Slabs of SnS archetype are outlined in Fig. 26.

(b) N is defined as the number of coordination pyramids on one side of the tightly-bonded double layer, diagonal across a slab of SnS archetype (Figs. 26 and 14).

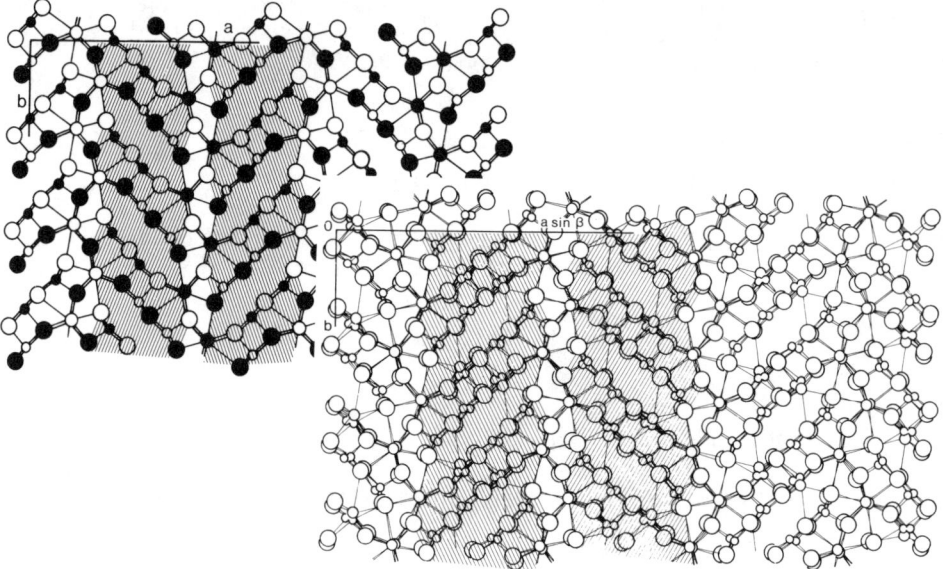

Figure 26. Slabs $(301)_{SnS}$ of SnS archetype (ruled) in the crystal structures of two simple sartorite homologues; (a) $PbAs_2S_4$ ($N = 3$) and (b) $Pb_{11.1}Tl_{0.9}As_{16}(As_{1.6}Ag_{2.3})S_{40}$. Compare with Figs 12-14.

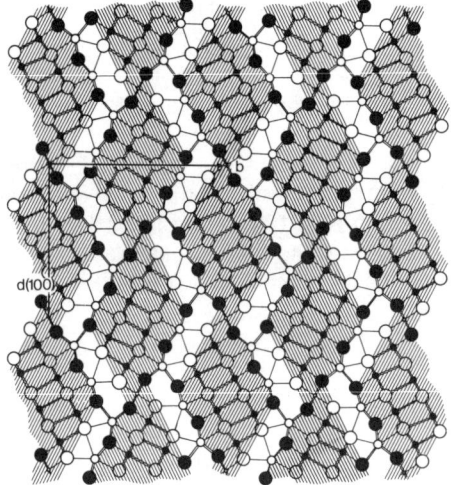

Figure 27. Definition of layers (indicated by ruling) for the crystal structure of jamesonite, $FePb_4Sb_6S_{14}$. Layers represent periodically twinned SnS archetype; the twinned and constricted portions are populated by octahedrally coordinated Fe; the non-commensurate interlayer spaces are unshaded. Compare with Figs. 23 and 24.

(c,d) Sartorite combinatorial series

$N_1;N_2$	Ag-free ideal composition	Ideal $\underline{a}$ or $\underline{d}_{100}$
(3;3)	$Pb_4As_8S_{16}$	19.62 Å
(3;3;4)	$Pb_{16}As_{24}S_{52}$	32.49 Å
(3;4)	$Pb_6As_8S_{18}$	22.68 Å
(3;4;4)	$Pb_{20}As_{24}S_{56}$	35.55 Å
(4;4)	$Pb_8As_8S_{20}$	25.74 Å

(e) Baumhauerite is a combined homologue (3;4) with $\underline{a}$ 22.8 Å and β 97.3 Å (Fig. 14).

Problem 4

(a) $(YS)_{1.23}NbS_2$, $(SmS)_{1.19}TaS_2$, $(CeS)_{1.14}TaS_2$ and $(LaS)_{1.13}TaS_2$.

(b) Cylindrite homologues ($7\underline{b}_Q \approx 11\underline{b}_H$)

Subcell match $m\underline{c}_Q/n\underline{c}_H$	Q component 7x	H component 11x	Composition $Me_Q/Me_H/S$	Me_Q/Me_H	Me/S
12/11	$12Me_4S_4$	$11Me_2S_4$	2.777/2/6.777	1.388	0.7048
13/12	$13Me_4S_4$	$12Me_2S_4$	2.758/2/6.758	1.379	0.7040
14/13	$14Me_4S_4$	$13Me_2S_4$	2.741/2/6.741	1.371	0.7033
15/14	$15Me_4S_4$	$14Me_2S_4$	2.727/2/6.727	1.364	0.7027
16/15	$16Me_4S_4$	$15Me_2S_4$	2.715/2/6.715	1.358	0.7022

Problem 5

Weibullite: alternation of two-planes-thick Q layers and double-octahedral H layers, respectively 5 1/2 Q subcells and 3 1/2 H subcells broad; interlayer match 6Q / 3 1/2 H; composition/structure non-conservative boundaries are composed of trigonal prisms represent glide planes in the layer pile.

$Pb_4In_3Bi_7S_{18}$: alternation of four-planes-thick Q layers and single-octahedral H layers, respectively 4Q and 3H subcells broad; interlayer match 3Q/2H. The composition- and structure non-conservative boundaries are based on a combination of 'vertical' and 'horizontal' trigonal prisms, respectively MeS_{6+2} and MeS_{6+1}, and represent out-of-phase boundaries.

$PbBi_2S_4$ **(galenobismutite)** represents the lowermost homologue for both of the above compounds; these relationships are schematically shown in Fig. 21 $\underline{a}$ and $\underline{b}$.

Junoite: alternation of two-planes thick Q layers and single H layers, the layer fragments are respectively 3Q and 2 1/2 H subcells broad; interlayer match is 3 1/2 Q / 2H subcell; (out-of-phase) shear planes are almost perpendicular to layers, they display doubling of the H layer and Cu atoms in the Q layer.

Cosalite ~$Pb_2Bi_2S_5$: a continuous layer of PbS archetype; building blocks are rods with cross-sections 4Q x 1 1/2 H, with interconnecting layer intervals that have octahedral, H character. Surfaces: respectively 3Q subcells and 2H subcells broad; interlayer match 3Q/2H.

Jamesonite $FePb_4Sb_6S_{14}$: a continuous layer of (in constricted portions periodically twinned) SnS archetype (Fig. 27); building blocks are 3Q x 1 1/2 (sheared) H rods that are in contact via the Fe octahedra in the constricted portions; rod surfaces and interlayer match: 2Q subcells / 1 1/2 (sheared) H subcells.

9. References

Andersson, S. & Hyde, B.G. (1974). 'Twinning on the unit cell level as a structure-building operation in the solid state', J. Solid State Chem. **9**, 92-101.

Bakker, M. & Hyde, B.G. (1978). 'A preliminary electron microscope study of chemical twinning in the system MnS + Y_2S_3, an anlogue of the mineral system PbS + Bi_2S_3 (galena + bismuthinite)', Phil. Magazine, **A38**, 615-628.

Bovin, J.O. & Andersson, S. (1977). 'Swinging twinning on the unit cell level as a structure-building operation in the solid state', J. Solid State Chem. **20**, 127-133.

Carré, D. & Laruelle, P. (1973). 'Structure cristalline du sulfure d'erbium et de lanthane, $Er_9La_{10}S_{27}$', Acta Cryst. **B29**, 70-73.

Engel, P. & Nowacki, W. (1969). 'Kristallstruktur von Baumhauerit', Z. Kristallogr. **129**, 178-202.

Euler, R. & Hellner, E. (1960). 'Über komplex zusammengesetzte sulfidische Erze VI. Zur Kristallstruktur des Meneghinits, $CuPb_{13}Sb_7S_{24}$', Z. Kristallogr. **113**, 345-372.

Harris, D.C. & Chen, T.T. (1975). 'Gustavite - two Canadian occurrences', Can. Mineral. **13**, 411-414.

Hyde, B.G., Bagshaw, A.N., Andersson, S. & O'Keeffe, M.O. (1974). 'Some defect structures in crystalline solids', Ann. Rev. Mat. Sci. **4**, 43-92.

Hyde, B.G., Andersson, S., Bakker, M., Plug, C.M. & O'Keeffe, M. (1979). 'The (twin) composition plane as an extended defect and structure-building entity in crystals', Progr. Solid State Chem. **12**, 273-327.

Iitaka, Y. & Nowacki, W. (1961). 'Refinement of the pseudo crystal structure of scleroclase, $PbAs_2S_4$', Acta Cryst. **14**, 1291-1292.

Iitaka, Y. & Nowacki, W. (1962). 'A redetermination of the crystal structure of galenobismutite, $PbBi_2S_4$', Acta Cryst. **15**, 691-698.

Kohatsu, I. & Wuensch, B.J. (1971). ' The crystal structure of aikinite, $PbCuBiS_3$'. Acta Cryst. **B27**, 1245-1252.

Krämer, V. & Reis, I. (1986). 'Lead indium bismuth chalcogenides. II. Structure of $Pb_4In_3Bi_7S_{18}$', Acta Cryst. **C42**, 249-251.

Kupčík, V. & Wendshuh, M. (1982). 'The structure of antimony bismuth tin sulphide $Bi_xSb_{2-x}Sn_2S_5$', Acta Cryst. **B38**, 3070-3071.

Kupčík, V. & Steins, M. (1991). 'Verfeinerung der Kristallstruktur von Gustavit $(Pb_{1.5}Ag_{0.9}Bi_{2.5}Sb_{0.1}S_6)$' Berichte Deutsch. Mineral. Gesellschaft 1990/2, 151.

Lima de Faria, J., Hellner, E., Liebau, F., Makovicky, E. & Parthé, E. (1990). 'Nomenclature of inorganic structure types'. Report of the IUCr Commission on Crystallographic Nomenclature, Subcommittee on the Nomenclature of Inorganic Structure Types. Acta Cryst. **A46**, 1-11.

Makovicky, E. (1974). 'Mineralogical data on cylindrite and incaite', N. Jahrb. Mineral. Abh. **126**, 304-326.

Makovicky, E. (1981). 'The building principles and classification of bismuth-lead sulphosalts and related compounds', Fortschr. Mineral. **59**, 137-190.

Makovicky, E. (1985). 'The building principles and classification of sulphosalts based on the SnS archetype', Fortschr. Mineral. **63**, 45-89.

Makovicky, E. (1989). 'Modular classification of sulphosalts - current status. Definition and application of homologous series', N. Jahrb. Miner. Abh. **160**, 269-297.

Makovicky, E. & Hyde, B.G. (1981). 'Non-commensurate (misfit) layer structures', Structure & Bonding **46**, 101-170.

Makovicky, E. & Hyde, B.G. (1992). 'Incommensurate, two-layer structures with complex crystal chemistry: minerals and related synthetics'. In "Incommensurate Misfit Sandwiched Layered Compounds", ed. Meerschaut, A., Trans.Tech Publ. Ltd., in press.

Makovicky E. & Karup-Møller, S. (1977). 'Chemistry and crystallography of the lillianite homologous series. I. General properties and definitions', N. Jahrb. Miner. Abh. **130**, 264-287.

Makovicky, E. & Mumme, W.G. (1983) 'The crystal structure of ramdohrite, $Pb_6Sb_{11}Ag_3S_{24}$ and its implications for the andorite group and zinckenite', N. Jahrb. Mineral. Abh. **147**, 58-79.

Makovicky, E., Mumme, W.G. & Watts, J.A. (1977). 'The crystal structure of synthetic pavonite, $AgBi_3S_5$ and the definition of the pavonite homologous series', Can. Mineral. **15**, 339-348.

Makovicky, E., Mumme, W.G. & Madsen, I.C. (1993). 'The crystal structure of vikingite', N. Jahrb. Mineral. Abh., in press.

Makovicky, E. & Mumme, W.G. (1979). 'The crystal structure of benjaminite $Cu_{0.50}Pb_{0.40}Ag_{2.30}Bi_{6.80}S_{12}$', Can. Mineral. **17**, 607-618.

Marumo, F. & Nowacki, W. (1965). 'The crystal structure of rathite - I', Z. Kristallogr. **122**, 433-456.

Matzat, E. (1979). 'Cannizzarite', Acta Cryst. **B35**, 133-136.

Moëlo, Y., Makovicky, E. & Karup-Møller, S. (1988). 'Sulfures complexes plombo-argentifères: Minéralogie et cristallochimie de la série andorite-fizelyite (Pb, Mn,Fe, Cd,Sn)$_{3-2x}$ (Ag,Cu)$_x$(Sb,Bi,As)$_{2+x}$(S,Se)$_6$', Documents BRGM (Orléans) **167**, 107 pp.

Mumme, W.G. (1975). 'Junoite, $Cu_2Pb_3Bi_8(S,Se)_{16}$, a new sulfosalt from Tennant Creek, Australia: Its crystal structure, and relationship with other bismuth sulfosalts. Amer. Mineral. **60**, 548-558.

Mumme, W.G. (1980). 'Weibullite, $Ag_{0.32}Pb_{5.02}Bi_{8.55}Se_{6.08}S_{11.92}$ from Falun, Sweden. A higher homologue of galenobismutite'. Can. Mineral. **18**, 1-18.

Niizeki, W. & Buerger, M.J. (1957). 'The crystal structure of jamesonite, $FePb_4Sb_6S_{14}$', Z. Kristallogr. **109**, 161-183.

Ohmasa, M. & Nowacki, W. (1973). 'A redetermination of the crystal structure of aikinite [BiS_2 | S | $Cu^{IV}Pb^{VIII}$]', Z. Kristallogr. **137**, 422-432.

Otto, H.H. & Strunz, H. (1968). ' Zur Kristallchemie synthetischer Blei-Wismut-Spiessglanze', N. Jahrb. Mineral. Abh. **108**, 1-19.

Srikrishnan, T. & Nowacki, W. (1974). 'A redetermination of the crystal structure of cosalite, $Pb_2Bi_2S_5$', Z. Kristallogr. **140**, 114-136.

Skowron, A. & Tilley, R.J.D. (1990). 'Chemically twinned phases in the Ag_2S-PbS-Bi_2S_3 syst em. Part 1. Electron microscope study'. J. Solid State Chemistry **85**, 235-250.

Takagi, J. & Takeuchi, Y. (1972). 'The crystal structure of lillianite', Acta Cryst. **B28**, 649-651.

Takeuchi, Y. (1978). 'Tropochemical twinning: A mechanism of building complex structures', Recent Progress Nat. Sci. Japan **3**, 153-181.

Thompson, J.B., Jr. (1978). 'Biopyriboles and polysomatic series', Am. Mineral. **58**, 239-249.

Veblen, D.R. (1991). 'Polysomatism and polysomatic series: A review and applications", Amer. Mineral. **76**, 801-826.

THE BOND VALENCE METHOD IN CRYSTAL CHEMISTRY

MICHAEL O'KEEFFE
Department of Chemistry
Arizona State University
Tempe, Arizona 85287
USA

ABSTRACT. The concept of bond valence is defined and the dependence of bond lengths on valences described. The use of these correlations for the prediction and interpretation of bond lengths in crystals is then discussed. The crystal-chemical implications of rules for predicting bond valences of crystals are illustrated.

1. The Concept of Bond Valence

The idea that a bond can be assigned a valence has proved enormously fruitful in chemistry. It may be considered as a generalization of the concept of bond order (single and double bonds *etc.*). An important observation is that bond lengths and bond valences are correlated. This leads to two complementary topics that are discussed in this article: the *prediction* and the *interpretation* of bond lengths. Rules for the prediction of bond valences, in particular Brown's *equal valence rule* allow some generalizations to be made about the stability of structure types. Although the method is some 40 years old, and was discussed extensively in the 70's, it is only now becoming the preferred method of discussing bond lengths in molecules and crystals.

In the interest of economy, references are not provided to crystal structures used as examples where these are to be found in standard reference sources such as *Structure Reports*. On the other hand a fairly extensive bibliography for applications of the bond valence method in crystal chemistry is provided.

1.1. DEFINITION OF BOND VALENCE

The valence of a bond between two atoms i and j is v_{ij}. Bond valences are defined so that the sum of all the bond valences of the bonds formed by a given atom i is the atom valence V_i.

$$\Sigma_j v_{ij} = V_i \qquad (1)$$

Thus, for example, if an Al atom with valence 3 forms four equal bonds to other atoms the bond valences are 3/4.

1.2. DEPENDENCE OF BOND LENGTH ON BOND VALENCE

Many empirical expressions have been proposed to express the correlation of bond length d_{ij} with valence (see *e.g.* Donnay & Donnay, 1973, Brown, 1981). Probably the best of these is that originally used (Pauling, 1947):

$$d_{ij} = R_{ij} - b \ln v_{ij} \qquad (2)$$

In this expression R_{ij} is (at least in the formal sense) the length of a single bond ($v_{ij} = 1$) and is a parameter characteristic of bonds between atoms i and j. It transpires (Brown & Altermatt, 1985) that, to a good approximation, b may be considered a "universal" parameter equal to 0.37 Å. This value is used here.

1.3. BOND VALENCE PARAMETERS

From equations (1) and (2) one has for an atom i forming k bonds with atoms of type j and of length d_k:

$$R_{ij} = b \ln[V/\Sigma \exp(-d_k/b)] \qquad (3)$$

Thus values of R_{ij} can be determined for pairs of atoms in a variety of coordinations in experimental crystal structures, and are generally found to be nearly constant for a given pair. Particularly in the case of oxides, for which a large data base exists, R_{ij} values are known rather accurately (Brown & Altermatt, 1985). Bond valence parameters for a wide variety of cation - anion bonds were subsequently determined by Brese & O'Keeffe (1991) using an interpolation technique. Parameters for anion-anion bonds in solids have also been determined (O'Keeffe & Brese, 1992).

Approximate bond valence parameters can be estimated for virtually any bond using a method of O'Keeffe & Brese (1991) who showed that for some 600 atom pairs involving both homopolar and heteropolar bonds it is a good approximation to write

$$R_{ij} = r_i + r_j - [r_i r_j (\sqrt{c_i} - \sqrt{c_j})^2]/(c_i r_i + c_j r_j) \qquad (4)$$

Here r and c are parameters characteristic of a given atom. c was found to be closely correlated with electronegativity. The last term on the right of equation (4) is very small except for bonds between very electronegative and electropositive elements (it is about 10% of the total for Cs-F bonds). As R_{ij} is, in a formal sense, a single bond length, the right hand side of equation (4) can be considered as a sum of single bond radii with an electronegativity-difference correction.

It was also found that although no data for metal-metal bonds were used in determining the parameters r and c, the values of R_{ij} obtained from equation (4) for metal-metal bonds, when used in equation (2) accurately predicted the bond lengths in metals with unambiguous valences (non-transition metals).

It should be remarked however, that the bond valence method is generally less suitable for "metallic" bonding than for "covalent" and "ionic" bonds and some caution is required in the interpretation of metal-metal distances. Particularly for the more-electropositive elements, it is often found that interatomic distances decrease substantially on forming a compound. To take a fairly typical and familiar example, the Mg-Mg distance in MgO is 2.97 Å, in the elemental metal it is 3.20 Å. In the former case it is generally conceded that there is not Mg-Mg bonding, but there must be in the latter.

2. Prediction of Bond Valences and Bond Lengths in Crystals

Normally one has an experimentally determined crystal structure and one wishes to interpret the experimental bond lengths. However there are occasions where one wishes to predict expected bond lengths in crystals (see *e.g.* O'Keeffe, 1991a; Brown, 1992). Perhaps more important, the fact that the prediction of bond lengths is successful suggests that the method is soundly based, and leads to useful crystal chemical generalizations.

2.1. THE VALENCE CONSTRAINT AND CONNECTIVITY MATRICES

We start by calculating bond lengths in a simple structure: that of B_2O_3 in its high-pressure form with B in 4-coordination. Denoting coordination numbers by prefixed Roman numerals, the coordination numbers are $^{iv}B_2{}^{ii}O(1)^{iii}O(2)_2$. The topology of the bonding is conveniently set out in a *connectivity matrix*:

	O(1)	2 O(2)
2 B	2α	6β

Here the numbers preceding the the element symbols are the numbers per formula unit and the entries in the matrix are the numbers of bonds and symbols (α and β) for their valences.

The bond valence sums are easily read off the matrix:

horizontally: $\quad 2\alpha + 6\beta = 2V_B = 6$
vertically: $\quad 2\alpha = V_O = 2 \quad\quad 6\beta = 2V_O = 4$

With the solutions $\alpha = 1$, $\beta = 2/3$. Using $R_{BO} = 1.37$ Å, we calculate that each B forms one bond of length 1.37 Å to O(1) and three bonds of length 1.52 Å to O(2) in accord with the experimentally described structure.

This almost trivial example illustrates two important points. (a) If there are *m* cations and *n* anions there are *m+n*-1 independent valence sums (note that the first equation given above is the sum of the second and third). (b) We assume that the bonds from O(1) will be equal in valence (length) and likewise that the bonds from O(2) will be equal in valence (we have implicitly used an equal valence rule).

Bond valence sum constraints dictate that in this instance the coordination of B by O will be irregular. Note the superiority of the bond valence method over a sum-of-radii method which would predict four equal B-O bond lengths.

2.2. THE BROWN EQUAL VALENCE RULE

We now consider a somewhat more complex structure: that of $SrBe_3O_4$. This compound may be written $SrBe(1)_2Be(2)O(1)O(2)_3$ emphasizing the five different kinds of atom in the structure. The connectivity matrix is set out below:

	O(1)	3 O(2)
Sr	3α	6β
2 Be(1)	2γ	6δ
Be(2)	-	3ε

From the connectivity matrix, one can see that there are 5 kinds of bonds (with valences indicated as α, β, γ, δ and ε). However with 5 kinds of atom there are only 5-1 = 4 independent bond valence sums and another condition is required to determine the five bond valence parameters. This is provided by Brown's *equal valence rule*. This rule may be stated as requiring that, subject to the bond valence constraint, bonds between like pairs of atoms have valences as nearly equal as possible.

This constraint can be shown (Brown, 1992) to be equivalent to, in this instance, requiring that the difference between the valences of the Sr-O(1) and Sr-O(2) bonds is equal to the differences between the valences of the Be(1)-O(1) and Be(1)-O(2) bonds (note that the valence, ε, of the Be(2)-O(2) bonds is given directly by the bond valence sum constraint $3\varepsilon = 2$). In the notation given above:

$$\alpha - \beta = \gamma - \delta. \tag{5}$$

This is the required fifth equation which allows the five expected valences to be determined by solving five simultaneous linear equations.

It has been shown (O'Keeffe, 1989) that the necessary extra equations can always be read off the connectivity matrix and that the resulting equations always have a unique solution. A more complicated example illustrating the procedure is given in § 2.3 below.

The equal valence rule, as described above, gives undue weight to weak (low valence) bonds, and a minor modification which gives generally better predictions, assigns weights to bonds that are proportional to their average valences. Thus letting s_{SrO} be the average valence of the Sr-O bonds and s_{BeO} be the average valence of the Be(1)-O bonds, equation (5) is replaced by:

$$(\alpha - \beta)/s_{SrO} = (\gamma - \delta)/s_{BeO}. \tag{6}$$

Note that s_{SrO} (= 2/9) and s_{BeO} (= 2/4) are equal to the Pauling (1929) electrostatic bond strengths of the Sr-O and Be(1)-O bonds.

The table below shows calculated bond lengths using equations (5) and (6) respectively for $SrBe_3O_4$. The differences are rather small but eq. (6) is slightly better. Most important is the observation that the lengths of the Sr-O and Be(1)-O bonds are correctly ordered, and indeed calculated to be close to those observed.

bond	Eq. (5)	Eq. (6)	observed
Sr - O(1)	2.58	2.61	2.65
Sr - O(2)	2.73	2.71	2.71
Be(1)-O(1)	1.59	1.57	1.60
Be(1)-O(2)	1.66	1.66	1.66
Be(2)-O(2)	1.53	1.53	1.53

2.3. COMPUTING ALGORITHMS

I now treat a slightly more complex example to show how predicted bond valences are determined in practice. Below is the connectivity matrix for β-Ga_2O_3 = Ga(1)Ga(2)O(1)O(2)O(3).

	O(1)	O(2)	O(3)
Ga(1)	α	2β	γ
Ga(2)	2δ	ε	3ζ

The bond valence sums (of which only four are independent) are:

horizontal $\quad\alpha + 2\beta + \gamma = 3$ $\qquad$ (i)
$\qquad\qquad\qquad 2\delta + \varepsilon + 3\zeta = 3$ $\qquad$ (ii)
vertical $\qquad\quad \alpha + 2\delta = 2$ $\qquad$ (iii)
$\qquad\qquad\qquad 2\beta + \varepsilon = 2$ $\qquad$ (iv)
$\qquad\qquad\qquad \gamma + 3\zeta = 2$ $\qquad$ (v)

Note that (v) = (i) + (ii) - (iii) - (iv) just expresses the overall satisfaction of valences (*i.e.* sum of cation valences = sum of anion valences).

The equal valence rule translates into sums around circuits on the connectivity matrix. There are three four-circuits yielding three equations, of which only two are independent. These are given here without weights for simplicity:

$\qquad\qquad\qquad \alpha - \beta + \varepsilon - \delta = 0$ $\qquad$ (vi)
$\qquad\qquad\qquad \beta - \gamma + \zeta - \varepsilon = 0$ $\qquad$ (vii)
$\qquad\qquad\qquad \alpha - \gamma + \zeta - \delta = 0$ $\qquad$ (viii)

Note that (viii) = (vi) + (vii) so that one has six independent equations to solve for the six bond valences. Again one finds predicted and observed bond lengths in good agreement with experiment (O'Keeffe, 1989).

This example is already a little tedious to solve by hand and it is a good idea to resort to computer methods especially for more complex structures. I (O'Keeffe, 1990a) find it easiest to program the computer to read off all equations from the connectivity matrix and to use an algorithm that solves linear equations that include redundancies. It should be noted that one never has an over-determined set of equations.

For a more complicated examples, including the use of six-circuits to express the equal valence rule, see Wagner & O'Keeffe (1988) and O'Keeffe (1990a).

A related algorithm for calculating bond valences in crystals has recently been described (Rutherford, 1990).

2.4. APPLICATION TO DEFECTS AND INTERFACES

The discussion so far has focussed on the calculation of valences (and bond lengths) in periodic crystals, but the bond valence method is equally applicable to aperiodic structures and is potentially very useful for predicting relaxations around defects in crystals (Brown, 1988) and at surfaces and interfaces (O'Keeffe, 1991b).

2.5. COMPARISON WITH USE OF RADII

Although the bond valence method has a long history (see § 5), it has only recently begun to supplant the use of sums of radii to predict bond lengths in crystals. One reason is possibly that one needs only to add two numbers to sum radii, whereas on needs to solve linear equations to predict valences (§ 2.3) and then take a logarithm to calculate bond lengths using the bond valence method. Although the increase in complexity is rather modest, it is worth demonstrating how the bond valence method is superior.

In § 2.1 we showed how the bond valence method correctly accounted for the irregular coordination tetrahedron around B in high-pressure B_2O_3 and remarked that a sum-of-radii method would predict a regular tetrahedron. In fact most compilations of radii ascribe most, if not all, of the variation of bond length with valence to a variation of cation radius with coordination number (e.g. Shannon, 1976). Apart from the fact that one would not expect a cation "size" (however defined) to vary significantly with coordination number, this approach can lead to some wrong conclusions as is now shown.

In the crystal structure of β-LaOF there are two kinds of anion, each forming four bonds to La. In the original structural work it was not possible to determine which anion was O and which was F. As the ionic radius (Shannon, 1976) of F is smaller than that of O, it was generally assumed that the shorter bonds were to F and the longer bonds were to O, but this is incorrect, as accurate structural work subsequently showed. The La-F bonds have a valence of 1/4 and the La-O bonds have valence 1/2 and in fact the weaker bonds are the longer ones. Thus compare the predictions of bond length of the two methods with the observed values:

	La-O	La-F
in crystal	2.42	2.60
bond valence method	2.43	2.57
sum of radii (^{viii}La)	2.54	2.47
sum of radii (^{vi}La and ^{xii}La)	2.41	2.67

Note that the bond valence method (using parameters from Brese & O'Keeffe, 1991) predicts close to the observed values, but the sums of ionic radii (using the radius for 8-coordinated La, Shannon, 1976) is substantially in error. The last row of the table shows the sums of ionic radii assuming a coordination number of 6 for La in La-O bonds (corresponding to a bond valence of 3/6 = 1/2) and a coordination number of 12 for La in the La-F bonds (corresponding to a bond valence of 3/12 = 1/4). Now the sums of radii are approximately correct, illustrating that it is bond valence rather than coordination number that determines bond length.

3. The Interpretation of Bond Lengths in Crystals

3.1. VERIFYING CRYSTAL STRUCTURES

A fairly obvious use of the bond valence method is to verify experimentally-determined crystal structures. In a good structure the bond valence sums should come close to the expected atom valences. However, particularly in compounds rich in large electropositive elements, sums less than anticipated may occur (see § 3.3).

3.2. DETERMINING ATOMIC VALENCES AND LOCATING ATOMS

Bond valence sums can be very useful in determining atom valences in crystals in cases where there is ambiguity. A simple example (O'Keeffe, 1989) to illustrate the point is afforded by the structure of ilmenite which may be formulated Fe(II)Ti(IV)O_3 or Fe(III)Ti(III)O_3. From the observed bond lengths and known bond valence parameters

(Brese & O'Keeffe, 1991) we can test the two hypotheses (note that in this instance we use bond valence parameters that depend on oxidation state):

hypothesis	V_{Fe}	V_{Ti}
Fe(II), Ti(IV)	2.08	3.99
Fe(III), Ti(III)	2.22	3.74

Clearly Fe(II) and Ti(IV) are indicated.

The currently topical question of valences in copper oxides is less straight-forward. For a discussion of these compounds in terms of bond valence sums see (O'Keeffe & Hansen, 1988, Brown, 1989). A complementary approach using Madelung potentials is described by O'Keeffe (1990b).

A related use of bond valences (Donnay & Allmann, 1970) is in distinguishing O^{2-}, OH^- and H_2O in crystals where the H atoms have not been located (as frequently occurs). The bond valence sums (other than for O-H bonds) should be close to 2 for O^{2-}, 1 for OH^- and 0 for H_2O. A related application is in distinguishing atoms close in atomic number, but different in valence, such as Al(III) and Si(IV) in aluminosilicates (Brown, 1978). In § 2.5 above we discussed the case of distinguishing between O and F in LaOF. Let the two anions be $X(1)$ and $X(2)$ and compare bond valence sums for the two possible assignments of anions:

hypothesis	V_{La}	$V_{X(1)}$	$V_{X(2)}$
$X(1) = O, X(2) = F$	3.00	2.04	0.96
$X(1) = F, X(2) = O$	2.80	1.50	1.30

Clearly the first hypothesis is indicated as correct.

It has also been shown that likely positions for light atoms such as Li in crystal structures can be identified by locating sites where the bond valence sum is the correct value (Waltersson, 1978).

3.3. INDICATIONS OF NON-BONDED INTERACTIONS

In corundum (= $^{vi}Al_2{}^{iv}O_3$) there is only one kind of Al and one kind of O atom, so with R_{AlO} = 1.65 Å we would expect for a valence of 3/6, that the Al-O bond lengths would be 1.91 Å. In fact there are two different bond lengths, 1.85 Å (3x) and 1.97 Å (3x). In the structure, pairs of AlO_6 octahedra share faces and the Al-Al repulsions across the face stretch one half of the Al-O bonds; to maintain the correct bond valence sums at Al and O, the other half of the bonds are correspondingly shortened. If one wishes to model this effect non-bonded repulsions must be included (O'Keeffe, 1991a). One then minimizes the repulsion energy with the bond valence sums constrained to be correct. It transpires that both O-O and Al-Al repulsions are required to model the crystal structure correctly.

A more complicated example is that of garnet $^{viii}A_3{}^{vi}B_2{}^{iv}C_3{}^{iv}O_{12}$. Here there are just four kinds of atom: A, B, C and O, but two independent A-O bonds which are generally of different lengths and which again can be successfully modeled using atom-atom repulsions. In this example if one only constrains the bond valence sum at O, it is found (O'Keeffe, 1991a) that the experimental bond valence sum at the individual cations can be reproduced. Two striking examples of the effect of non-bonded repulsions on bond lengths (and hence on bond valence sums) is given for two garnets $A_3Al_2Si_3O_{12}$ (A = Ca, Mg) in the table below in which the valences are the sums determined from the experimental bond lengths.

A	V_A	V_{Al}	V_{Si}	V_O
Mg	1.72	3.17	3.90	1.93
Ca	2.51	2.87	3.78	2.05

When a "small" atom such as O is highly coordinated by "large" atoms such as the more electropositive metals, the effect of non-bonded repulsions can be very severe (McGuire & O'Keeffe, 1984) and it is natural to ascribe the effect to cation-cation non-bonded repulsions. The stretching of bonds due to this effect is also mirrored in anomalously low bond energies (O'Keeffe & Hyde, 1984). A particularly revealing example is that of the A-type rare earth oxide structure exemplified by $La_2O_3 =$ $^{vii}La_2{}^{vi}O(1){}^{iv}O(2)_2$. Here the apparent valences from bond valence sums are:

V_{La}	$V_{O(1)}$	$V_{O(2)}$
2.89	1.33	2.22

Note the very low value of the sum at the 6-coordinated O(1) and the compensatingly high value at the 4-coordinated O(2). The observed La-O(1) distance is 2.73 Å compared to the distance of 2.58 Å calculated for a valence of 1/3. The fact that this elongation of the bond is attributable to crowding around O(1) can be verified by examining the La-O bond length in $La(OH)_3 = {}^{ix}La^iH_3{}^{iv}O_3$, here again the valence of the La-O bond should be 1/3. The bond lengths are 2.59 (6x) and 2.55 (3x) *i.e.* on average 2.58 Å as predicted.

4. Some Crystal Chemical Principles

4.1. IMPLICATIONS OF THE EQUAL VALENCE RULE

The equal valence rule has some simple but far-reaching implications in crystal chemistry. We may infer from its general applicability that low energy configurations of solids are those in which valences are as nearly as equal as possible. Thus we might expect an important factor in the determination of coordination numbers to be the ability of a structure with given coordination numbers to accommodate equal valences. Some simple examples, taken from oxide and nitride chemistry, illustrate this point.

About 30 sesquioxides M_2O_3 are known. Of these only the (metastable) high pressure form of B_2O_3 has all the metal atoms in 4-coordination. Why is this? 4-coordination of the metal atoms requires coordination numbers $^{iv}M_2{}^{ii}O_2{}^{iii}O$ which in turn requires bonds from O to have valences 2/3 and 1 and a valence range of 1/3 (compare § 2.1). In contrast $^{vi}M_2{}^{iv}O_3$ will have all bonds with valence 1/2 and the great majority of sesquioxides have metal atoms six-coordinated (in either the corundum or bixbyite structures), although a few oxides are known with coordinations $^{vii}M_2{}^{vi}O^{iv}O_2$ with valences of 1/2 and 1/3 (valence range 1/6).

Si is generally 4-coordinated in oxides, but is known in 6-coordination in phosphates such as $^{vi}Si^{iv}P_2{}^{ii}O_7$ and in mixed coordination in compounds as in $Si_5P_6O_{25} = {}^{vi}Si_3{}^{iv}Si_2{}^{iv}P_6{}^{ii}O_{25}$. In these compounds the O atoms are all two-coordinated; if the Si atoms were all in four-coordination, some of the O atoms would be bonded to only one atom.

Many striking examples of unusual coordination numbers that can be understood in terms of stoichiometry constraints and the equal valence rule can be found in nitride chemistry. Examples are $^{V}Ca_2^{II}Zn^{VI}N_2$ and $^{V}Ca_3^{III}Cr^{VI}N_3$. For these and other examples see (Brese & O'Keeffe, 1992).

It is a general rule (O'Keeffe & Hyde, 1984) that increasing anion/cation ratio increases cation coordination number, but there are some exceptions readily explicable by the requirements of equal valences. Thus compare $^{VI}Al_2^{IV}O_3$ (already discussed) with $^{IV}Al^{IV}P^{II}O_4$. Equally striking is the contrast in the compounds $^{VI}Na^{III}N^{III}O_3$ and $^{IV}Na_3^{IV}N^{IV}O_4$.

4.2. EXCEPTIONS TO THE EQUAL VALENCE RULE

In oxide chemistry at least, there are three classes of "cation" that often have local environments that are in gross violation of the equal valence rule. These are:

(a) "Lone pair cations" with outer electron configurations ns^2 [*e.g.* with $n = 6$: Tl(I), Pb(II) and Bi(III)] usually in environments such as distorted octahedra with three short and three long bonds. For a good account of the crystal chemistry of these compounds see Hyde & Andersson (1989).

(b) Transition metals with d^4 and d^9 configurations. Mn(III) and Cu(II) are conspicuous examples. The occurrence of 5- or 6-coordination with four short and one or two long bonds is well documented and understood. For discussions of compounds of these cations within the framework of the bond valence method see (O'Keeffe & Hansen, 1988; Brown, 1989; O'Keeffe,1990a).

(c) Transition elements with a d^0 configuration. This refers particularly to early transition elements in their maximum oxidation states: *e.g.* Ti(IV), Nb(V) and W(VI). Thus in contrast to the simple cubic structure of ReO_3 [Re(VI) is d^1] with six equal Re-O bonds, WO_3 with the same topology of bonding has a complex triclinic structure at room temperature with a wide range of W-O bond lengths. The fact that such cations can tolerate (prefer?) irregular coordination in oxides provides a rationalization of some important phenomena. For example virtually all the important ferroelectric oxides (typified by $BaTiO_3$) involve irregular octahedral coordination of d^0 cations. The almost incredibly rich and complex oxide chemistry involving these elements is attributable to the same cause. As an example in the series Ti_nO_{2n-1} with Ti in 6-coordination, oxygen must be in 3- and 4-coordination with a consequently wide range of bond valences not normally found (compare § 4.1). Likewise reduced tungsten oxides WO_{3-x} with W in 6-coordination necessitate oxygen having 2- and 3-coordination.

5. A Historical Note

Although the bond valence method has come into prominence only rather recently, it is worth recalling that it has a long history. Readers of the early papers cited here may care to speculate on why it has taken so long for the method to be generally accepted.

Pauling (1929) introduced the concept of *electrostatic bond strength* of bonds formed in "ionic" crystals. He subsequently (Pauling, 1947) used the concept of bond order or *bond number* particularly in discussion of bond lengths and valences in metals and gave an explicit expression for the dependence of bond length on bond number [equation (2)]. Shortly thereafter Byström & Wilhelmli (1951) applied the same method to a discussion of Cr-O bond lengths in crystals, and a number of papers followed in which bond lengths in oxides were discussed in similar terms (*e.g.* Zachariasen, 1954, Evans, 1960, Kihlborg, 1964). A particularly good account of bond valences (which he called *valence strength*) was given by Zachariasen (1963). In this paper equation (1) was given explicitly.

These papers were mainly concerned with *interpreting* observed bond lengths in crystals. An important step was taken by Brown (1977) who first proposed a method for *predicting* bond valences (and hence lengths) to be expected in crystals. This work introduced the equal valence rule for the first time.

Tables of parameters relating bond length to valence were published by Allmann (1975), by Brown & Wu (1976) and by Zachariasen (1978).

6. References

Allmann, R. (1975). 'Beziehungen zwischen Bindungslängen und Bindungstärken in Oxidstrukturen', Monatsh. Chem. **106**, 779-793.

Brese, N. E. & O'Keeffe, M. (1991). 'Bond Valence Parameters for Solids'. Acta Crystallogr. B**47**, 192-197.

Brese, N. E. & O'Keeffe, M. (1992). 'Crystal Chemistry of Inorganic Nitrides'. Structure and Bonding **79** in press.

Brown, I. D. (1977). 'Predicting Bond Lengths in Inorganic Crystals' Acta Crystallogr. B**37**, 1305-1310.

Brown, I. D. (1978). 'Bond Valences—A Simple Structural Model for Inorganic Chemistry'. Chem. Soc. Revs. **7**, 359-376.

Brown, I. D. (1981). 'The Bond-Valence Method: An Empirical Approach to Chemical Structure and Bonding' in *Structure and Bonding in Crystals* (M. O'Keeffe & A.Navrotsky, eds.) New York, Academic Press. II, 1-30.

Brown, I. D. (1988). 'Chemical Modelling of Frenkel Defects in Fluorite'. Solid State Ionics **31**, 203-208.

Brown, I. D. (1989). 'A Determination of Oxidation States and Internal Stresses in $Ba_2YCu_3O_x$, x = 6-7 Using Bond Valences'. J. Solid State Chem. **82**, 122-131.

Brown, I. D. (1992). 'Modelling the Structures of La_2NiO_4'. Z. Kristallogr. **199**, 255-272.

Brown, I. D. & Altermatt, D. (1985). 'Bond Valence Parameters Obtained from a Systematic Analysis of the Inorganic Crystal Structure Database'. Acta Crystallogr. B**41**, 244-247.

Brown, I. D. & Wu, K. K. (1976). 'Empirical Parameters for Calculating Cation-Oxygen Bond Valences'. Acta Crystallogr. B**32**, 1957-1959.

Byström, A. & Wilhelmli, K.-A. (1951). 'The Crystal Structure of $(NH_4)_2Cr_2O_7$ with a Discussion of the Relation between Bond Number and Interatomic Distances'. Acta Chemica Scand. **5**, 1003-1010.

Donnay, G. & Allmann, R. (1970). 'How to Recognize O^{2-}, OH^- and H_2O in Crystal Structures Determined by X-rays'. Amer. Mineral. **55**, 1003-1015

Donnay, G. & Donnay, J. H. D. (1973). 'Bond Valence Summations for Borates'. Acta Crystallogr. B**29**, 1417-1425.

Evans, H. T. (1960). 'Crystal Structure Refinement and Vanadium Bonding in the

Metavanadates KVO_3, NH_4VO_3 and $KVO_3.H_2O$'. Z. Kristallogr. 11?, 257-277.
Hyde, B. G. & Andersson, S. (1989) *Inorganic Crystal Structures*, New York: Wiley.
Kihlborg, L. (1964) 'The Structural Chemistry of the Higher Molybdenum Oxides'. Arkiv för Kemi, **21**, 471-495.
McGuire, N. K. & O'Keeffe, M. (1984). 'Bond Lengths in Alkali Metal Oxides'. J. Solid State Chem. **54**, 49-53.
O'Keeffe, M. (1989). 'The Prediction and Interpretation of Bond Lengths in Solids'. Structure and Bonding **71**, 161-190.
O'Keeffe, M. (1990a). 'A Method for Computing Bond Valences in Crystals' Acta Crystallogr. **A46**, 138-142.
O'Keeffe, M. (1990b). ' Site Potentials and Valences in Copper and Silver Oxides'. J. Solid State Chem. **85**, 108-116.
O'Keeffe, M. (1991a). 'Empirical Methods in Oxide Crystal Chemistry' in *Chemistry of Electronic Ceramic Materials* (Davies, P. K. & Roth, R. S., eds.) Washington: NIST Spec. Publ. 804, 485-496.
O'Keeffe, M. (1991b). 'Application of the Bond Valence Method to $Si/NiSi_2$ Interfaces'. J. Mats. Res. **6**, 2371-2374.
O'Keeffe, M. & Brese, N. E. (1991) 'Atom Sizes and Bond Lengths in Molecules and Crystals'. J. Am. Chem. Soc. **113**, 3226-3229.
O'Keeffe, M. & Brese, N. E. (1992) 'Bond Valence parameters for Anion-Anion Bonds in Solids' Acta Crystallogr. **B48**, in press.
O'Keeffe, M. & Hyde, B. G. (1984). 'The Role of Stoichiometry in Determining the Structure and Stability of Inorganic Solids'. Nature, **309**, 411-414.
O'Keeffe, M. & Hansen, S. (1988). 'The Crystal Chemistry of $Ba_2YCu_3O_7$ and Related Compounds'. J. Am. Chem. Soc. **110**, 1506-1510.
Pauling, L. (1929). 'The Principles Determining the Structure of Complex Ionic Crystals' J. Am. Chem. Soc. **51**, 1010-1026.
Pauling, L. (1947). 'Atomic Radii and Interatomic Distances in Metals' J. Am. Chem. Soc. **69**, 542-553.
Rutherford, J. S. (1990). 'Theoretical Prediction of Bond-Valence Networks'. Acta Crystallogr. **B46**, 289-292.
Shannon, R. D. (1976). 'Revised Effective Ionic Radii and Systematic Studies of Interatomic Distances in Halides and Chalcogenides' Acta Crystallogr. **A34**, 751-767.
Wagner, T. R. & O'Keeffe, M. (1988). 'Bond Lengths and Valences in Aluminates with the Magnetoplumbite and β-Alumina Structures'. J. Solid State Chem. **73**, 211-216.
Waltersson, K. (1978). 'A Method Based upon "Bond-Strength" Calculations for Finding Probable Lithium Sites in Crystal Structures'. Acta Crystallogr. **A34**, 901-905.
Zachariasen, W. H. (1954). 'Crystal Chemical Studies of the 5f-series of Elements XXIII. On the Crystal Chemistry of Uranyl Compounds and of Related Compounds of Transuranic Elements.' Acta Crystallogr. **7**, 795-799.
Zachariasen, W. H. (1963). 'The Crystal Structure of Monoclinic Metaboric Acid'. Acta Crystallogr. **16**, 385-389.
Zachariasen, W.H. (1978). 'Bond Lengths in Oxygen and Halogen Compounds of d and f Elements'. J. Less-Common Metals, **62**, 1-7.

7. Problems

Problem 1. The connectivity matrix for $Si_5P_6O_{25}$ [H. Meyer, Monatsh. Chem. **105**, 46-54 (1974)] is:

	O(1)	6 O(2)	6 O(3)	6 O(4)	6 O(5)
Si(1)	0	0	6	0	0
2 Si(2)	0	6	0	0	6
2 Si(3)	2	0	0	6	0
6 P	0	6	6	6	6

(i) Determine the coordinations of each atom.
(ii) Predict the P-O bond lengths using $R_{SiO} = 1.62$ Å, $R_{PO} = 1.60$ Å.

Problem 2. In the structure [K. K. Wu & I. D. Brown, Mats Res. Bull. **8**, 593 (1975)] of $CaCrF_5 = CaCrF(1)F(2)_2F(3)_2$ the bond lengths (in Å) are:

Ca-F(1) 2.49
Ca-F(2) 2.29 (X2), 2.39 (X2)
Ca-F(3) 2.21 (X2)
Cr-F(1) 1.94 (X2)
Cr-F(2) 1.92 (X2)
Cr-F(3) 1.85 (X2)

(i) Calculate R_{CaF} and R_{CrF}.
(ii) Write out the connectivity matrix for $CaCrF_5$.
(iii) Explain why the bonds to F(3) are shorter than the other bonds.
(iv) Use the equal valence rule to predict bond valences and hence bond lengths [using the parameters from (i)].

Problem 3. $CoMnO_3$ is isostructural with $FeTiO_3$ (ilmenite) and each cation forms six bonds to O of length 1.881 (3x) and 2.124 (3x) Å. Test the hypotheses $Co(II)Mn(IV)O_3$ and $Co(III)Mn(III)O_3$ using the bond valence parameters (in Å) Co(II)-O 1.692; Co(III)-O 1.701; Mn(III)-O 1.760; Mn(IV)-O 1.753.

7.1 SOLUTIONS TO THE PROBLEMS

Problem 1.

(i) $^{vi}Si(1)^{vi}Si(2)_2{}^{iv}Si(3)_2{}^{iv}P_5{}^{ii}O(1)^{ii}O(2)_2{}^{ii}O(3)_2{}^{ii}O(4)_2{}^{ii}O(5)_2$

(ii) predicted valences and bond lengths (observed in parentheses):

	v = 4/3	d = 1.50 (1.52)
P-O(2)	4/3	1.50 (1.52)
P-O(3)	4/3	1.50 (1.51)
P-O(4)	1	1.60 (1.58)
P-O(5)	4/3	1.50 (1.50)

Problem 2.

(i) $R_{CaF} = 1.85$ Å, $R_{CrF} = 1.65$ Å.

(ii)
	F(1)	2 F(2)	2 F(3)
Ca	1	4	2
Cr	2	2	2

(iii) If all the bonds from each metal atom were of equal valence, the bond valence sums at the metal atoms would be:

at F(1) 2/7 + 2(3/6) = 1.28
at F(2) 2(2/7) + (3/6) = 1.07
at F(3) 2/7 + 3/6 = 0.78

thus F(3) would be underbonded. Accordingly the valence of bonds to F(3) must be increased (length decreased) to make the bond valence sum equal to 1.0.

(iv) Calculation of valences (without weights) and lengths with results from (i):

bond	valence	calc. (obs.) length
Ca-F(1)	0.185	2.47 (2.49)
Ca-F(2)	0.259	2.35 (2.34 average)
Ca-F(3)	0.389	2.20 (2.19)
Cr-F(1)	0.407	1.98 (1.94)
Cr-F(2)	0.481	1.92 (1.92)
Cr-F(3)	0.611	1.83 (1.85)

Problem 3.

Bond valence sums at the cations are:

hypothesis	V_{Co}	V_{Mn}	
Co(II),Mn(IV)	2.74	3.23	
Co(III),Mn(III)	2.80	3.29	this is more likely!

VALENCE ELECTRON RULES FOR COMPOUNDS WITH TETRAHEDRAL STRUCTURES AND ANIONIC TETRAHEDRON COMPLEXES

ERWIN PARTHÉ
Laboratoire de Cristallographie aux Rayons X
Université de Genève
24, Quai Ernest Ansermet
CH - 1211, Geneva 4
Switzerland

ABSTRACT. For compounds with tetrahedral structure or anionic tetrahedron complex two valence electron concentration rules can be formulated which correlate the number of available valence electrons with particular features of the crystal structure. These two rules are known as the tetrahedral structure equation where the total valence electron concentration, VEC, is used as parameter and the generalized 8 - N rule where the parameter of interest is the partial valence electron concentration in respect to the anion, VEC_A. From the tetrahedral structure equation one can calculate the average number of non-bonding orbitals per atom and, in the case of non-cyclic molecular tetrahedral structures, the number of atoms in the molecule. An application of the generalized 8 - N rule allows the derivation of the average number of anion - anion bonds per anion or the number of valence electrons which remain with the cation to be used for cation - cation bonds and/or lone electron pairs. These rules have been used not only to predict probable structural features of unknown compounds but also to point out possible errors in composition or structure of known compounds.

1. Tetrahedral Structure Compounds

Tetrahedral structure compounds form a subset of the general valence compounds where each atom has at most four neighbours which are positioned at the corners of a surrounding tetrahedron. The tetrahedral structures are found with iono-covalent compounds which can be considered either as covalent or as ionic. For each hypothetical bonding state a particular valence electron concentration rule can be formulated which allows certain structural features to be predicted.

1.1. TETRAHEDRAL STRUCTURES INTERPRETED BY AN OVERLAP OF sp^3 HYBRIDIZED BONDING ORBITALS (COVALENT BONDING STATE)

Each atom forms four tetrahedral sp^3 hybrid orbitals which overlap with the sp^3 orbitals of the neighbouring atoms. To form a sp^3 hybrid each atom needs four valence electrons. Each orbital not used for bonding obtains a second electron and becomes a non-bonding orbital. Non-bonding orbitals can be seen indirectly in a structure by the absence of an expected tetrahedral neighbour atom. For a compound of composition $C_m A_n$ one can formulate the following equation :

$$m \cdot e_C + n \cdot e_A = (m + n) \cdot 4 + (m + n) \cdot N_{NBO} \qquad (1)$$

where e_C and e_A are the valence electron numbers of the cations and anions in the non-ionized state (which corresponds for non-transition elements to their group numbers in the Periodic Table) and N_{NBO} is the average number of non-bonding orbitals ($\equiv$ lone electron pairs) per atom.

Introducing as parameter the total valence electron concentration, VEC, defined as

$$VEC = (m \cdot e_C + n \cdot e_A) / (m + n), \qquad (2)$$

equation (1) can be rewritten as

$$VEC = 4 + N_{NBO}. \qquad (3)$$

This is the *tetrahedral structure equation* (Parthé, 1972).

Calculation of VEC allows one to find out if a compound might have a tetrahedral structure and what kind it should be. One distinguishes between normal tetrahedral structures, where each atom in the structure has four tetrahedral neighbours, and defect tetrahedral structures where some atoms have less than four neighbour atoms.

If $VEC < 4$: A tetrahedral structure can not be formed.
If $VEC = 4$: *Normal tetrahedral structure* with $N_{NBO} = 0$.
If $VEC > 4$: *Defect tetrahedral structure* with $N_{NBO} > 0$.

Note that the tetrahedral structure equation describes a necessary but not sufficient condition for the occurrence of a tetrahedral structure. For a discussion of the other conditions, which can not always be formulated in a precise mathematical manner, see Parthé (1990).

1.2. TETRAHEDRAL STRUCTURES INTERPRETED BY CATION - ANION INTERACTIONS (IONIC BONDING STATE)

The valence electrons of the cations are used to complete the octets of the anions. For a compound of composition $C_m A_n$ one can formulate

$$m \cdot (e_C - CC) = n \cdot (8 - e_A - AA) \qquad (4)$$

where CC is the average number of valence electrons per cation which are not transferred to the anions but remain with the cation, either used for a cation - cation bond or for a lone electron pair. AA is the average number of valence electrons per anion used for covalent anion - anion bonds. For each bond which an anion extends to another anion it needs one electron less to complete its octet.

Introducing as new parameter the partial valence electron concentration in respect to the anion, VEC_A, defined as

$$VEC_A = (m \cdot e_C + n \cdot e_A) / n, \qquad (5)$$

equation (4) can be rewritten as

$$VEC_A = 8 + CC/(n/m) - AA. \qquad (6)$$

This is the *generalized 8 - N rule* in the formulation of Parthé (1973).

Calculation of VEC_A allows to classify a compound as a polyanionic, normal or polycationic valence compound. Polyanionic valence compounds are characterized by anion - anion bonds. In the normal valence compounds there are neither anion - anion nor cation - cation bonds and in the polycationic valence compounds some valence electrons are used for cation - cation bonds and/or lone electron pairs on the cations.

If $VEC_A < 8$: *Polyanionic valence compound* with $AA > 0$, that is

$$AA = 8 - VEC_A \qquad \text{if } CC = 0. \tag{7}$$

If $VEC_A = 8$: *Normal valence compound* with $CC = AA = 0$.
If $VEC_A > 8$: *Polycationic valence compound* with $CC > 0$, that is

$$CC = (n/m) \cdot (VEC_A - 8) \quad \text{if } AA = 0. \tag{8}$$

The experimental evidence has shown that the limiting conditions, that is $CC = 0$ for polyanionic compounds and $AA = 0$ for polycationic compounds, can be taken for granted. But note, that in general these equations can be applied only if all bonds are two electron bonds and additional electrons remain inactively in the non-bonding orbitals of the cations or anions. Thus the compounds must be semiconductors. If the compounds are metallic, the applicability of these equations is not guaranteed.

1.3. EXAMPLES OF TETRAHEDRAL STRUCTURES

1.3.1. *Crystal chemical formulae.* To describe the predicted and observed structural features we shall use crystal chemical formulae (Lima-de-Faria, Hellner, Liebau, Makovicky & Parthé, 1990) where one can denote in codified form
the linkage type of the overall structure, such as

$${}_{\infty}^{3}C \text{ (diamond)}, \quad {}_{\infty}^{2}GaSe, \quad {}_{\infty}^{1}Se, \quad \circledast[S_8],$$

the linkage type of a partial structure

$$Zn \, {}_{\infty}^{1}P_2, \quad Cs_2{}_{\infty}^{\Lambda}[S_6], \quad Na_4 \oslash [Si_4],$$

the hetero- and homonuclear coordination of an atom and its coordination polyhedron

$$Fe^{[6o;]}S_2^{[(3;1)t]}$$

and the lone electron pair assigned to a particular atom, as for example

Ga<u>Se</u>.

For more details on crystal chemical formulae, particulary for the meaning of the symbols and the commas, semi-colons and small letters within the superscripted square brackets used in the crystal chemical formulae, see the contribution of Lima-de-Faria in this book.

1.3.2. *Examples for normal valence compounds ($VEC_A = 8$) with normal and defect tetrahedral structures.* There are known many normal valence compounds with normal and defect tetrahedral structure. A subgroup of this family is formed by the adamantane structure compounds characterized by structures which are related to zinc blende or wurtzite as substitution and/or vacancy variants (for details, see Parthé, 1990). As example for a normal adamantane structure we shall present in Figure 1 the zinc blende structure and, as example for a defect adamantane structure, the $CdIn_2Se_4$ structure, a vacancy and substitution variant of zinc blende.

ZnS : VEC = 4, thus $N_{NBO} = 0$; $VEC_A = 8$, thus AA = CC = 0.
Crystal chemical formula :

$$Zn^{[4t;]}S^{[4t;]}$$

The same crystal chemical formula is also valid for wurtzite, a stacking variant of zinc blende.

$CdIn_2Se_4$: VEC = 4.571, thus $N_{NBO} = 4/7$; $VEC_A = 8$, thus AA = CC = 0.
Crystal chemical formula :

$$Cd^{[,4t;]}In_2^{[,4t;]}\underline{Se}_4^{[(1,2)n;]}$$

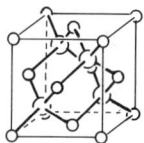   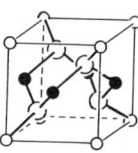

ZnS $CdIn_2Se_4$

Figure 1 : One unit cell of zinc blende and of $CdIn_2Se_4$, a ternary ordered vacancy variant of zinc blende. Small circles represent cations, of those the open circles Zn or Cd and the full circles In atoms. Large open circles correspond to S or Se atoms.

As compared to the zinc blende structure there is one vacant Zn site per unit cell of $CdIn_2Se_4$. One vacancy, ☐, corresponds to four non-bonding orbitals (one each on the four surrounding Se atoms). In one formula unit of $CdIn_2Se_4$, consisting of 7 atoms, there are four non-bonding orbitals, that means $N_{NBO} = 4/7$.

1.3.3. *Examples for polyanionic valence compounds ($VEC_A < 8$) with normal and defect tetrahedral structures.* We shall present red ZnP_2 as an example for a normal and Cu_2P_7 as an example for a defect tetrahedral structure.

ZnP_2 : VEC = 4, thus $N_{NBO} = 0$; $VEC_A = 6$, thus AA = 2. With AA = 2 one may find an infinite anion chain or a closed anion ring. Simplest crystal chemical formula assuming an infinite chain :

$$Zn^{[4t;]}{}_\infty^1 P_2^{[(2;2)t]}$$

The same crystal chemical formula applies also to isoelectronic α-CdP$_2$ and ZnAs$_2$, shown in Figure 2, all being characterized by infinite anion chains. In α-CdP$_2$ the chains are parallel, in ZnAs$_2$ antiparallel and in red ZnP$_2$ mutually perpendicular. The valence electron rules do not permit the prediction of which type of anion chain will occur in the structure.

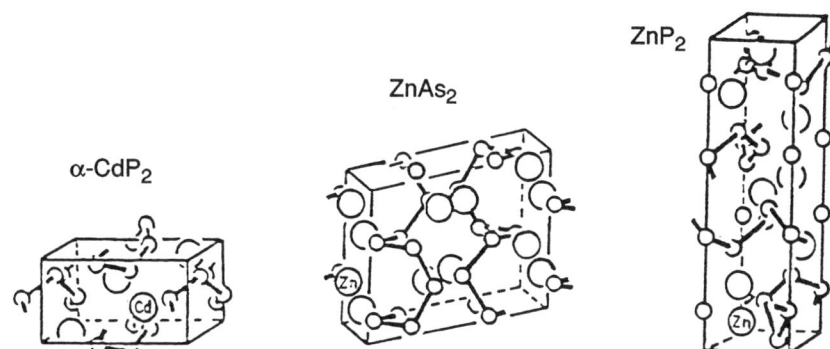

Figure 2 : Structures of red ZnP$_2$ and isoelectronic compounds with same crystal chemical formula.

Cu$_2$P$_7$: VEC = 4.111, thus N$_{NBO}$ = 1/9; VEC$_A$ = 5.286, thus AA = 19/7. One can formulate two equivalent crystal chemical formulae which differ only in the single non-bonding orbital being either on a P atom with two or on a P atom with three homonuclear bonds :

$$Cu_2^{[4t;]}\underline{P}^{[(1;2)n]}P^{[(2;2)t]}P_5^{[(1;3)t]} \quad \text{or} \quad Cu_2^{[4t;]}P_2^{[(2;2)t]}\underline{P}^{[;3n]}P_4^{[(1;3)t]}$$

The real structure, as determined by Möller & Jeitschko (1982), corresponds to the second formula.

Other examples are Ag$_3$P$_{11}$ (Möller & Jeitschko, 1981) and CuTe$_2$Cl (Fenner, 1976).

1.3.4. *Examples for polycationic valence compounds (VEC$_A$ > 8) with defect tetrahedral structures.* Except for the high pressure form of B$_2$O with diamond-like structure (Endo, Sato & Shimada, 1987), no polycationic valence compounds with normal tetrahedral structure are known. As examples for compounds with defect tetrahedral structure we discuss here GaSe and "In$_5$S$_4$". The latter compound served as a test case for the validity of these valence electron rules.

GaSe : VEC = 4.5, thus N$_{NBO}$ = 1/2; VEC$_A$ = 9, thus CC = 1.
Crystal chemical formula :

$$[Ga_2^{[(3;1)t]}\underline{Se}_2^{[3n;]}]$$

GaSe, shown in Figure 3, has a layer structure with a very pronounced cleavage between the layers. The Ga atoms form dumb-bells. The non-bonding orbitals on the Se atoms are directed towards the space between the layers.

Figure 3 : Atom arrangement in $^2_\infty$ GaSe.

"In_5S_4" : The original crystal structure determination (Wadsten, Arnberg & Berg, 1980) revealed a defect tetrahedral structure which, based on the reported atom coordinations and atom contacts, can be described by the following crystal chemical formula :

$$In_4^{[(3;1)t]}In^{[;4t]}\underline{S}_4^{[3n;]}$$

To this formula correspond the values : $N_{NBO} = 4/9$ and $CC = 8/5$. However, if one uses the valence electron rules and bases the calculation on the reported composition "In_5S_4" different values for N_{NBO} and CC are obtained :
"In_5S_4" : $VEC = 4.333$, thus $N_{NBO} = 1/3$ instead of 4/9; further $VEC_A = 9.75$, thus $CC = 7/5$ instead of 8/5.

Assuming that the valence electron rules are correct, the error had to be looked for either in the crystal structure or in the composition of the compound. Since "In_5S_4" had been prepared in a bath of liquid tin the possibility could not be excluded, that some Sn atoms had been incorporated in the crystals and that the true chemical composition might be different. Taking instead of "In_5S_4" as composition In_4SnS_4 one calculates the following values :
In_4SnS_4 : $VEC = 4.444$, thus $N_{NBO} = 4/9$; $VEC_A = 10$, thus $CC = 8/5$, considering the Sn atoms here as cations. The corresponding crystal chemical formula would be :

$$In_4^{[(1,3)t;]}Sn^{[4t,;]}\underline{S}_4^{[3n,;]}$$

One notes, that the structural features corresponding to this crystal chemical formula do agree with those observed experimentally. The problem, which remained unsolved for 10 years, has now finally been settled. A recent reexamination (Deiseroth & Pfeifer, 1991) has proven that the compound can not exist without Sn, its true composition being In_4SnS_4.

Other examples (free of problems) for polycationic valence compounds are Si_2Te_3 and Al_7Te_{10} (Nesper & Curda, 1987).

1.4. MOLECULAR TETRAHEDRAL STRUCTURES

One can write the valence electron equation for tetrahedral structures as follows :

$$4 = 8 - VEC + N_{NBO} \qquad (9)$$

| total number of orbitals / atom | number of bonding orbitals / atom | number of non-bonding orbitals / atom |

If for a compound VEC > 6 it follows that the average number of bonding orbitals / atom is < 2, which means that the structure consists of a non-cyclic molecule with a finite number of atoms. Let us denote by $N_{A/M}$ the number of atoms in the non-cyclic molecule. A molecule with $N_{A/M}$ atoms has $N_{A/M}$ - 1 bonds. The ratio of the number of bonds to the number of atoms can have the values 0/1, 1/2, 2/3 or in general $(N_{A/M} - 1) / N_{A/M}$. Because each bond is formed by the overlapping of two orbitals, the average number of bonding orbitals / atom is the double, that means $2 \cdot (N_{A/M} - 1) / N_{A/M}$. Thus it follows from (9) that

$$8 - VEC = 2 \cdot (N_{A/M} - 1) / N_{A/M} \qquad (10)$$

which can be transformed into

$$N_{A/M} = 2 / (VEC - 6) \quad \text{for VEC} > 6. \qquad (11)$$

This equation serves to calculate the number of atoms in a non-cyclic tetrahedral structure molecule. Simple solutions of (11) are given in Table 1.

TABLE 1. The VEC value of a compound and the corresponding number of atoms in the non-cyclic tetrahedral structure molecule.

VEC	6.2	6.222	6.25	6.286	6.333	6.4	6.5	6.667	7
$N_{A/M}$	10	9	8	7	6	5	4	3	2

1.4.1. *Examples for non-cyclic molecular tetrahedral structures.* Molecular tetrahedral structures are found with normal and polycationic valence compounds. Of the examples given in Figure 4 the compound SnI_4 is a normal valence compound, all others are polycationic valence compounds. The values of N_{NBO}, $N_{A/M}$ and AA or CC, calculated with (3), (11) and (7) or (8), correspond to the crystal chemical formulae of the molecules, presented on the last line of the text blocks.

For simplicity the non-bonding orbitals are often omitted in the crystal chemical formulae or they are shown only for certain kinds of atoms. In the case of a tetrahedral structure the number of non-bonding orbitals assigned to an atom can easily be obtained by noting that the sum of the number of covalent bonds, given within the superscripted square brackets, and the number of non-bonding orbitals must be four.

Other examples for non-cyclic molecular tetrahedral structures are SCl_2 ($N_{A/M}$ = 3) and $SiCl_3$ ($N_{A/M}$ = 8).

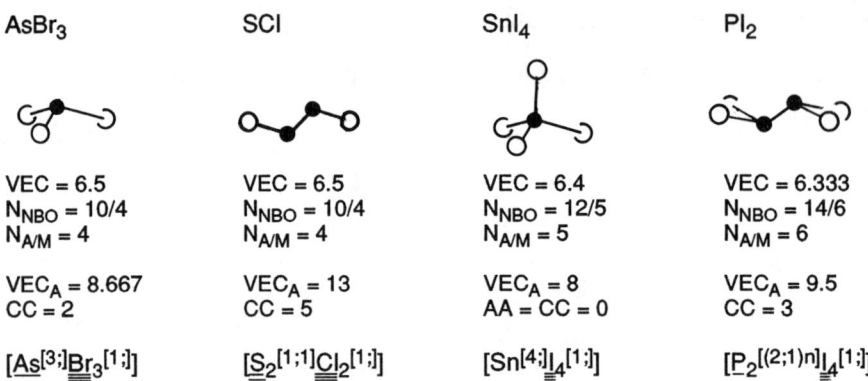

Figure 4: Non-cyclic tetrahedral structure molecules with 4 (twice), 5 and 6 atoms per molecule in the structures of $AsBr_3$, SCl, SnI_4 and PI_2.

2. Structures with Tetrahedral Anion Partial Structures

There are known many iono-covalent compounds where only the anions adopt a tetrahedral structure. In a formalistic approach one may assume that the cations transfer all the valence electrons to the anions. The tetrahedral structure equation can then be applied to the charged anion partial structure. We shall use primed parameters, such as VEC', N'_{NBO} and $N'_{A/M}$, to indicate that we refer to a charged anion partial structure.

In the case of a binary compound of composition $C_m A_n$ the valence electron concentration of the charged anion partial structure, VEC', is defined as

$$VEC' = (m.e_C + n.e_A) / n \equiv VEC_A \quad \text{for } C_m A_n. \tag{12}$$

The parameter VEC' is here identical to VEC_A, formulated before in (5).

In the case of a ternary compound of composition $C_m C'_{m'} A_n$ there may occur an anionic tetrahedron complex formed by a central atom C', tetrahedrally surrounded by anions A (see below). Here VEC' and VEC_A are different because VEC' refers to all atoms of the charged anionic tetrahedron complex, but VEC_A only to the anions. We thus have to distinguish for $C_m C'_{m'} A_n$ between

$$VEC' = (m.e_C + m'.e_{C'} + n.e_A) / (m' + n) \tag{13}$$

and

$$VEC_A = (m.e_C + m'.e_{C'} + n.e_A) / n. \tag{14}$$

In analogy to (3) and (11) one can now write the tetrahedral structure equation as applied to the charged tetrahedral anion partial structure (Parthé, 1989) as

$$VEC' = 4 + N'_{NBO} \qquad (15)$$

and the corresponding derivative equation for a molecular tetrahedral anion partial structure as

$$N'_{A/M} = 2 / (VEC'- 6) \quad \text{for VEC' > 6.} \qquad (16)$$

2.1. EXAMPLES OF STRUCTURES WITH TETRAHEDRAL ANION PARTIAL STRUCTURES

As example for a binary compound with a tetrahedral anion partial structure we shall present in Figure 5 Sr_3As_4 (Deller & Eisenmann, 1977) and as example for an anionic tetrahedron complex the compound $Na_6Ge_2Se_7$ (Eisenmann, Hansa & Schäfer, 1986). The drawings are complemented with text blocks where are listed all the parameters which can be calculated with the equations discussed above. In the last line are given the crystal chemical formulae of the tetrahedral anion partial structure or the charged anionic tetrahedron complex which can be derived from these parameters and which agree with the observed crystal structures.

Sr_3As_4 　　VEC = 26/7 < 4　　　　$Na_6Ge_2Se_7$　　VEC = 56/15 < 4

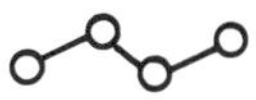

　　　　　　　VEC' = 26/4 > 6　　　　　　　　　　　　　VEC' = 56/9 > 6
　　　　　　　N'_{NBO} = 10/4　　　　　　　　　　　　　N'_{NBO} = 20/9
　　　　　　　$N'_{A/M}$ = 4　　　　　　　　　　　　　　$N'_{A/M}$ = 9

　　　　　　　VEC_A = 26/4 < 8　　　　　　　　　　　VEC_A = 56/7 = 8
　　　　　　　AA = 6/4　　　　　　　　　　　　　　　AA = C'C' = CC = 0

$^A_4[As_2^{[;1]}As_2^{[;2]}]^{6-}$ or $^A_3[As_3^{[;1]}As^{[;3]}]^{6-}$ 　　　$[Ge_2^{[4t;]}Se_6^{[1;]}Se^{[2;]}]^{6-}$

Figure 5 : The molecular tetrahedral anion partial structures of Sr_3As_4 and $Na_6Ge_2Se_7$.

For both compounds a tetrahedral structure including all atoms is not possible because VEC < 4. However, both can have tetrahedral anion partial structures, which in the particular case, since for both VEC' > 6, should consist of non-cyclic molecules. For Sr_3As_4 there exist two possibilities, that is an unbranched chain of 4 As atoms, as shown in the drawing and actually found in Sr_3As_4, or a chain of 3 anions which is branched, the anion in the middle of the chain being linked to a forth anion. Note, that in both the crystal chemical formulae and the drawings the non-bonding orbitals have not been indicated, however, it should not be difficult for the reader to verify that the listed values of N'_{NBO} are correct.

3. Structures with Anionic Tetrahedron Complexes

In the compounds of composition $C_m C'_{m'} A_n$ the tetrahedron complex, as stated above, is formed with all the C' (central atoms) and A atoms (anions). The C atoms (cations) are outside the complex and are assumed to transfer their valence electrons to the tetrahedron complex which is then negatively charged.

A problem might exist to recognize which atoms in a ternary compound participate on an anionic tetrahedron complex and which do not. Experimental evidence has shown that the alkali elements Na, K, Rb and Cs, the alkaline earth elements Ca, Sr and Ba and the rare earth elements never have a tetrahedral coordination. They function as cations C only which transfer their valence electrons to the remaining atoms forming the anionic tetrahedron complex. There are, however, other elements such as Al which in some compounds participate on the complex but in others not. To simplify matters will shall in the following examples consider only compounds where the recognition of the elements which function as cations C causes no problems. If for a compound VEC < 4 it is evident that all atoms can not participate on a tetrahedral structure. But also if VEC ≥ 4, where in principle a tetrahedral structure involving all elements would be possible, the above cited elements will never participate on the anionic tetrahedron complex.

In the most simple case the anionic tetrahedron complex consists of a simple isolated $C'A_4$ tetrahedron. However, there have to be considered four different kinds of variants of this tetrahedron, shown in Figure 6, three of them allowing for a linkage of tetrahedra and two of them involving the removal of an anion neighbour of a central atom (Parthé & Chabot, 1990).

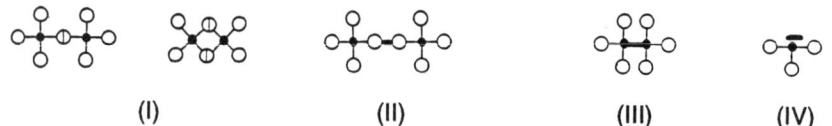

(I)   (II)   (III)   (IV)

Figure 6 : The four different kinds of variants of the simple $C'A_4$ tetrahedron. For simplicity there has been chosen a planar graph presentation of the tetrahedra. The central atoms are indicated by filled circles and the anions by larger open circles. For tetrahedron variants (I), (II) and (III) are shown here two linked tetrahedra.

(I) The anion of a tetrahedron is shared with another tetrahedron. Depending on how many anions of one tetrahedron are shared with one other tetrahedron one may further distinguish between corner and edge shared tetrahedra.

(II) The anion of a tetrahedron forms a covalent bond with an anion of another tetrahedron (A - A bond).

(III) The central atom of a tetrahedron forms a covalent bond with the central atom of another tetrahedron (replacement of a C' - A bond by a C' - C' bond).

(IV) An anion neighbour of a central atom is replaced by a lone electron pair (formation of a so-called psi-tetrahedron).

The tetrahedron linkage described by (I) can be calculated from the composition of the compound, to be exact, from the anion to central atom ratio (n/m'). The linkage type (II), (III) and the formation of a psi-tetrahedron (IV) can be derived from a modified generalized 8 - N rule, i .e. one needs to calculate the VEC_A value of the compound. The anion sharing according to (I) is independent of the VEC_A value and can occur in addition to case (II), (III) and/or (IV).

3.1. THE MODIFIED VALENCE ELECTRON RULES FOR STRUCTURES WITH ANIONIC TETRAHEDRON COMPLEXES.

3.1.1. *The modified tetrahedral structure equation.* The tetrahedral structure equation and its derivative equation for non-cyclic molecular tetrahedral structures can be applied only to the charged anionic tetrahedron complex in the form of (15) and (16). The first equation allows the calculation of the average number of non-bonding orbitals per atom of the charged complex, N'_{NBO}, and the second, applicable only when $VEC' > 6$, allows the determination of the number of atoms in the non-cyclic molecular anionic tetrahedron complex, $N'_{A/M}$.

A demonstration of the calculation of the VEC', N'_{NBO} and $N'_{A/M}$ values for $Na_6Ge_2Se_7$ had been given above in Figure 5. For all compounds treated in Figures 7 to 10 the values of these three parameters are listed in separate text blocks.

3.1.2. *The modified generalized 8 - N rule.* The generalized 8 - N rule shall be used in the modified form

$$VEC_A = 8 + C'C'/(n/m') - AA \qquad \text{for } C_m C'_{m'} A_n \qquad (17)$$

where VEC_A is a given by (14), n/m' is the ratio of the number of anions to the number of central atoms, C'C' is the average number of C' - C' bonds per central atom and/or the average number of valence electrons per central atom which rest inactively with the central atom as a lone electron pair and, finally, AA is the average number of A - A bonds per anion. It was stated above, that it is assumed that there are no bonds between the non-tetrahedral cations (CC = 0); they transfer all their valence electrons to the central atoms (C') and anions (A).

As above one can, depending on the VEC_A value, distinguish between polyanionic, normal and polycationic valence compounds.

If $VEC_A < 8$: Polyanionic valence compound with $AA > 0$ and $C'C' = 0$. The value of AA can be calculated using (7). Tetrahedra are linked by covalent A - A bonds (tetrahedron variant (II)) and, depending on the composition, also by sharing anions (tetrahedron variant (I)). The parameter of interest is the average number of A - A bonds per tetrahedron. Since there are n anions for every m' central atoms, its value is given by

$$AA \cdot (n/m') = (n/m') \cdot (8 - VEC_A). \qquad (18)$$

If $VEC_A = 8$: Normal valence compound with $AA = C'C' = 0$. Depending on the composition there are either isolated $C'A_4$ tetrahedra or tetrahedra which are linked by sharing anions (tetrahedron variant (I)).

If $VEC_A > 8$: Polycationic valence compound with $C'C' > 0$ and $AA = 0$. A tetrahedron variant (III) and/or (IV) occurs and, depending on the composition, also variant (I). The parameter of interest is the average number of C' - C' bonds per tetrahedron and/or the number of electrons on the central atom per tetrahedron used for lone electron pairs. Since each tetrahedron of the anionic tetrahedron complex has only one central atom, this value is identical to $C'C'$, that is

$$C'C' = (n/m') \cdot (VEC_A - 8). \tag{19}$$

Experimentally one finds with more than a 90% probability that when $C'C' = 1$ there is a C' - C' bond between the central atoms of two tetrahedra and when $C'C' = 2$ there is a lone electron pair on the central atom of each tetrahedron.

3.1.3. Examples for polyanionic, normal and polycationic valence compounds with anionic tetrahedron complexes. In Figures 7 to 10 are given examples for the various kinds of compounds with anionic tetrahedron complexes. In the schematic drawings a planar graph presentation of the tetrahedra has been used. In the case of Na_8SnSb_4 (Figure 8) and Na_3AsS_3 (Figure 10) the anionic tetrahedron complex is constituted by a single isolated $C'A_4$ or a psi-tetrahedron, respectively. In all other examples the anionic tetrahedron complex consists of more tetrahedra which are linked with themselves. In these drawings there is not presented the complete anionic tetrahedron complex but only a so-called base tetrahedron which when linked with itself can be used to construct the anionic tetrahedron complex found in the compound.

In these base tetrahedron drawings the central atoms are shown by filled small circles and the anions by larger open circles. Anions shared with an other tetrahedron are presented by half circles. A bond between anions or between central atoms is indicated by a short heavy line extending from the open or filled circle, respectively. In the case of a psi tetrahedron the lone electron pair on the central atom is shown by a heavy bar alongside the filled circle.

The drawings are complemented with text blocks detailing the numerical values of the different parameters which can be calculated from the valence electron equations discussed above. On top is given the total valence electron concentration, VEC. If VEC < 4 a tetrahedral structure involving all atoms is impossible. The parameters listed below VEC are derived from the valence electron concentration of the charged anion partial structure, VEC', and the next one from the partial valence electron concentration in respect to the anion, VEC_A. The parameter C'AC', to be discussed in the next paragraph, refers to the sharing of the anions and can be calculated from the composition of the compound. Finally, on the last row one finds a classification code for the base tetrahedron, also to be discussed later on.

For all the compounds discussed in Figures 7 to 10 the observed structural features, in particular the features of the base tetrahedron which can be used to construct the complete anionic tetrahedron complex, are in perfect agreement with the predictions based on the valence electron rules.

Literature references for all the compounds mentioned here can be found in the paper by Parthé & Chabot (1990).

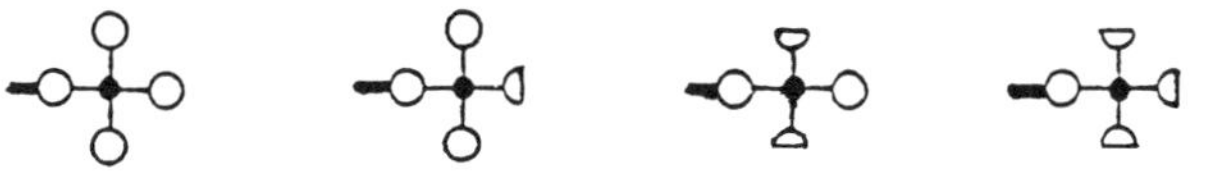

Na$_3$SiSe$_4$	Cs$_4$Sn$_2$Te$_7$	Ca$_5$Ga$_2$As$_6$	Na$_7$Al$_2$Sb$_5$

VEC = 3.875   VEC = 4.154   VEC = 3.538   VEC = 2.714

VEC' = 6.2   VEC' = 6   VEC' = 5.75   VEC' = 5.429
N'$_{NBO}$ = 22/10   N'$_{NBO}$ = 18/9   N'$_{NBO}$ = 14/8   N'$_{NBO}$ = 10/7
N'$_{A/M}$ = 10

VEC$_A$ = 7.75   VEC$_A$ = 7.714   VEC$_A$ = 7.667   VEC$_A$ = 7.6
AA = 1/4   AA = 2/7   AA = 1/3   AA = 2/5
n/m' = 4   n/m' = 7/2   n/m' = 6/2   n/m' = 5/2
AA·(n/m') = 1   AA·(n/m') = 1   AA·(n/m') = 1   AA·(n/m') = 1

C'AC' = 0   C'AC' = 1   C'AC' = 2   C'AC' = 3

^{1}7.75/0 ^{1}7.714/1 ^{1}7.667/2 ^{1}7.6/3

Figure 7 : Examples for polyanionic valence compounds with an anionic tetrahedron complex where each tetrahedron extends one A - A bond to another tetrahedron.

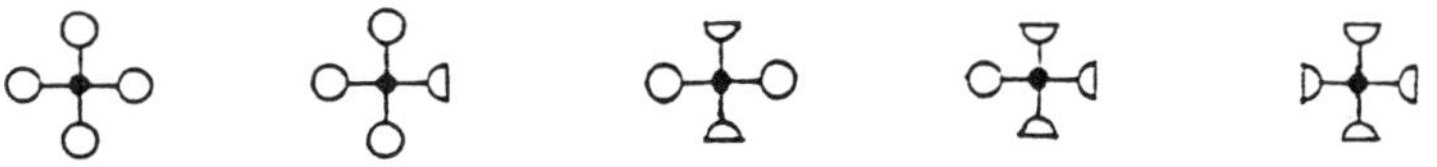

Na$_8$SnSb$_4$	Ba$_4$Ga$_2$S$_7$	Na$_5$GeP$_3$	K$_2$Si$_2$Te$_5$	K$_2$SiP$_2$

VEC = 2.462   VEC = 4.308   VEC = 2.667   VEC = 4.444   VEC = 3.2

VEC' = 6.4   VEC' = 6.222   VEC' = 6   VEC' = 5.714   VEC' = 5.333
N'$_{NBO}$ = 12/5   N'$_{NBO}$ = 20/9   N'$_{NBO}$ = 8/4   N'$_{NBO}$ = 12/7   N'$_{NBO}$ = 4/3
N'$_{A/M}$ = 5   N'$_{A/M}$ = 9

VEC$_A$ = 8   VEC$_A$ = 8   VEC$_A$ = 8   VEC$_A$ = 8   VEC$_A$ = 8
AA = C'C' = 0   AA = C'C' = 0   AA = C'C' = 0   AA = C'C' = 0   AA = C'C' = 0

C'AC' = 0   C'AC' = 1   C'AC' = 2   C'AC' = 3   C'AC' = 4

^{0}8/0 ^{0}8/1 ^{0}8/2 ^{0}8/3 ^{0}8/4

Figure 8 : Examples for normal valence compounds with anionic tetrahedron complexes where AA = C'C' = 0.

K_3GeTe_3	$Ba_2Ge_2Te_5$	$Sr_3Si_2As_4$	$Na_2Ga_2Se_3$
VEC = 3.571	VEC = 4.667	VEC = 3.778	VEC = 3.714
VEC' = 6.25	VEC' = 6	VEC' = 5.667	VEC' = 5.2
N'_{NBO} = 18/8	N'_{NBO} = 14/7	N'_{NBO} = 5/3	N'_{NBO} = 6/5
$N'_{A/M}$ = 8			
VEC_A = 8.333	VEC_A = 8.4	VEC_A = 8.5	VEC_A = 8.667
n/m' = 3	n/m' = 5/2	n/m' = 4/2	n/m' = 3/2
C'C' = 1	C'C' = 1	C'C' = 1	C'C' = 1
C'AC' = 0	C'AC' = 1	C'AC' = 2	C'AC' = 3
^{1}8.333/0	^{1}8.4/1	^{1}8.5/2	^{1}8.667/3

Figure 9 : Examples for polycationic valence compounds with anionic tetrahedron complexes where C'C' = 1 with a covalent C' - C' bond between two central atoms.

Na_3AsS_3	$MgTe_2O_5$	$KAsSe_2$
VEC = 3.714	VEC = 5.5	VEC = 4.5
VEC' = 6.5	VEC' = 6.286	VEC' = 6
N'_{NBO} = 10/4	N'_{NBO} = 16/7	N'_{NBO} = 6/3
$N'_{A/M}$ = 4	$N'_{A/M}$ = 7	
VEC_A = 8.667	VEC_A = 8.8	VEC_A = 9
n/m' = 3	n/m' = 5/2	n/m' = 2
C'C' = 2	C'C' = 2	C'C' = 2
C'AC' = 0	C'AC' = 1	C'AC' = 2
^{2}8.667/0	^{2}8.8/1	^{2}9/2

Figure 10 : Examples for polycationic valence compounds with anionic tetrahedron complexes where C'C' = 2, characterized by a psi tetrahedron, that means an anion neighbour of the central atom has been replaced by a lone electron pair.

3.2. TETRAHEDRA WITH SHARED ANIONS AND THE NUMBER OF C' - A - C' LINKS OF A TETRAHEDRON.

If the tetrahedra are linked by sharing anions one can express this formally by specifying the parameter C'AC', the number of C' - A - C' links which originate from a tetrahedron. We shall present at first a general equation to calculate C'AC' and then two simplified equations which can be used with a more than 90% probability of being correct.

As discussed in detail by Parthé & Chabot (1990) and Parthé (1990) the general equation to calculate the value of C'AC' from the anion to central atom ratio n/m' is given by

$$C'AC' = 2 \cdot [4 - (n/m')] - (1 + x) \cdot C'C' \quad \text{for } 4(3) \geq n/m' \geq 2(1). \quad (20)$$

This equation is applicable in the case that each anion is either unshared or shared with only one other tetrahedron. The x parameter refers to the electrons which remain with the central atom in the case of polycationic valence compounds. It expresses the ratio of the number of valence electrons used for C' - C' bonds between the central atoms to the total number of electrons which rest with the central atom to be used for C' - C' bonds (tetrahedron variant (III)) and lone electron pairs (tetrahedron variant (IV)). When $x = 1$ the electrons are used only for C' - C' bonds, and when $x = 0$ only for lone pairs. We shall find out below that for the most common cases one does not need to consider the x value.

In the case of polyanionic and normal valence compounds where C'C' = 0, equation (20) simplifies to

$$C'AC' = 2 \cdot [4 - (n/m')] \quad \text{for } 4 \geq n/m' \geq 2. \quad (21)$$

The number of shared anions of a tetrahedron depends thus in a simple manner on the n/m' ratio. As examples one can study the compounds given in Figures 7 and 8. The calculated C'AC' values are listed in the second last row. The value of C'AC' corresponds in the drawings to the number of anions which are presented as half circles.

The main classification of the silicates, all being normal valence compounds, is based on the average number of shared O atoms per SiO_4 tetrahedron. As shown in Table 2 this can conveniently be expressed by means of the C'AC' and corresponding n/m' values.

TABLE 2. The main classification of the silicates based on C'AC' and n/m' values.

Subdivision	C'AC'	n/m'	Examples
Nesosilicates	0	4	$Zr[SiO_4]$
Sorosilicates	1	7/2	$Sc_2[Si_2O_7]$
Inosilicates			
Pyroxenes	2	3	$Mg(SiO_3)$
Amphiboles	5/2	11/4	$Ca_2Mg_5(OH)_2(Si_4O_{11})_2$
Phyllosilicates	3	5/2	$Al_2(OH)_4(Si_2O_5)$
Tectosilicates	4	2	$Na(AlSi_3O_8)$

For polycationic valence compounds where $C'C' \neq 0$ one may rewrite (20) in the form

$$C'AC' + x \cdot C'C' = 2 \cdot [4 - (n/m')] - C'C' \quad \text{for } 3 \geq n/m' \geq 1. \tag{22}$$

The right hand part of (22) can be calculated without any assumptions (no x value involved) and has a definite value for a given compound. It is equal to the sum of the number of $C' - A - C'$ links of a tetrahedron and the number of its $C' - C'$ bonds. To obtain the actual value of $C'AC'$ one has then to make assumptions on the value of x.

However, there exists a simple solution for some compounds. The experimental evidence has shown that when $C'C' = 1$ there are with more than 90% probability only $C' - C'$ bonds, that means $x = 1$, and, further, when $C'C' = 2$ only psi tetrahedra, that means $x = 0$. This leads to the following simplified formula for the calculation of the $C'AC'$ values of polycationic valence compounds whenever $C'C' = 1$ or 2

$$C'AC' = 2 \cdot [3 - (n/m')] \quad \text{for } 3 \geq n/m' \geq 1. \tag{23}$$

The $C'AC'$ values calculated with (23) for the polycationic valence compounds with $C'C' = 1$ and 2 given in Figures 9 and 10 are listed there in the second last row of the text blocks. These values correspond to the number of anion half circles in each drawing.

3.3. ANIONIC TETRAHEDRON COMPLEXES BUILT UP WITH BASE TETRAHEDRA

3.3.1. Definition of a base tetrahedron. In the examples given above we have demonstrated how one can use the two valence electron rules to predict certain structural features of a compound assuming that an anionic tetrahedron complex is formed. In particular, for each compound we had given a list of calculated parameters and had presented a corresponding graph of a base tetrahedron with which the anionic tetrahedron complex of the compound can be constructed.

One denotes as base tetrahedron the isolated $C'A_4$ tetrahedron and all other single tetrahedra which differ by having an unshared anion (or more) replaced by a shared anion or by a $C' - C'$ bond or a lone electron pair or where the anion extends a bond to another anion of a different tetrahedron. Since reference is made to a single tetrahedron the $C' - C'$ and $A - A$ bonds are considered as tangling bonds and the anions shared with another tetrahedron are counted only as half an anion.

In Figure 11 are presented the 32 possible base tetrahedra where each central atom has at least two anion neighbours and where each anion is either unshared or shared with only one other tetrahedron. The shared anions are shown as half circles and the tangling $A - A$ and $C' - C'$ bonds by short heavy lines extending from the open or filled circles, respectively. A lone electron pair is indicated by a heavy bar on a filled circle.

The base tetrahedra are arranged vertically according to decreasing n/m' value, which is equivalent to increasing $C'AC'$ value and horizontally to decreasing number of tangling $A - A$ bonds per tetrahedron, that means decreasing $AA \cdot (n/m')$ value or, on the right hand side, to increasing number of electrons used for tangling $C' - C'$ bonds and/or lone electron pairs, that means increasing $C'C'$ value. For completeness there are given two columns for $C'C' = 2$, the first for $x = 0$ (the most common case with a psi tetrahedron) and the second for $x = 1$.

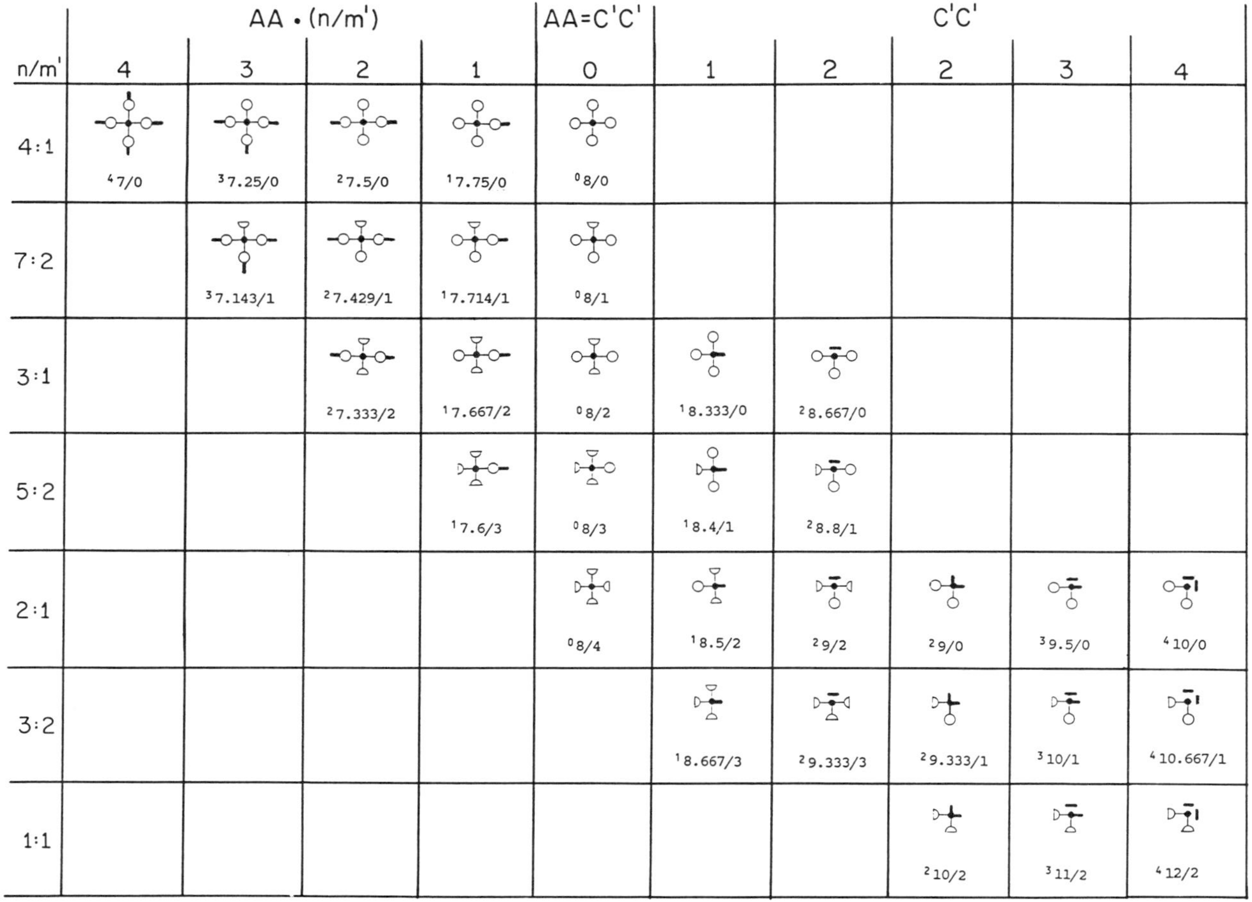

	AA · (n/m')				AA=C'C'			C'C'			
n/m'	4	3	2	1	0	1	2	2	3	4
4:1	⁴7/0	³7.25/0	²7.5/0	¹7.75/0	⁰8/0					
7:2		³7.143/1	²7.429/1	¹7.714/1	⁰8/1					
3:1			²7.333/2	¹7.667/2	⁰8/2	¹8.333/0	²8.667/0			
5:2				¹7.6/3	⁰8/3	¹8.4/1	²8.8/1			
2:1					⁰8/4	¹8.5/2	²9/2	²9/0	³9.5/0	⁴10/0
3:2						¹8.667/3	²9.333/3	²9.333/1	³10/1	⁴10.667/1
1:1								²10/2	³11/2	⁴12/2

Figure 11 : Graphs of 32 base tetrahedra with unshared anions or shared between two tetrahedra only.

For each base tetrahedron can be specified a particular VEC_A and corresponding VEC' value which a compound must have if its anionic tetrahedron complex is to be constructed with this particular base tetrahedron. The VEC_A value of a base tetrahedron, based alone on information which can be found in the graph, can be calculated by means of the generalized 8 - N rule, that is

$$VEC_A = 8 - [AA \cdot (n/m')] / (n/m') \tag{24}$$

or

$$VEC_A = 8 + C'C' / (n/m') \tag{25}$$

and the corresponding VEC' value with

$$VEC' = VEC_A \cdot n / (m' + n) \tag{26}$$

For the base tetrahedra in Figure 11 with tangling A - A bonds one calculates VEC_A < 8. These base tetrahedra can be used for the construction of the anionic tetrahedron complexes in polyanionic valence compounds. For all base tetrahedra where AA = C'C' = 0 one finds that VEC_A = 8. These are the base tetrahedra which are important for the anionic tetrahedron complexes in normal valence compounds. Finally, all base tetrahedra with C'C' > 0 have VEC_A > 8. These are the base tetrahedra which build up the anionic tetrahedron complexes in polycationic valence compounds.

Note, that AA·(n/m') and C'C' (for our case also 2·n/m') can have integral values only. The number of possible base tetrahedra is therefore limited. Equally limited is the list of specific VEC_A values which compounds must have for them to be built up of one kind of base tetrahedron only.

3.3.2. The classification codes for base tetrahedra. It was found desirable to design for the base tetrahedra a classification code which contains all the information necessary to draw a graph and includes also the VEC_A value which a compound must have for its anionic tetrahedron complex to be built up of this base tetrahedron (Parthé & Chabot, 1990). We shall use for the classification code the values of AA·(n/m') or C'C', VEC_A and C'AC', which, except for VEC_A, can be obtained directly from a study of the drawing of the base tetrahedron. Depending on the VEC_A value the classification code of a base tetrahedron will be written as

$$^{AA \cdot (n/m')}VEC_A/C'AC' \quad \text{or} \quad ^{0}8/C'AC' \quad \text{or} \quad ^{C'C'}VEC_A/C'AC' \tag{27}$$

If the value of VEC_A < 8, then the superscript indicates the AA·(n/m') value and if VEC_A > 8 the C'C' value. In the case that VEC_A = 8 one uses as superscript 0. Following the VEC_A value, and separated by a slash, is given the C'AC' value.

In Figures 7 to 10 the classification codes had been listed in the last rows of the text blocks. As seen in Figure 11, where the classification code is written below each base tetrahedron graph, all 32 different base tetrahedra have different codes and are thus unambiguously identified.

3.3.3. The classification codes for compounds with anionic tetrahedron complexes. For the classification code of a compound one uses the same expressions as (27). However, VEC_A has here to be calculated from the composition and the number of available valence electrons according to (14). The value obtained for VEC_A is further used to calculate AA·(n/m') using (18) or C'C' by means of (19). The most probable value of C'AC' is obtained by (21) or (23), respectively.

Depending on the classification code of a compound two cases have to be distinquished to find the most probable base tetrahedron(a) :
- The classification code of the compound is identical with the classification code of a base tetrahedron. Here, one expects as most simple solution that the anionic tetrahedron complex is constructed alone with this base tetrahedron. This is the case for all the examples treated in Figures 7 to 10.
- The classification code of the compound is not identical with the classification code of a base tetrahedron. Whenever the C'AC' or the C'C' or AA·(n/m') values of the classification code of a compound are not integers, the tetrahedron complex is built up of more than one kind of base tetrahedron. As simple solution one selects from Figure 11 a pair of base tetrahedra with similar classification codes. As a guide to find their proper porportions one may use the following equations which relate the classification code and the n/m' ratio of a tetrahedron complex built up of different base tetrahedra to the classification codes and n/m' ratios of the base tetrahedra involved.

$$AA \cdot (n/m') = <AA \cdot (n/m')> \tag{28a}$$

$$C'C' = <C'C'> \tag{28b}$$

$$VEC_A = <VEC_A \cdot (n/m')> / <n/m'> \tag{28c}$$

$$C'AC' = <C'AC'> \tag{28d}$$

$$n/m' = <n/m'> \tag{28e}$$

where the angled brackets indicate that the enclosed data are the average values of the base tetrahedra involved.

3.3.4. Examples for anionic tetrahedron complexes constructed of two kinds of base tetrahedra. As a simple example for a compound where the anionic tetrahedron complex is built up of two kinds of base tetrahedra one may consider the amphiboles, listed in Table 2. Based on their classification code $^0 8/2.5$ one can conclude, considering (28d), that the anionic tetrahedron complex is constructed of equal amounts of $^0 8/2$ and $^0 8/3$ base tetrahedra. From (28e) one finds that in this case n/m' = 11/4, which agrees with the compositions of the amphiboles.

Two more examples of normal valence compounds with two different kinds of base tetrahedra are presented in Figure 12. In the case of the ultraphosphate NdP_5O_{14} with classification code $^0 8/2.4$ one expects, in agreement with (28d), that for every three $^0 8/2$ base tetrahedra there are two $^0 8/3$ base tetrahedra. This agrees with the experimentally determined anionic tetrahedron complex of NdP_5O_{14}, shown in schematic form on the left hand side.

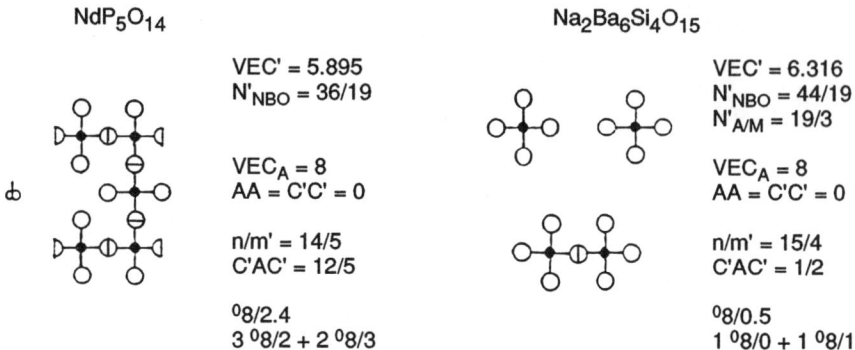

Figure 12 : The observed anionic tetrahedron complexes of two normal valence compounds where the anionic tetrahedron complexes are built up of two kinds of base tetrahedra. In the second last row of the text columns are listed the classification codes of the compounds and in the last row the classification codes of the base tetrahedra involved and their proportions.

In the case of $Na_2Ba_6Si_4O_{15}$ with classification code $^08/0.5$ one expects equal amounts of $^08/0$ and $^08/1$ base tetrahedra. To the code $^08/0$ corresponds an isolated $C'A_4$ tetrahedron. However, the $^08/1$ base tetrahedron with an anion half circle in its graph presentation must be linked with a second one and forms an isolated C'_2A_7 double tetrahedron. Equal amounts of $^08/0$ and $^08/1$ means thus that the anionic tetrahedron complex of $Na_2Ba_6Si_4O_{15}$ should consist of two isolated SiO_4 tetrahedra with 5 atoms each for every isolated Si_2O_7 double tetrahedron with 9 atoms. The average number of atoms per molecule is 19/3 in agreement with the calculated $N'_{A/M}$ value. All these predictions agree with the experimentally determined anionic tetrahedron complex, shown in a graph presentation on the right hand side of Figure 12.

3.3.5. Structural features which can not be predicted with valence elctron rules. The valence elctron rules allow the prediction of a probable base tetrahedron for a compound with an anionic tetrahedron complex, however not the details how this base tetrahedron is linked with itself. In addition, instead of an expected single base tetrahedron there may occur two different base tetrahedra but with the restriction that the average of their codes, calculated with (28), corresponds to the classification code of the compound.

As example one may consider the observed anionic tetrahedron complexes of the four normal valence compounds with the same classification code $^08/3$, shown in Figure 13. The first three tetrahedron complexes are constructed alone of $^08/3$ base tetrahedra. In $Na_2Ge_2Se_5$ and $Na_2Ge_2S_5$ the base tetrahedra are only corner linked, however in $Rb_4In_2S_5$ corner and edge linked. Even if there are only corner linked base tetrahedra the anionic tetrahedron complex may have the shape of a two-dimensional layer or a cyclic molecule. Finally, the anionic tetrahedron complex in $Cs_4Ga_2Se_5$, a multicyclic molecule, is constructed, instead of $^08/3$ base tetrahedra, of $^08/2$ and $^08/4$ base tetrahedra in equal proportions.

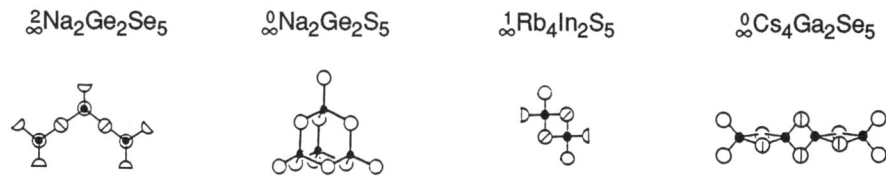

Figure 13 : The different anionic tetrahedron complexes of four normal valence compounds with same classification code $^0 8/3$.

The interchange of corner and edge linked tetrahedra or the substitution of one base tetrahedron by two corresponding ones has been observed even with different modifications of one compound. An example is $Ca_3Si_2O_7$ for which the classification code is $^0 8/1$. In the rankinite modification a double tetrahedron of two linked $^0 8/1$ base tetrahedra occurs as shown for $Na_6Ge_2Se_7$ in Figure 5. However, in the kilchoanite modification there are two kinds of molecules, that is an isolated SiO_4 tetrahedron ($^0 8/0$) and an isolated molecule of composition Si_3O_{10} which consists of a chain of three corner linked tetrahedra ($^0 8/1 + {}^0 8/2 + {}^0 8/1$). Note, that the average value of the codes of the base tetrahedra involved and the average number of the atoms in the molecules correspond to the code of the compound and the calculated $N'_{A/M}$ value, respectively.

3.3.6. Limits for the application of the valence electron rules. For the fruitful application of these valence electron rules certain assumptions have to be made which can be summarized as follows :
- One has to assume that an anionic tetrahedron complex is formed and one has to make also assumptions which of the elements are the cations C, the central atoms C' and the anions A. The parameters used in the Mooser - Pearson diagrams to separate compounds with adamantane structure from those having no adamantane structures are also here of some relevance. If the central atom and the anion is from the second period of the Periodic Table then a planar triangular anion complex is preferred to a tetrahedron complex, as for example with $(BO_3)^{3-}$, $(CO_3)^{2-}$ or $(NO_3)^{1-}$. Tetrahedral complexes are also rarely found if the central atom is from a very high period. Central atoms from periods inbetween can be expected to form tetrahedral complexes, but there are exceptions.
- The valence electron contribution is evident in most cases, but there are exceptions such as the frequently encountered cations Tl^{1+} and Pb^{2+}. In the case of transition elements their valence electron contribution can only be calculated from the observed structural features of the anionic tetrahedron complex. This may in case be verified by magnetic measurements.
- Equation (23) represents a simplified expression which is correct in only 90 % of the structures. In the most general case one is obliged to make assumptions how the electrons on the central atoms are used for C' - C' bonds or lone electron pairs. The reader is advised to look up the original publication of Parthé & Chabot (1990) to find out how the C'AC' values can be calculated for less common cases of anionic tetrahedron complexes.

Inspite of certain limitations for the application of the valence electron rules, they have been found very useful to systematize and classify structures with anionic tetrahedron complexes. They have been of help to point out inconsistencies between reported structure and composition and they offer an aid in the synthesis of new compounds with particular structural features.

4. References

Deiseroth, H.J. & Pfeifer, H. (1991). 'In$_5$S$_4$ = SnIn$_4$S$_4$: Eine Korrektur I.' Z.Kristallogr. **196**, 197 - 205.

Deller, K. & Eisenmann, B. (1977). 'Die Kristallstruktur des Sr$_3$As$_4$.' Z.Naturforsch. **32b**, 1368 - 1370.

Eisenmann, B., Hansa, J. & Schäfer, H. (1986). 'Na$_6$Ge$_2$Se$_7$, das erste Selenidosorogermanat(IV).' Revue Chim. Minér. **23**, 8 - 13.

Endo, T., Sato, T. & Shimada, M. (1987). 'High-Pressure Synthesis of B$_2$O with Diamond-Like Structure.' J.Mater.Science Lettr. **6**, 683 - 685.

Fenner, J. (1976). 'The Crystal Structure of CuTe$_2$Cl.' Acta Cryst. **B32**, 3084 - 3086.

Lima-de-Faria, J., Hellner, E., Liebau, F., Makovicky, E. & Parthé, E. (1990). 'Nomenclature of Inorganic Structure Types. Report of the International Union of Crystallography Commission on Crystallographic Nomenclature Subcommittee on the Nomenclature of Inorganic Structure Types.' Acta Cryst. **A46**, 1 - 11.

Nesper, R. & Curda, J. (1987). 'Al$_7$Te$_{10}$ - das erste Chalcogenid mit zweiwertigem Aluminium.' Z. Naturforsch. **42b**, 557 - 564.

Möller, M.H. & Jeitschko, W. (1981). 'Preparation and Crystal Structure of Trisilver Undecaphosphide, Ag$_3$P$_{11}$, an Unusual Defect Tetrahedral Compound.' Inorg.Chem. **20**, 828 - 833.

Möller, M.H. & Jeitschko, W. (1982). 'Darstellung, Eigenschaften und Kristallstruktur von Cu$_2$P$_7$ und Strukturverfeinerungen von CuP$_2$ und AgP$_2$.' Z. Anorg. Allg. Chem. **491**, 225 - 236.

Parthé, E. (1972). "Cristallochimie des Structures Tétraédriques." 349 pages. Paris : Gordon & Breach.

Parthé, E. (1973). 'Valence Electron Concentration Rules and Diagrams for Diamagnetic, Non - Metallic Iono - Covalent Compounds with Tetrahedrally Coordinated Anions.' Acta Cryst. **B29**, 2808 - 2815.

Parthé, E. (1989). 'Comments on the Valence Rule by Nesper and von Schnering.' Z.Kristallogr. **189**, 101 - 107.

Parthé, E. (1990). "Elements of Inorganic Structural Chemistry, a Course on Selected Topics." 100 pages. Petit-Lancy (Switzerland), 49 Chemin du Gué : K. Sutter Parthé Publisher.

Parthé, E. & Chabot, B. (1990). 'Classification of Structures with Anionic Tetrahedron Complexes using Valence - Electron Criteria.' Acta Cryst. **B46**, 7 - 23.

Wadsten, T., Arnberg, L. & Berg, J.E. (1980). 'The Structure of Pentaindium Tetrasulfide.' Acta Cryst. **B36**, 2220 - 2223.

5. Appendix : The Periodic Table of the Elements

H	A			T									B					He
1	2											3	4	5	6	7	8	
Li	Be											B	C	N	O	F	Ne	
Na	Mg	3	4	5	6	7	8	9	10	1	2	Al	Si	P	S	Cl	Ar	
K	Ca	Sc	Ti	V	Cr	Mn	Fe	Co	Ni	Cu	Zn	Ga	Ge	As	Se	Br	Kr	
Rb	Sr	Y	Zr	Nb	Mo	Tc	Ru	Rh	Pd	Ag	Cd	In	Sn	Sb	Te	I	Xe	
Cs	Ba	L*	Hf	Ta	W	Re	Os	Ir	Pt	Au	Hg	Tl	Pb	Bi	Po	At	Rn	
Fr	Ra	A*	Rf	Ha														

L*	La	Ce	Pr	Nd	Pm	Sm	Eu	Gd	Tb	Dy	Ho	Er	Tm	Yb	Lu
A*	Ac	Th	Pa	U	Np	Pu	Am	Cm	Bk	Cf	Es	Fm	Md	No	Lr

6. Problems

Problem 1 : Derive possible crystal chemical formulae for Ag_3P_{11} and Al_7Te_{10}, assuming that the two compounds crystallize with tetrahedral structures.

Problem 2 : The following binary compounds are characterized by tetrahedral anion partial structures. Find out what the probable features of these charged anion partial structures might be and denote them by means of crystal chemical formulae.

SrP $LaAs_2$ Ca_2As_3 LiAs BaP_3

Problem 3: The compounds listed below are characterized by an anionic tetrahedron complex. Write down the classification codes of the compounds and make graph drawings of the probable base tetrahedron(a) involved in the construction of the anionic tetrahedron complexes.

K_2GeTe_4 KSO_3 $LiGeTe_2$ $BaSnS_2$ $Na_3Mg_2P_5O_{16}$

6.1. SOLUTIONS OF THE PROBLEMS

Problem 1 : The values for VEC, N_{NBO}, VEC_A and AA or CC are obtained from (2), (3), (5) and (7) or (8), respectively.

Ag_3P_{11}

VEC = 4.142
N_{NBO} = 2/14

VEC_A = 5.273
AA = 30/11

$Ag_3^{[4t;]}P_3^{[(2;2)t]}\underline{P}_2^{[;3n]}\underline{P}_6^{[(1;3)t]}$

Al_7Te_{10}

VEC = 4.765
N_{NBO} = 13/17

VEC_A = 8.1
CC = 1/7

$Al_6^{[4t;]}Al^{[(3;1)t]}\underline{Te}_7^{[3n;]}\underline{Te}_3^{[2n;]}$

The formula given for Ag_3P_{11} corresponds to the case where the two lone electron pairs are both associated with P atoms forming three homonuclear P - P bonds. The valence electron rules specify only the number of lone electron pairs per formula unit. Some or all of the lone electron pairs might therefore be on the P atoms having two homonuclear bonds.

Problem 2 : The values of VEC', $N'_{A/M}$ or AA have been calculated by means of (12), (16) and (7), respectively.

SrP	LaAs$_2$	Ca$_2$As$_3$	LiAs	BaP$_3$
VEC' = 7	VEC' = 6.5	VEC' = 6.333	VEC' = 6	VEC' = 5.667
$N'_{A/M}$ = 2	$N'_{A/M}$ = 4	$N'_{A/M}$ = 6	AA = 2	AA = 7/3
$^A_2[P_2]^{4-}$	$^A_4[As_4]^{6-}$	$^A_6[As_6]^{8-}$	$^1_\infty(\underline{As}^{[2]})^{1-}$	$(\underline{P}_2^{[2]}\underline{P}^{[3]})^{2-}$

In the low temperature modification of LaAs$_2$ occur the expected finite chains of four As atoms. However, in the high temperature modification there are two kinds of chains, one with three and the second with five atoms.

In the actual structure of Ca$_2$As$_3$ one does not find the expected finite As chain of six As atoms, but one with four and one with eight As atoms.

Problem 3 : To obtain the values of VEC', $N'_{A/M}$, VEC_A, AA or C'C' and C'AC' use was made of (13), (16), (14), (7) or (19) and (21) or (23), respectively.

The anionic tetrahedron complex in $Na_3Mg_2P_5O_{16}$ should consist of a molecule with 21 atoms which, in the most simple case, is constructed of two $^08/1$ and three $^08/2$ base tetrahedra. This corresponds to a finite chain of five corner linked tetrahedra which is actually observed.

The actual anionic tetrahedron complex of LiGeTe$_2$ is built up not of the expected $^18.5/2$ base tetrahedra, but of equal amounts of $^18.4/1$ and $^18.667/3$ base tetrahedra. Using (28b), (28c) and (28e) one can verify that the average corresponds to $^18.5/2$, the classification code of the compound.

	K_2GeTe_4	KSO_3	$LiGeTe_2$	$BaSnS_2$	$Na_3Mg_2P_5O_{16}$
VEC'	6	6.25	5.667	6	6.095
$N'_{A/M}$	-	8	-	-	21
VEC_A	7.5	8.333	8.5	9	8
AA	0.5	-	-	-	0
C'C'	-	1	1	2	0
n/m'	4	3	2	2	16/5
C'AC'	0	0	2	2	8/5
	$^2 7.5/0$	$^1 8.333/0$	$^1 8.5/2$	$^2 9/2$	$^0 8/1.6 \equiv$ $2\,^0 8/1 + 3\,^0 8/2$

STRUCTURAL PRINCIPLES OF SILICATES, PHOSPHATES, SULPHATES AND RELATED STRUCTURES

DIMITRI Y. PUSHCHAROVSKY
Department of Geology
Moscow State University
119899 Moscow
Russia

ABSTRACT. The review summarizes the results of recent structural investigations of silicates and other groups of compounds with anionic tetrahedron complexes. The different approaches to their crystal chemical classifications are considered. A description of new types of anionic tetrahedron complexes, found in silicates and their analogues during the last years, is given. The concept emphasizing the dominating role of the nontetrahedral cations is useful for the interpretation of structure transformations of silicates under high pressure and temperature. Several problems and answers are given in the last section.

1. Structural Diversity of Silicates and the Si - O Bond

Silicates are compounds containing silicon and oxygen. Their main specific structural feature is the presence of Si atoms in a tetrahedral oxygen atom coordination, which is favored by the ratio $R_{Si} : R_O = 0.28$ (Table 1). Bond lengths and bond angles in the Si,O tetrahedra tend to be close to their mean values : $d(Si - O) = 1.62$Å; $\angle O - Si - O = 109.47°$ and $\angle Si - O - Si = 140°$ (Liebau, 1985). Si atoms in 6-fold coordination are found in a relatively small group of essentially high-pressure silicates and Si atoms in 5-fold coordination in some of the organosilicon compounds.

TABLE 1. The different coordination numbers (CN) and coordination figures as a function of the ratio of ionic radii

CN	Coordination	Ratio of ionic radii
2	dumb-bells	≤ 0.15
3	triangle	0.15 - 0.22
4	tetrahedron	0.22 - 0.41
6	octahedron	0.41 - 0.73
8	cube	0.73 - 1.37

The structures of silicates and related compounds contain anionic tetrahedron complexes of different configuration (single tetrahedra, pyrogroups, rings, chains, layers, frameworks etc.). Their comparative analysis considerably contributes to the scientific ideas on the structures of crystalline materials in general. The interest in silicates is related to their widespread abundance in rocks (Table 2) and the exceptional variability of the anionic tetrahedron complexes in their structures.

TABLE 2. Bulk content of silicates in the continental crust

Feldspars	64%	
Pyroxenes + amphibols	9%	
Quartz	18%	total 96.5%
Biotite	4%	
Olivine	1.5%	

If one takes into account the most specific features of a Si,O tetrahedron complex, such as the number of tetrahedra in the chain period, or the types of rings which can be recognized in different layers, it is possible to estimate the total number of anionic tetrahedron complexes in the silicates as one hundred.

One might be tempted to suppose, that this diversity of the silicates should have a common root with the great number of compounds of carbon, which is situated above silicon in the fourth main group of the Periodic Table. Like silicates, which are the most abundant inorganic compounds, carbon compounds dominate in organic chemistry. However, according to Bragg and Liebau, the definite distinctive features can be revealed if one compares the nature of the bonds formed by silicon and carbon with other elements.

1) Si atoms have $3s^2 3p^2$ valence electrons, but C atoms $2s^2 2p^2$ electrons. Thus, the distances between C atoms are much shorter and double bonds $C = C$ can be formed, while the Si - Si bonds are much weaker (bond energy for C - C is 346 kJ/mol, but for Si - Si 222 kJ/mol).

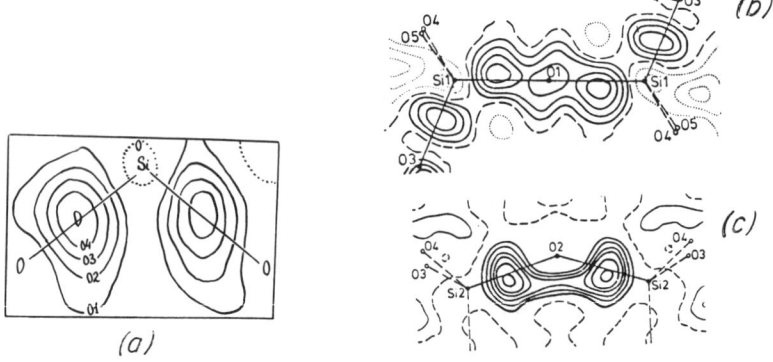

Figure 1 : Difference Fourier maps of andalusite (a) and coesite (b, c). In (b) the angle $\angle Si - O - Si = 180°$, but in (c) less than $180°$ (Ross, 1980).

2) Differently from the carbon atoms, the Si atoms form very strong Si - O bonds (bond energy for C - O is 358 kJ/mol, but for Si - O 452 kJ/mol). In the Si atoms, due to the higher nuclear charge, the energy of the empty d orbitals is decreased and

energetically they become closer to the 2p orbitals of the O atoms. Therefore, additional $d\pi - p\pi$ components reinforce the bonds between silicon and oxygen and, consequently, $d(Si - O)_{exp} = 1.626Å < d(Si - O)_{calc} = 1.760Å$. Deviations of the peaks from the lines connecting the silicon and oxygen atoms (example coesite, SiO_2) as well as their nonspherical symmetry on the deformation electron density maps (example andalusite, Al_2SiO_5) can also be attributed to the π component of the Si - O bonds (Figure 1).

2. Main Concept of Silicate Structures

In the silicate structures the Si,O anionic tetrahedron complex will be adjusted to the cation-oxygen polyhedron fragments (Belov, 1959). In particular, this principle was confirmed with the structures having big cations (K, Na, Ca, R = rare earth elements), where the Si,O anionic tetrahedron complex - in the form of pyrogroups $[Si_2O_7]$ - is commensurable with the edges of the cation-oxygen polyhedra.

Several quantitative correlations actually complete this qualitative approach. For example, there are more than 60 compounds with general formula $M_a[T_2O_7]_b$, where T = As, Be, Cr, Ge, P, S, Si (Clark & Morley, 1976). They can be subdivided into two groups : a) thortveitite-like with angle $\angle T - O - T > 140°$ and b) bichromate-like with angle $\angle T - O - T < 140°$. Figure 2 illustrates that the frontier between both groups corresponds to the equation : $r_T = 1.5 \cdot r_M - 1.1$, where r_T is the radius of the tetrahedral (T) and r_M of the nontetrahedral (M) cation, respectively.

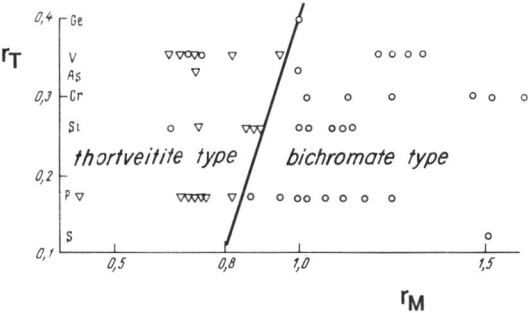

Figure 2 : Correlation between the stability of thortveitite - like and bichromate - like structures $M_a[T_2O_7]_b$ and the radii of the M and T cations (Clark & Morley, 1976).

3. Crystal Chemical Classifications of Silicates

Belov's concept enabled the division of the silicate structures into two groups (or chapters). However, there are several different approaches to their systematics. The common feature of these studies is the analysis of Si,O anionic tetrahedron complex. The first steps in this direction were made by Machatschki, Náray-Szabó and Bragg.

Liebau (1971) elaborated a very detailed classification on the basis of 7 aspects :
- 1) the coordination number of Si
- 2) the number of oxygens positioned between two adjacent SiO_n polyhedra (linkedness)
- 3) the number of SiO_n polyhedra with which a given Si,O polyhedron is connected (connectedness)
- 4) branchedness : a) an unbranched complex contains SiO_n polyhedra, which are isolated or are in contact with 1 or 2 others; b) a branched complex has some SiO_n polyhedra with 3 or more neighbours
- 5) dimensionality
- 6) multiplicity and
- 7) periodicity of a chain or a ring of SiO_n polyhedra.

Kostov (1975) based his classification partially on the crystal structure. He considered the ratio of Si atoms to nontetrahedral atoms as well as the crystal morphology, and the association of chemical elements in minerals.

Pushcharovsky (1986) considered the composition of the anionic tetrahedron complex $[T_mO_n]$ as the most important aspect in the classification of silicates, phosphates and germanates. He showed that the number of observed different Si : O ratios in silicates (28) is greater than the different P : O ratios in phosphates (10) and Ge : O ratios in germanates (9). One distinguishes between bridging oxygen atoms O_{br} and nonbridging atoms O_{nbr}. The average number of O_{br} per tetrahedron, expressed by K, is related with the composition of the anionic tetrahedron complex $[T_mO_n]$ according to

$$K = 8 - 2 \cdot (n/m).$$

The identical formula was derived by Parthé & Engel (1986) for the modified tetrahedron sharing coefficient TT, which corresponds to the average number of (central atom - bridging atom - central atom) bonds per tetrahedron in a structure with anionic tetrahedron complexes. If one restricts oneself to the case where the bridging atom forms a bridge between only two central atoms, that which is observed in all silicates, then TT = K.

About 20 compounds with different kinds of anionic tetrahedron complexes in the same structure are considered in a special subdivision (Table 3). These structures do not obey Pauling's fifth rule. Belov's concept, emphasizing the dominating role of the cations, can be used for the interpretation of these structures with different tetrahedron complexes (Table 4).

For the purposes of comparative crystal chemistry, the recognition of tetrahedron fragments is very useful if the tetrahedra share corners. If a condensation of tetrahedra does not occur, the analysis of the structures and their classification is more effective in the frame of the concept of mixed complexes. The predominant structural fragments are cation-oxygen polyhedra - where the bond strengths are comparable with those of the bonds between silicon and oxygen - and Si,O tetrahedra. This approach was recently extended to other kinds of compounds and has been used during the last years for the classification of carbonates, sulphates, tellurates and so on. Figure 3 illustrates Hawthorn's classification (Hawthorn, 1986) of carbonates with mixed chain complexes.

The structural classification of sulphates on the basis of this concept proves that the variability of mixed complexes in sulphates (Raszvetaieva & Pushcharovsky, 1989) is comparable with the diversity of the anionic tetrahedron complexes in silicates (Figure 4).

TABLE 3. Common occurrence of different anionic tetrahedron complexes in the same structure.

First anion complex	Second anion complex
orthotetrahedron	pyrogroup triple tetrahedron fivefold tetrahedron 4-membered ring chain layer
pyrogroup	triple tetrahedron 4-membered ring 12-membered ring chain band
triple tetrahedron	band
3-membered ring	9-membered ring
chain	band
double chain	triple chain
band	layer
layer	double layer

TABLE 4. Alkali Ta phosphates with two different kinds of anionic tetrahedron complexes in the same structure (Nickolaev, 1983).

Compound	r_M(Å)	N_M/N_{Ta}	complex I	complex II
$CsTa_2[P_3O_{10}][PO_4]_2$	1.69	1 : 2	P_3O_{10}	PO_4
$Rb_2Ta_2[P_5O_{16}][PO_4]_2$	1.48	2 : 3	P_5O_{16}	PO_4
$KTa[PO_3]_2[P_2O_7]$	1.33	1 : 1	PO_3	P_2O_7

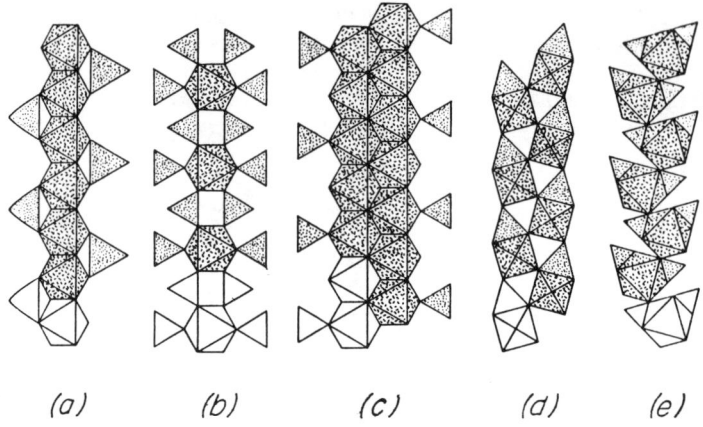

Figure 3 : Mixed chain complexes in carbonate structures : (a) dundasite $Pb[Al(CO_3)(OH)_2]_2 \cdot H_2O$; (b) sahamalite $R_2[(Mg,Fe)(CO_3)_4]$; (c) artinite $[Mg_2(CO_3)(OH)_2(H_2O)_3]$; (d) nesquehonite $[Mg(CO_3)(H_2O)_2]H_2O$; (e) chalconatronite $Na_2[Cu(CO_3)_2H_2O] \cdot 3H_2O$.

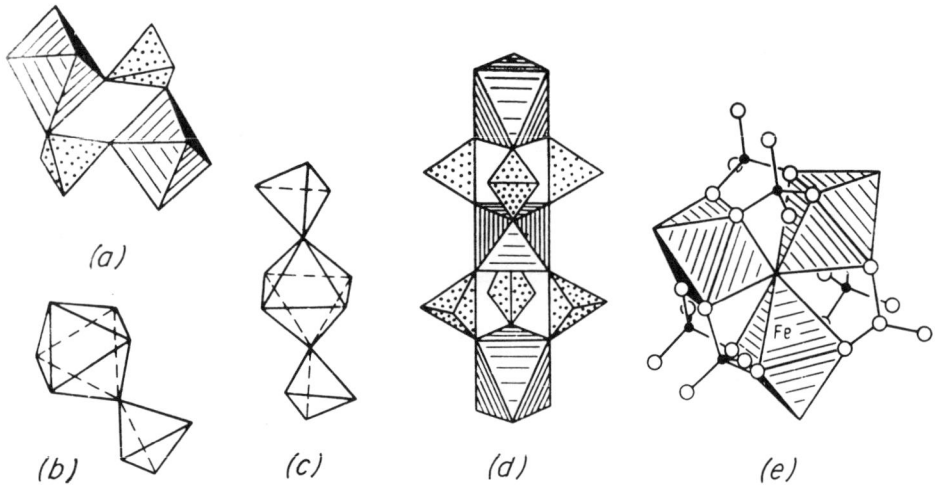

Figure 4 : Mixed insular complexes in sulphate structures : (a) starkeyite $MgSO_4 \cdot 4H_2O$; (b) $VOSO_4 \cdot 5H_2O$; (c) astrakhanite $Na_2Mg(SO_4)_2 \cdot 4H_2O$; (d) $Fe_2(SO_4)_3 \cdot 9H_2O$; (e) Mause's salts $A_5Fe_3O(SO_4)_6 \cdot nH_2O$, where A = Li, Na, K, Rb, Cs, NH_4, Tl and $n = 5 - 10$.

4. New Types of Anionic Tetrahedron Complexes in Silicates and their Analogues

Several new anionic tetrahedron complexes, all belonging to the main subdivisions, have been revealed during the last years.

4.1. INSULAR ANIONIC TETRAHEDRON COMPLEXES IN SILICATES

The biggest insular tetrahedron complex, built up of 48 Si,O tetrahedra, was found as the main structural unit of ashcroftine $K_{10}Na_{10}(Y,Ca)_{24}(OH)_4(CO_3)_{16}[Si_{56}O_{140}]\cdot 16H_2O$.

Pyrophosphate groups $[P_2O_7]$ were discovered in canaphite, the only mineral with linked P,O tetrahedra.

At the present time the number of known isolated linear tetrahedron complexes has been considerably increased. The first known complexes of this type were triple tetrahedra $[Si_3O_{10}]$ in ardenite $Mn_4Al_6[(As,V)O_4][SiO_4]_2[Si_3O_{10}](OH)_6$ and Na, Cd silicates (Figure 5). In $Ag_{10}[Si_4O_{13}]$ were found complexes of 4 tetrahedra and in $Na_4Sn_2[Si_5O_{16}]\cdot H_2O$ complexes of 5 tetrahedra. The phosphate analogues of these tetrahedron complexes are also known. Several Mg, Sc silicates, synthesized in the MgO - Sc_2O_3 - SiO_2 diagram, contain linear tetrahedron complexes, formed by 8, 9 and even 10 Si,O tetrahedra. There are indications of oligophosphates with tetrahedron complexes built of more than 5 P,O tetrahedra (Averbuch - Pouchot & Durif, 1991a). In some structures the linear anion complexes are built of chemically different tetrahedra. The association of a triple tetrahedron $[Si_3O_{10}]$ with an As,O tetrahedron was found in tyragalloite, $Mn_4[AsSi_3O_{12}(OH)]$ and of a complex of 5 Si,O tetrahedra linked to one V,O tetrahedron in medaite $HMn_6[VSi_5O_{16}]O_3$ (Figure 5).

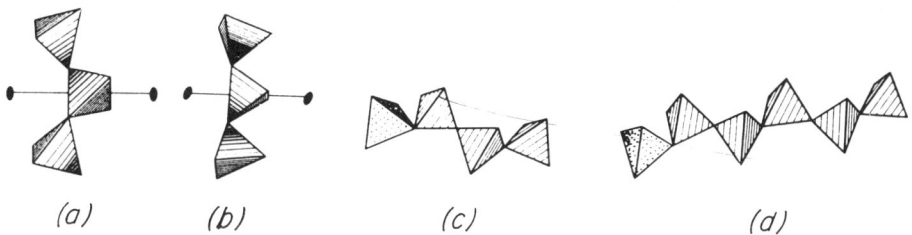

(a)   *(b)*   *(c)*   *(d)*

Figure 5 : Triple tetrahedra in (a) $Na_4Cd_2[Si_3O_{10}]$ and (b) $Na_2Cd_3[Si_3O_{10}]$. Linear anionic tetrahedron complexes in (c) tiragalloite and (d) medaite. The As,O tetrahedron (tiragalloite) and the V,O tetrahedron (medaite) are dotted.

4.2. ANIONIC TETRAHEDRON COMPLEXES IN THE FORM OF RINGS AND CHAINS

There are nine types of tetrahedron rings in silicate, phosphate and germanate structures (Table 5). The biggest 18-membered ring (Figure 6) were found in 1990 in the natural K,Na silicate $KNa_8[Si_9O_{18}(OH)_9]\cdot 19H_2O$. Unlike the simple tetrahedron rings $[SiO_3]_n$, the double rings have the composition $[Si_2O_5]_n$. The 3-membered and 4-membered double rings are rather rare, while 6-membered double rings, first discovered in the milarite structure $KCa_2Be_2Al[Si_{12}O_{30}]\cdot H_2O$, were later found in quite a large group of silicates.

TABLE 5. Tetrahedron rings observed in silicates, phosphates and germanates

Type of ring	Silicates	Phosphates	Germanates
3-membered	+	+	+
4-membered	+	+	-
5-membered	-	+	-
6-membered	+	+	-
8-membered	+	+	-
9-membered	+	-	-
10-membered	-	+	-
12-membered	+	-	+
18-membered	+	-	-

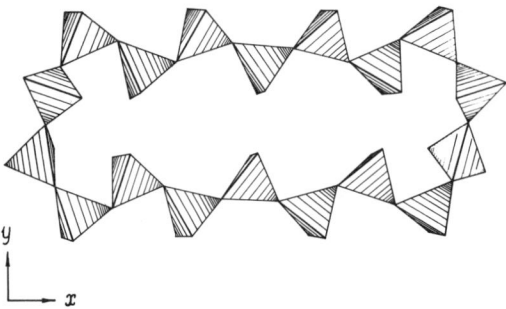

Figure 6 : 18-membered tetrahedron ring of $KNa_8[Si_9O_{18}(OH)_9] \cdot 19H_2O$.

Chakoumakos, Hill & Gibbs (1981) proposed to estimate the s component $\delta(\%)$ of the Si - O_{br} - Si bonds in silicate rings by the formula:

$$\delta(\%) = 100/(1 + \lambda^2),$$

where $\lambda^2 = -1/\cos \angle Si - O_{br} - Si$. Correspondingly, the higher the value of δ, the shorter is the distance Si - O : $d(Si - O) = -3.5 \cdot 10^{-3} \cdot \delta + 1.78$.

On the basis of 28 accurate structure determinations Averbuch - Pouchot & Durif (1991b) examined the main characteristics of 6 - membered P,O rings. The general average P - P = 2.927Å and $\angle P - O - P = 132.4°$ are comparable with those previously reported for phosphoric anions. However, the average angles P - P - P of 111.9° for internal ring symmetry P-1 and 109.2° for -3 and 3, deviate significantly from their ideal value of 120° in contrast with the cyclotri- and cyclotetraphosphates.

15 different types of tetrahedron chains are known to this day (Table 6). Recently, in 1989, a new type of spiral chain with 16 tetrahedra in the period was found in $KEr[PO_3]_4$ - VII. Figure 7 demonstrates different configurations of tetrahedron chains.

TABLE 6. Tetrahedron chains in silicates, phosphates and germanates.

Tetrahedron(a) in period	Silicates	Phosphates	Germanates
1	-	-	+
2	+	+	+
3	+	+	+
4	+	+	-
5	+	-	+
6	+	+	+
7	+	-	-
8	-	+	-
9	+	-	-
10	-	+	-
12	+	-	-
14	-	+	-
16	-	+	-
22	+	-	-
24	+	-	-

TABLE 7. Influence of the valence and the electron affinity of the cations on the shape of the [PO_3] chains in metaphosphate structures.

Compound	Period	Stretching factor	Cation valence	Electron affinity of cation (kJ/mol)
$Rb[PO_3]$	2	0.835	1	213.5
$K[PO_3]$	2	0.934	1	221.9
$Na[PO_3]$ - A	4	0.633	1	263.8
$Al[PO_3]_3$	6	0.626	3	1620.4
$Nd[PO_3]_3$	6	0.493	3	1101.2
$Bi[PO_3]_3$	6	0.466	3	1578.5
$Zr[PO_3]_4$	8	0.708	4	1817.2

TABLE 8. Influence of the cation sizes on the shape of the [PO_3] chains in metaphosphate structures.

Compound	Period	Stretching factor	Av. valence	Difference of cation sizes (Å)
III - $KNd[PO_3]_4$	4	0.747	2	0.34
III - $KEr[PO_3]_4$	4	0.739	2	0.46
$CdBa[PO_3]_4$	4	0.733	2	0.52
$CsPr[PO_3]_4$	8	0.466	2	0.64
VI - $CsNd[PO_3]_4$	8	0.464	2	0.66
$CsTb[PO_3]_4$	8	0.462	2	0.74

According to Liebau & Pallas (1981) two main parameters can be used for their description :
- 1) the number of tetrahedra in their repeat unit and
- 2) the stretching factor $f_s = l / \Sigma l_i$, where l is the length of the repeat unit in Å and l_i is the length of the edge between two bridge oxygen atoms in the tetrahedra, participating in the chain period (Figure 8). The stretching coefficients of the tetrahedral chains decrease with an increase of the electronegativity, the valence or the difference in size of the nontetrahedral cations (Tables 7 and 8).

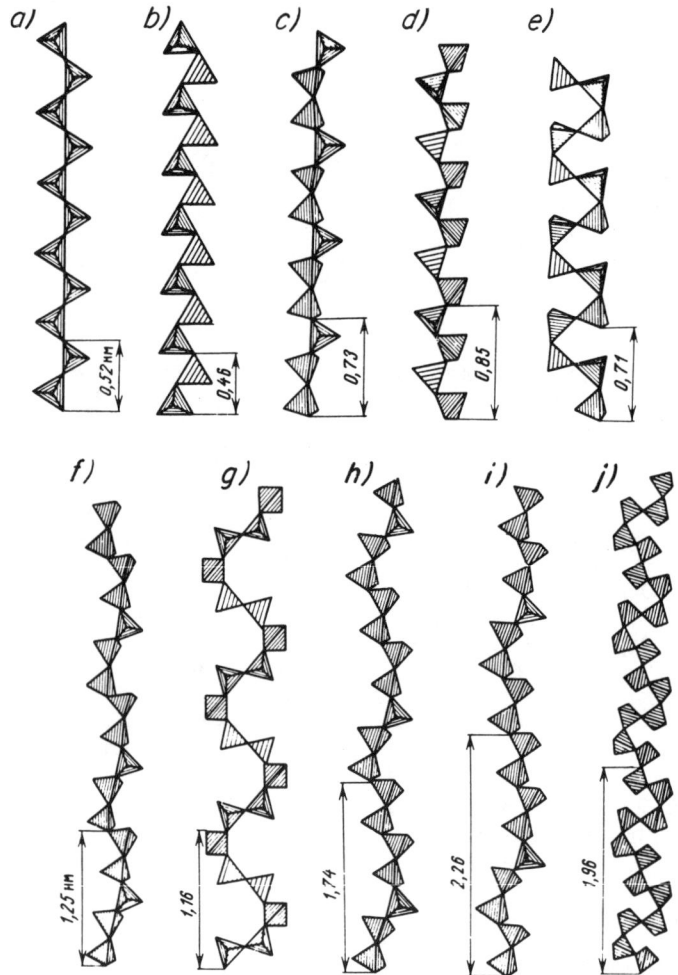

Figure 7 : Different types of chains in silicate structures : (a) enstatite $Mg_2[Si_2O_6]$; (b) $Ba_2[Si_2O_6]$ - HT; (c) wollastonite $Ca[SiO_3]$; (d) krauskopfite $H_4Ba_2[Si_4O_{12}]$; (e) haradaite $Sr_2V_2[Si_4O_{12}]O_2$; (f) rhodonite $CaMn_4[Si_5O_{15}]$; (g) stokesite $Ca_2Sn_2[Si_6O_{18}]\cdot 4H_2O$; (h) pyroxferroite $(Fe,Ca)_7[Si_7O_{21}]$; (i) ferrosilite $Fe[SiO_3]$ - III; (j) alamosite $Pb[SiO_3]$ (Liebau, 1985).

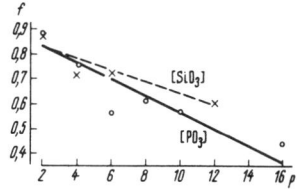

Figure 8 : Values of the stretching factors f for different chain silicates and phosphates as a function of their chain period P (Pushcharovsky, 1986).

The basis of many chain silicates can be considered as an association of octahedron bands and silicate chains, which consists of linked pairs of Si,O tetrahedra. The angle between their centers and the chain axis is equal to 30° (Figure 9). Several structures with such elements form together with the pyroxenes joint homological structure series.

The band's width may include two or three octahedra. If the width of a band is equal to 3 octahedra, the chain period is equal to 3 tetrahedra (wollastonite). If the width of an octahedral band is equal to 2 octahedra, the chain period is equal to 2 tetrahedra (pyroxenes). Several intermediate structures of the so called pyroxenoids are listed in Table 9.

TABLE 9. Topology of chain silicates.

Mineral		Width of octahedral band	Chain period
pyroxene	$CaMgSi_2O_6$	-2-	2
ferrosilite	$FeSiO_3$ III	-3-3-2-2-2-2-2-2-	9
pyroxmangite	$(Fe,Mn)_7Si_7O_{21}$	-3-3-2-2-2-2-	7
rhodonite	$CaMn_4Si_5O_{15}$	-3-3-2-2-	5
wollastonite	$CaSiO_3$	-3-	3

4.3. SILICATES WITH BRANCHED CHAINS OR RINGS

The ratio Si:O = 1:3 is also typical for a specific group of silicates, which, according to Liebau, belongs to the so-called "branched" chain or ring silicates. The basis of these complexes is formed by a tetrahedron chain or ring (Figure 10 and 11) linked with additional tetrahedron "branches". If one marks a pyrotetrahedron group by the number 2, and an orthotetrahedron group by the number 1, then the branched tetrahedron chain of astrophyllite can be represented by $(-2-)_\infty$, the aenigmatite complex by $(-2-2-1-1-)_\infty$, the surinamite complex by $(-2-1-1-1-)_\infty$ and the saneroite complex by $(-2-1-1-1-1-)_\infty$. The branched ring complexes are shown in Figure 11.

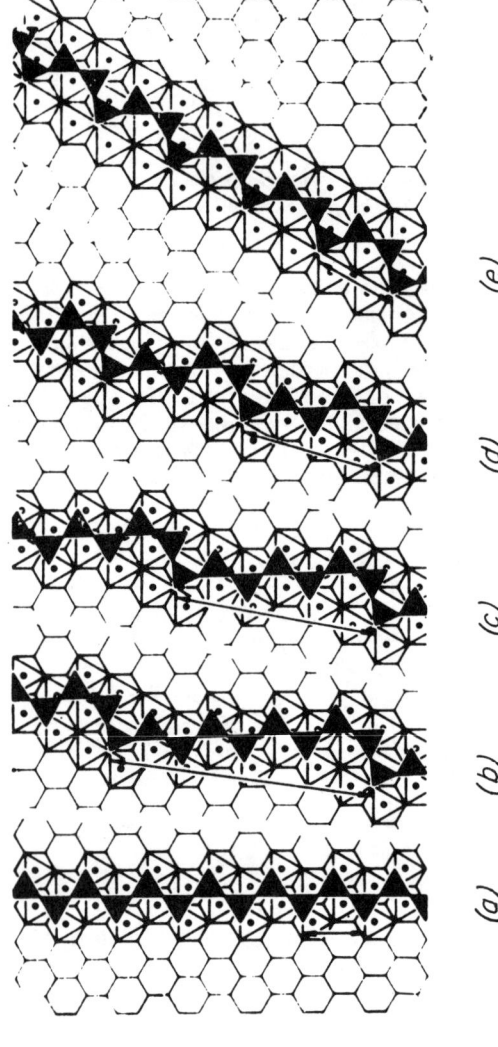

Figure 9 : The contact of the octahedral and the tetrahedral fragments in the structures of (a) pyroxenes; (b) ferrosilite III; (c) pyroxmangite; (d) rhodonite and (e) wollastonite (Liebau, 1985).

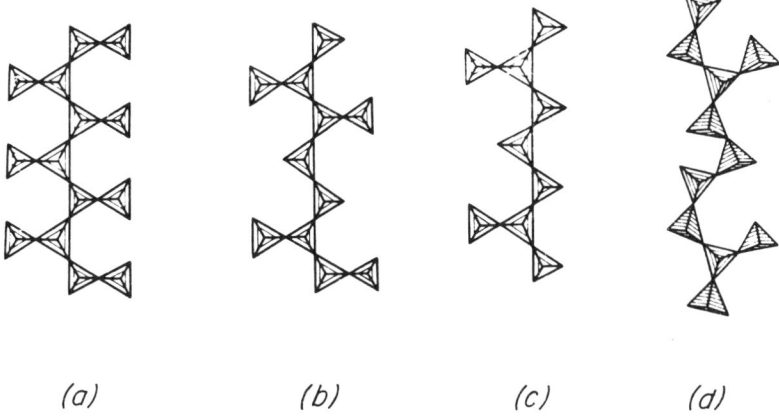

Figure 10: Branched chain silicate complexes in the structures of (a) astrophyllite $NaK_2Mg_2(Fe,Mn)_5Ti_2[Si_4O_{12}]_2(O,OH,F)$; (b) aenigmatite $Na_2Fe_5Ti[Si_6O_{18}]O_2$; (c) surinamite $Mg_3Al_4[BeSi_3O_{12}]O_4$ and (d) saneroite $HN_{1.15}Mn_5[(Si_{5.5}V_{0.5})O_{18}]OH$.

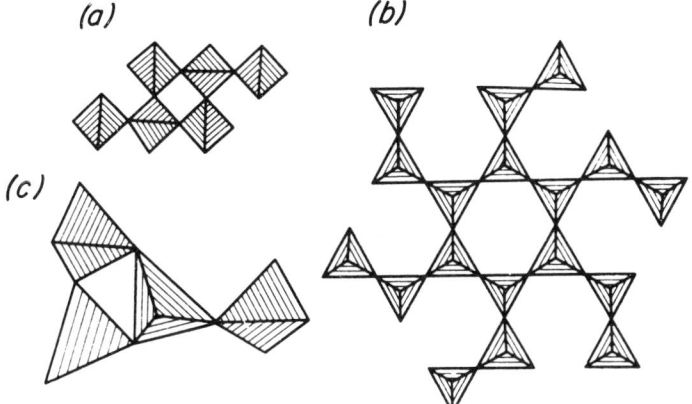

Figure 11. Branched ring silicate complexes in the structures of (a) eakerite $Ca_2SnAl_2[Si_6O_{18}](OH)_{22} \cdot H_2O$; (b) tienshanite $KNa_9Ca_2Ba_6(Mn,Fe)_6(Ti,Nb,Ta)_6B_{12}[Si_{18}O_{54}]_2O_{15}(OH)_2$; (c) branched $[B_4O_{12}]$ complex in the structure of uralborite $Ca_2[B_4O_4(OH)_4]$.

4.4. TETRAHEDRON BANDS IN SILICATE STRUCTURES

Some silicate structures contain the bands, formed by double chains. The structural connection between chain and band silicates is illustrated by Table 10. The different bands are shown in Figure 12.

TABLE 10. The topology of different Si,O bands.

Chain silicate		Period of chain	Band silicate	
$CuGeO_3$		1	sillimanite	$Al[AlSiO_5]$
pyroxene	$CaMg[Si_2O_6]$	2	vinogradovite	$Na_4(TiO)_4[Si_2O_6][Si_4O_{10}] \cdot H_2O$
wollastonite	$Ca[SiO_3]$	3*	epididymite	$Na_2Be_2[Si_6O_{15}] \cdot H_2O$
batisite	$Na_2Ba(TiO)_2[Si_4O_{12}]$	4	caysichite	$Y_4Ca_3R(OH)[Si_8O_{20}](CO_3)_6 \cdot 7H_2O$
rhodonite	$CaMn_4[Si_5O_{15}]$	5	inesite	$Ca_2Mn_7[Si_5O_{14}](OH)_2 \cdot 5H_2O$
stokesite	$CaSn[Si_3O_9] \cdot 2H_2O$	6	tuhualite	$(Na,K)_2Fe^{2+}{}_2Fe^{3+}{}_2[Si_{12}O_{30}] \cdot H_2O$

* The contact of two wollastonite chains via corners of the orthotetrahedron leads to the formation of the xonotlite $Ca_6Si_6O_{17}(OH)_2$ band with composition $[Si_6O_{17}]$.

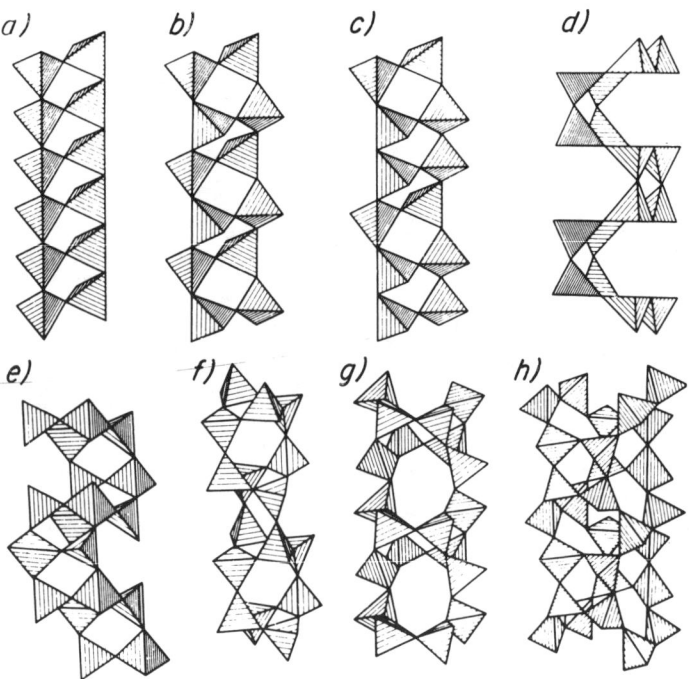

Figure 12 : Different types of [Si_2O_5] band silicate structures : (a) sillimanite; (b) vinogradovite; (c) epididymite; (d) caysichite; (e) tuhualite; (f) narsarsukite; (g) fenaksite and (h) canasite.

The new types of bands were found by high resolution electron microscopy (HREM) in biopyribols, which contain construction elements common also with pyroxenes, mica (biotite) and amphibols. There are bands, formed by 2, 3 and sometimes by even more than 10 pyroxene chains, which gradually become layers.

There is another group of bands, so called "tube-like", which contain sections with rings. Some of these structures with tube-like bands are listed in Table 11.

TABLE 11. Topology of tube-like bands

Silicate with tube-like bands		Section of band
stilvellite	$Ce[BSiO_5]$	3-membered ring
narsarsukite	$Na_2(TiO)[Si_4O_{10}]$	4-membered ring
fenaksite	$KNaFe[Si_4O_{10}]$	6-membered ring
agrellite	$NaCa_2[Si_4O_{10}]F$	6-membered ring
canasite	$(K,Na,Ca)_{11}[Si_{12}O_{30}](OH,F)_4$	8-membered ring
myserite	$KCa_5[Si_2O_7][Si_6O_{15}](OH)F$	8-membered ring

4.5. LAYER SILICATES

The most common tetrahedron layers found in clay minerals are polar and contain 6-membered rings. It is possible to consider them as the result of the condensation of pyroxene-like chains. However, there is a specific group of layer silicates, in which some chains are oriented in one direction, and some in the opposite direction. For example, in palygorskite such an inversion occurs after every band formed by two pyroxene-like chains and in sepyolite after every band formed by three pyroxene-like chains (Figure 13).

The difference in size between the Mg,O octahedral component and the Si,O tetrahedral component in serpentine minerals (antygorite and chrysotyle) leads to the deformation of their two layered blocks, which adopt a cylindric form with a diameter of 100Å and a Si,O tetrahedron layer, situated in the internal part of these cylinders.

Tetrahedral layers may contain 8- and 4-membered rings {apophyllite $KCa_4[Si_4O_{10}]_2F\cdot 8H_2O$}, 8-membered rings {$Na_2ZnSi_3O_8$}, 12-, 6- and 4-membered rings {manganpyrosmalite $Mn_8[Si_6O_{15}](OH)_9Cl$, 12-, 8- and 6-membered rings {$K_8Yb_3[Si_6O_{16}]_2(OH)$} etc. A new type of double layer $[T_5O_{12}]$ was recently found in ganophyllite $K_6Mn_{24}[(Si,Al)_{40}O_{96}](OH)_{16}\cdot 21H_2O$. Amphibol-like bands can be recognized in these layers. On their points of contact there are 5-, 6- and 7-membered rings.

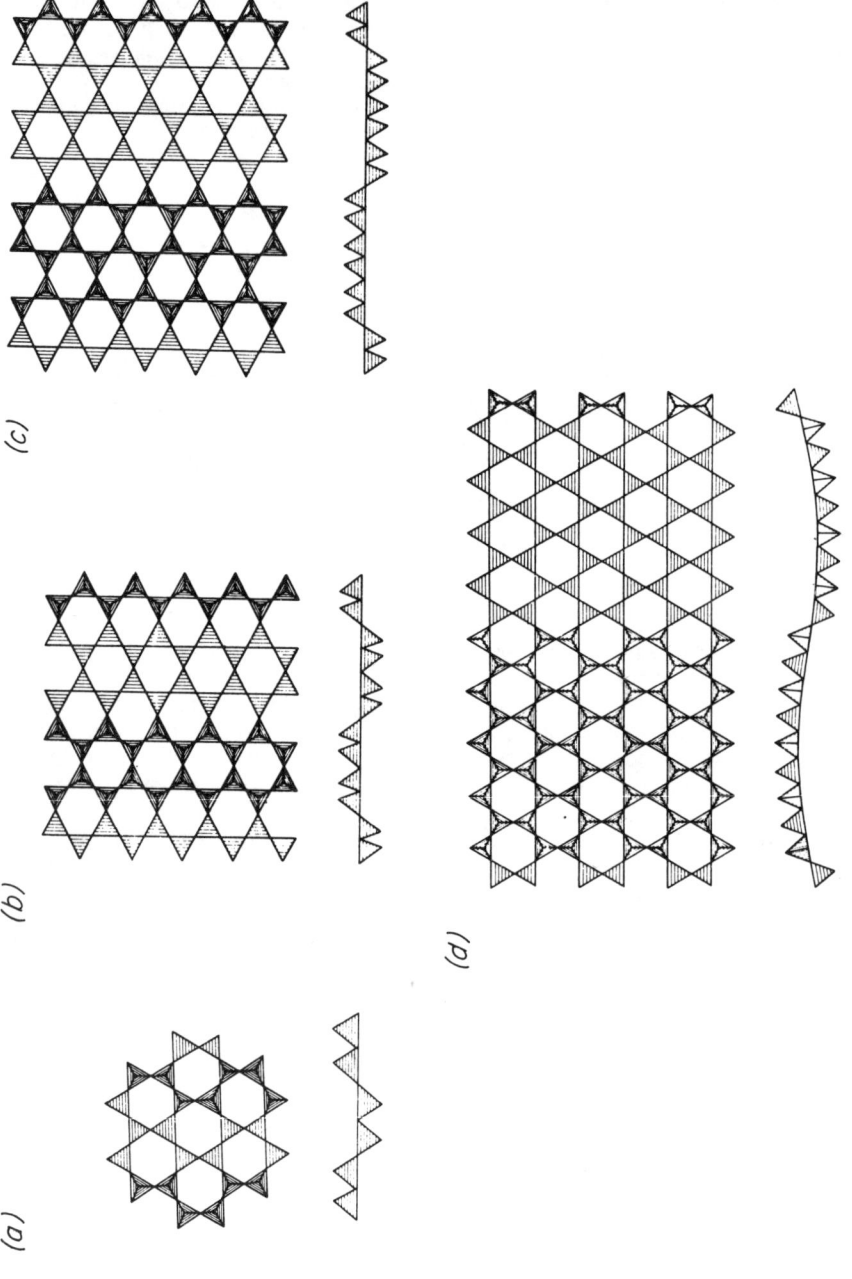

Figure 13 : Tetrahedron layers of composition Si_2O_5 in the structures of : (a) pentagonite $Ca(VO)[Si_4O_{10}]\cdot 4H_2O$; (b) palygorskite $Al_2Mg_2[Si_4O_{10}]_2(OH)_2\cdot 8H_2O$; (c) sepyolite $Mg_8[Si_4O_{10}]_3(O,OH)_4\cdot 8H_2O$ and (d) antygorite $Mg_6[Si_4O_{10}](OH)_8$.

4.6. FRAMEWORK SILICATES

A new type of tetrahedron framework was recently found in the structure of grumantite $Na[Si_2O_4(OH)] \cdot H_2O$. In connection with the grumantite structure it is worth noting that until 1960 many attempts had been made to correlate the composition of the silicates with the configuration of anionic tetrahedron complex. However, there are many examples of a better correlation between the shape of the Si,O complex and the average value of the negative charges on one Si,O tetrahedron (Pushcharovsky, 1986). In principle the framework of grumantite is not typical of the $[T_2O_5]$ complexes, which are mainly characterized by simple tetrahedron sheets. The presence of (OH) groups in the "grumantite" tetrahedron framework decreases the average negative charge of a Si,O tetrahedron and makes the formation of a framework more reasonable.

Among the framework silicates there is a specific zeolite group (47 mineralogical types + 120 synthetic compounds) with general formula : $M^{n+}{}_{x/n}[(AlO_2)^-{}_x(SiO_2)] \cdot zH_2O$, where M^{n+} is the cation, which balances the negative charge associated with the framework Al ions. Zeolite tetrahedron frameworks are characterized by low "framework density", d_f calculated according to

$$d_f = 1000 \cdot n_T / V_{cell},$$

where V_{cell} is the volume of the unit cell and n_T is the number of T atoms in the unit cell.

The values of d_f change from 12.7 in faujasite $(NaCa_{0.5})[Al_2Si_5O_{14}] \cdot 10H_2O$ (here $n_T = n_{Al} + n_{Si}$) up to 29.3 in coesite (high pressure form of SiO_2) (Liebau, 1985). Low framework density means that there are big voids and channels (3.5Å - 15Å) in the zeolite structures, which accomodate different cations and water molecules. The cations help to attract polar water molecules. Since the cations are only weakly bonded to the tetrahedron framework zeolites can be used for as ion exchangers.

Examples for the use of zeolites as ion exchangers :
- Water softening by zeolites which contain Na^+ involves the exchange reaction : $2Na \rightarrow Ca(Mg)$.
- Extraction of radioactive elements from nuclear waste by means of clinoptilolite $K_2Na_2Ca[AlSi_5O_{12}] \cdot 8H_2O$.

The voids in the zeolite structures permit them to be used as molecular sieves.

Examples for the use of zeolites as molecular sieves :
- Chabazite $(Ca,Na)_2[Al_2Si_4O_{12}] \cdot 6H_2O$ (Figure 14a) absorbs the vapors of small molecules such as water, formic acid, methyl and ethyl alcohols, but not large molecules such as aceton, ether, benzene or SO_2 from pollutants of smokestacks;
- The Ca form of Linde A (Figure 14b) with rather small voids (Table 12) removes from natural gasoline the undesirable hydrocarbons with straight molecular chains (these hydrocarbons burn more explosively). The presence of cyclic or branched molecules of hydrocarbons increases the quality of gasoline.
- Linde A can be also used for separating the air components, that is the O_2 and N_2 molecules, and to remove water from hydraulic brake systems (Kerr, 1989).

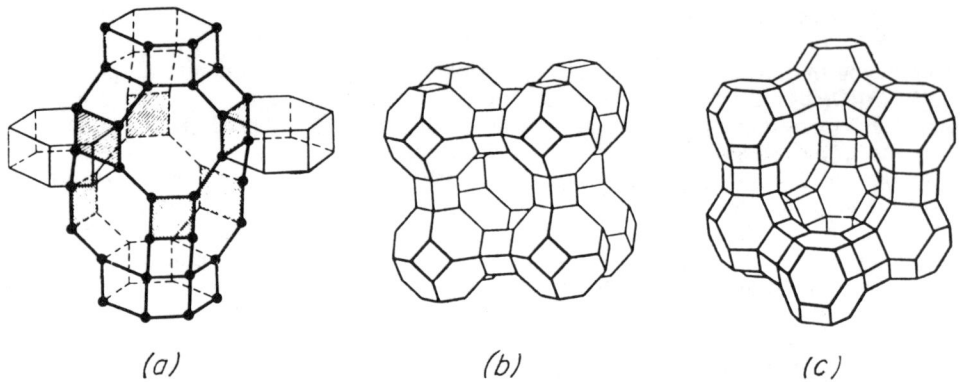

Figure 14 : Structural elements of zeolites : (a) chabazite; (b) synthetic crystal Linde A; (c) faujasite, synthetic Linde X and Linde Y, all with similar structure. Straight lines join the centers of neighbouring tetrahedra. Each vertex represents thus an aluminum or silicon atom.

TABLE 12. The composition and pore parameters of some zeolites (Maxwell, 1982).

Type	Unit cell composition	Void Vol.	Pore diam. (Å)	Thermal decomposition (°C)	Si/Al ratio
Linde A	$Na_{12}(AlO_2)_{12}(SiO_2)_{12}$	0.47	4.2	700	1.0
Linde X	$Na_{86}(AlO_2)_{86}(SiO_2)_{106}$	0.50	7.4	772	1.23
Linde Y	$Na_{56}(AlO_2)_{56}(SiO_2)_{136}$	0.48	7.4	793	2.43
mordenite	$Na_8(AlO_2)_8(SiO_2)_{40}$	0.28	6.7x7.0	1000	5.0

Of great technical importance is the use of the zeolites for the selective shape catalysis. The first zeolite catalist, Linde X, is a structural analogue of faujasite (Figure 14c). Its activation is connected with the substitution of Na^+ by Al^{3+} (crystals are bathed in an Al-rich solution). This exchange, important for their interaction with gas oil, converts the crystals into an acid : cations Al^{3+} detach $(OH)^-$ groups from water or gas oil, while H^+ cations occupy the channels. Such acidity leads to the catalytic cracking of large hydrocarbon molecules of gas oil.

Linde Y with a Si/Al ratio of 1.5 to 3 times higher, as compared with Linde X, is more stable to water at high temperatures. Its acidity can be achieved by substitution of Na^+ by NH^{4+}. The heating causes the reaction : $NH_4^+ \rightarrow NH_3 + H^+$. Steaming of this hydrogen form of zeolite drives a fraction of H^+ from the crystals and causes a part of Al^{3+} to leave their 4-fold coordinated sites and to occupy the structure channels. These crystals are catalysts and are used in the cracking of gas oil. The cracking of larger hydrocarbon molecules is possible with (Al,P)-zeolites where the channels have 12 to 13Å in diameter.

ZSM-5 and ZSM-12 with a Si/Al ratio of 35 and 90, respectively are not used in the cracking of gas oil because their elliptical channels (Figure 15) are too small (5.4 to 5.6Å in diameter). However this size is suitable for the catalysis of cyclic hydrocarbons.

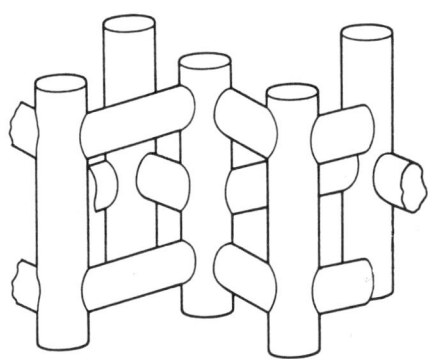

Figure 15 : A schematic view of the channel structure of ZSM-5.

Example for the use of ZSM-5 catalysts :
- Toluene $C_6H_5CH_3$, a by-product of petrolium refining, can be transformed by ZSM-5 into benzene and paraxylene $C_6H_4(CH_3)_2$. The small toluene molecules penetrate the ZSM-5 channels, pass through them and are heated to 520°C. Benzene and paraxylene escape from the crystals readily, but ortho- and meta-xylene are slowly converted into the para form.
- Using ZSM-5 catalysts methanol CH_3OH can be transformed into gasoline (MTG-process). This technology is attractive because methanol itself is toxic, very corrosive, contains up to 56% of water dissolved in it, and therefore has only half the energy equivalent of petrol. The first step involves the dehydration of methanol with an alumina catalyst and the formation of dimethylether (DME) : $2CH_3OH - H_2O \rightarrow CH_3OCH_3$. Then using the ZSM-5 catalyst DME is completely transformed into hydrocarbons, more than 80% of the reaction product being high quality gasoline. Using this technology the Synthetic Fuels Corporation (New Zealand) produces 1665 t petrol per day, which covers about 33% of New Zealand's needs.

The compressibility of the zeolites also depends on the size of the voids in their structures (Hazen & Finger, 1984). This results from high pressure experiments made with 4A-zeolite $12NaAlSiO_4 \cdot 27H_2O$ in a diamond-anvil X-ray camera. Different liquids were used as a hydrostatic pressure medium. When water was used the compressibility of zeolite was 6.5 times less as compared when glycerol $C_3O_3H_6$ or organofluorocarbon FC-75 $C_8F_{16}O$ was used as hydrostatic pressure medium. Different from the other liquids the relatively small water molecules can pass through the zeolite channels and increase the resistence of its structure to external pressure.

5. Comparative Crystal Chemistry of Silicates

One of the main modern crystal chemical problems is connected with the predictions of structure transformations at elevated temperatures and pressures. The diversity of the Si,O complexes, described in part 4, refers to silicates which are stable only at relatively low depth. The quite low coordination numbers of the oxygen atoms (2 to 3) in these structures make them unstable in the mantel. High pressure structure studies (Hazen & Finger, 1978) showed that under pressure the Si - O distances decrease until a "critical" value of 1.59Å is reached where then occurs a change from Si,O tetrahedra to Si,O octahedra. There is a certain advantage in such a transformation, because the Si,O octahedra with their longer Si - O distances can be linked not only by corners, but also by edges and even by faces. Thus the CN of oxygen becomes greater, and at the same time the distribution of Si atoms becomes more compact.

The "critical" value of 1.59Å is not accidental and is confirmed by the correlation between interatomic distances and angles in Si,O tetrahedra (Hill & Gibbs, 1979) :

$$\lg 2 \cdot [d(Si - O)_{av}] = \lg d(Si - Si) - \lg \sin(\angle Si - O - Si / 2)$$

With fixed Si - Si distances of 3.06Å and $\angle(Si - O - Si)_{max} \approx 180°$ one obtains $d(Si - O)_{min} = 1.59$Å. This value was obtained in the structure of forsterite Mg_2SiO_4 at a pressure of 14.9 GPa.

The stability of the Si,O tetrahedra depends on the properties of the nontetrahedral M cations. The presence of M cations with a high electronegativity value leads to a decrease in the CN of the O atoms. Consequently, under high pressures these structures transform into structures with Si,O octahedra more easily (Table 13).

In structures with more electronegative M cations the distances $Si - O_{br}$ and $Si - O_{nbr}$ become more equal and the average value of the Si - O distance in the Si,O tetrahedra decreases (Brown, 1978). In other words these cations have an influence on the structure just as high pressure.

The concept, emphasizing the dominating role of the cations for the formation of particular structures, can also be of use for the interpretation of structure transformations under high pressure. The compressibility of the cation-oxygen polyhedra (β) is inversely proportional to the polyhedral charge density z/d^3, where d is the distance between the central cation and the O atoms (Hazen & Prewitt, 1977). The behaviour of the Si,O tetrahedra under pressure depends thus also on the compressibility of the M,O polyhedra.

Example : Mg_2SiO_4, under high pressure, successively transforms from the α to the β and finally to the γ modifications with the $d(Mg - O)/d(Si - O)$ ratios decreasing from 1.291 (α form) to 1.260 (β form) and finally 1.251Å (γ form). The significant compressibility of the Mg,O octahedra during the transformation from the α to the γ form leads to an unexpected expansion of the Si,O tetrahedra : $d(Si - O)$ in $\alpha\ Mg_2SiO_4$ is 1.636Å but 1.655Å in $\gamma\ Mg_2SiO_4$.

It is noteworthy, that the structure changes in these compounds, as in some other silicates, results in an ordered oxygen atom packing (Table 14).

TABLE 13. The correlation between the electronegativity of the cations and the pressure needed to initiate a transformation of $Si^{[4t]} \rightarrow Si^{[6o]}$ in silicate and silicon containing structures.

Compound	M atom	E_M Electro-negativity	Pressure GPa	Temperature °C
$[(C_6H_4O_2)_3Si][C_5H_5NH]_2$	N,C,H	3.0; 2.5; 2.1	10^{-3}	
$Si(NH_4)_2[P_4O_{13}]$	N,H,P	3.0; 2.1; 2.1	10^{-4}	350
$Ca_3[Si(OH)_6][SO_4][CO_3] \cdot 12H_2O$ (thaumasite)	C,S,H	2.5; 2.44; 2.1	10^{-3}	
$Si[P_2O_7]$	P	2.1	10^{-3}	800-1000
$K_2Si^{[6]}[Si^{[4]}_3O_9]$	Si,K	1.8; 0.8	2-9	900-1200
SiO_2 (stishovite)	Si	1.8	9	200-1400
$In_2[Si_2O_7]$ (pyrochlore - type)	In	1.7	8-12	1300
$Mg_3(AlMg_{0.5}Si_{0.5})^{[6]}[SiO_4]_3$	Al,Mg	1.5; 1.2	10	900
$K(Al_{0.25}Si_{0.75})^{[6]}_4O_8$	Al,K	1.5; 0.8	12	900-1000
$Mn_3(MnSi)^{[6]}[SiO_4]_3$	Mn	1.5	12.5	800-1000
$NaAlSiO_4$	Na	0.9	24-30	1000-1200

TABLE 14. The decrease of the distortion (s)* of the cation-oxygen polyhedra and Si,O tetrahedra during α - β - γ transformations in olivine-like structures (Horiuchi & Sawamoto, 1981).

Polyhedron	Co_2SiO_4			Mg_2SiO_4	
	α	β	γ	α	β
Si,O tetrahedron	13.2	2.3	0	16.6	3.1
M(1),O octahedron	88.7	18.2		77.6	12.3
M(2),O octahedron	50.7	20.0	10.7	47.5	19.5
M(3),O octahedron		15.2			15.3
$\square_1$-octahedron	56.7	19.0		48.9	11.2
$\square_2$-octahedron	86.5	37.8	13.0	71.6	29.3
$\square_3$-octahedron		70.4			57.9

* $s = (\sigma/x)^2 \cdot 10^4$ and $\sigma^2 = \Sigma(x_j - x_{av})^2/(n - 1)$, where x_j and x_{av} are the definite and the average O - O distances in the polyhedron. $\square$ indicates an octahedron which is not centered by a cation.

According to Ringwood, the high pressure structures of silicates often correspond to the normal pressure structures of their germanate analogues. This can be understood if one takes into account that the ratio r_{cation}/r_{oxygen} should be increased under high pressure. At atmospheric pressure the radius of Ge^{4+} is by some 20% greater that the radius of Si^{4+}. Therefore, the structures of germanates at normal pressure reflect many specific features of the high pressure deformations in silicates. As an example,

one can compare the high pressure structure changes in quartz (Livien, Prewitt & Weidner, 1980) with the changes in its structure when Si atoms are partially substituted by Ge atoms (Pushcharovsky, Vyatkin, Yamnova & Sorokina, 1990). The investigated crystals of $(Si_{0.86}Ge_{0.14})O_2$ are unique, because until now the content of GeO_2 in monocrystals of quartz did not exceed 0.15 mol.%. The common features of pure quartz under high pressure and of quartz at ambiant pressure but with Si partially replaced by Ge are :
1) a decrease of the T - O - T angle,
2) a decrease of the intertetrahedral O - O distances,
3) a similar change of the tilt angles between the tetrahedra and
4) an increase of the tetrahedral distortion (Table 15).

TABLE 15. The comparison of structural parameters in quartz under high pressure and after partial substitution of Si by Ge.

Structural parameters	SiO_2 1 atm.	SiO_2 61.4kbar	$(Si_{.86}Ge_{.14})O_2$	GeO_2
$\angle$T - O - T	143.73	134.2	142.2	130.1
intertetrahedral O - O distances (Å)	3.331	2.925	3.304	3.024
	3.411	3.064	3.406	3.193
tilt angle	-16.37	-23.47	-17.41	-26.55
(O - T - O) angle dispersion	0.67	5.51	1.04	10.09

6. Summary of the Structural Principles of Silicates, Phosphates, Sulphates and Related Structures

1. The specific structural feature of the silicates is the presence of Si atoms in a 4-fold oyxgen atom coordination. The Si - O bonds are reinforced by $d\pi$-$p\pi$ components. Si,O tetrahedra form about 100 complexes of different configuration. The structural variability of silicates is significantly bigger in comparison with that of the phosphates (29 types of anionic tetrahedron complexes) and the germanates (14 types of anionic tetrahedron complexes).
2. Si,O tetrahedron complexes adjust to the cation-oxygen polyhedra. There are quantitative correlations between the properties of the nontetrahedral cations and the shape of the Si,O complexes.
3. Crystal chemical systematics allows the presentation of the structural data in a most compact form. The majority of the crystal chemical classifications of the silicates is based on the analysis of the Si,O complexes $[Si_mO_n]$. The average number of O_{br} per one tetrahedron is $K = 8 - 2 \cdot (n/m)$. The value of the average negative charge per tetrahedron controls the shape of the Si,O - complex.
4. For structures where the condensation of the anion complex (tetrahedra or triangles) is absent (phosphate minerals, sulphates, carbonates etc.), the concept of mixed complexes is more effective for the purpose of their systematics. The cation-oxygen polyhedra with bond strengths comparable to those found in T,O polyhedra, form in cooperation with the latter the predominate structural elements.

5. Several new types of anionic tetrahedron complexes were discovered during the last years : an isolated complex, built of 48 Si,O tetrahedra, pyrogroups [P_2O_7] in a phosphate mineral, a 18-membered [SiO_3] rings, a [PO_3] chain with 16 tetrahedra in the period, an interrupted [Si_2O_5] framework, etc. The new scientific ideas on the structures of the silicates and their analogues stimulate their applications in different technologies (ex.: shape selective catalysis with zeolites).

6. In silicate structures the participation of cations of high electronegativity reduces the pressure necessary for a transformation from tetrahedrally to octahedrally coordinated Si. The structural alteration under high pressure in some silicates results in a higher regularity of the oxygen atom packing. The structural changes of silicates under high pressure are similar with chemical deformations connected with the substitution of Si by Ge.

7. References

Averbuch-Pouchot, M.T. & Durif, A. (1991a). 'Crystal Chemistry of Oligophosphates'. Annu.Rev.Mater.Sci. **21**, 65 - 92.

Averbuch-Pouchot, M.T. & Durif, A. (1991b). 'Present State of Cyclohexaphosphate Crystal Chemistry'. Eur.J.Solid State Inorg.Chem. **28**, 9 - 22.

Belov, N.V. (1959). "Crystal Chemistry of Large Cation Silicates." Consultants Bureau : New York.

Brown, I.D. (1978). 'Bond Valences - A Simple Structural Model for Inorganic Chemistry.' Chem.Soc.Rev. **7**, 359 - 376.

Chakoumakos, B.C., Hill, R.J. & Gibbs, G.V. (1981). 'A Molecular Orbital Study of Rings in Silicates and Siloxanes.' Amer.Mineral. **66**, 1237 - 1249.

Clark, G.M. & Morley, R. (1976). 'Inorganic Pyrocompounds $M_a(X_2O_7)_b$'. Chem.Soc.Revs. **5**, 268 - 295.

Hawthorn, F.C. (1986). 'Structural Hierarchy in $^{VI}M_x^{III}T_yO_z$ Minerals.' Canad.Mineral. **24**, 625 - 642.

Hazen, R.M. & Finger, L.W. (1978). 'Crystal Chemistry of Silicon - Oxygen Bonds at High Pressure : Implications for the Earth's Mantle Mineralogy.' Science **201**, N4361, 1122 - 1123.

Hazen, R.M. & Finger, L.W. (1984). 'Compressibility of Zeolite 4Å is Dependent on the Molecular Size of the Hydrostatic Pressure Medium.' J.Appl.Phys. **56**, 1838 - 1840.

Hazen, R.M. & Prewitt, Ch.T. (1977)..'Effects of Temperature and Pressure on Interatomic Distances in Oxygen - Based Minerals.' Amer.Minerai. **62**, 309 - 315.

Hill, R.J. & Gibbs, G.V. (1979). 'Variation in d(T-O), d(T...T) and TOT in Silica and Silicate Minerals, Phosphates and Aluminates.' Acta.Crystallogr. **B35**, 25 - 30.

Horiuchi, H. & Sawamoto, H. (1981). 'β - Mg_2SiO_4 : Single - Crystal X - Ray Diffraction Study.' Amer.Mineral. **66**, 568 - 575.

Kerr, G.T. (1989). 'Synthetic Zeolites.' Scientific American, July, 82 - 87.

Kostov, I. (1975). 'Crystal Chemistry and Classification of the Silicate Mineralogy.' Geochem.Mineral.Petrol. **1**. Sofia, 5 - 41.

Liebau, F. (1985). "Structural Chemistry of Silicates." Springer Verlag : Berlin.

Liebau, F. & Pallas, I. (1981). 'The Influence of Cation Properties on the Shape of Silicate Chains.' Z.Kristallogr. **155**, 139 - 153.

Livien, L., Prewitt, Ch.T. & Weidner, D.J. (1980). 'Structure and Elastic Properties of Quartz at Pressure.' Amer.Mineral. **65**, 920 - 930.

Maxwell, I.E. (1982). 'Nonacid Catalysis with Zeolites.' Advances in Catalysis and Related Subjects **31**, 2 - 68.

Nickolaev, V.P. (1983). "Crystal Chemistry of Phosphates with Metals of Vanadium Subgroup". Ph.D. Thesis. Inst. of Inorg. Chem., Acad. Sci. USSR, Moscow (in Russian).

Parthé, E. & Engel, N. (1986). 'Relation between Tetrahedron Connections and Compositions for Structures with Anionic Tetrahedron Complexes'. Acta Cryst. **B42**, 538 - 544.

Pushcharovsky, D.Yu. (1986). "Structural Mineralogy of Silicates and their Synthetic Analogues." Nedra : Moscow (in Russian) and "Crystal Chemistry of Minerals and Synthetic Compounds with Tetrahedral Anions". D.Sc. Dissertation IGEM, Acad. Sci. USSR, Moscow (in Russian).

Pushcharovsky, D.Yu., Vyatkin, S.V., Yamnova, N.A. & Sorokina, S.L. (1990). 'Comparative Crystal Chemistry of Ge - Containing Synthetic Quartz.' Sov.Phys.Cryst. **35**, 689 - 691.

Raszvetaieva, R.K. & Pushcharovsky D.Yu. (1989). "Crystal Chemistry of Sulphates." VINITI Ser.Crystal Chemistry **V23** (in Russian).

Ross, F.K. (1980). 'Experimental Charge Density Studies in Minerals and Other Inorganic Compounds.' Trans.Amer.Cryst.Assoc. **16**, 79 - 95.

8. Problems

Problem 1 : Compare the anionic tetrahedron complexes of the two layer structures shown in Figure 16a and 16b. Describe their common and distinctive structural features. Determine the chemical formulae and the charges of the two tetrahedron complexes.

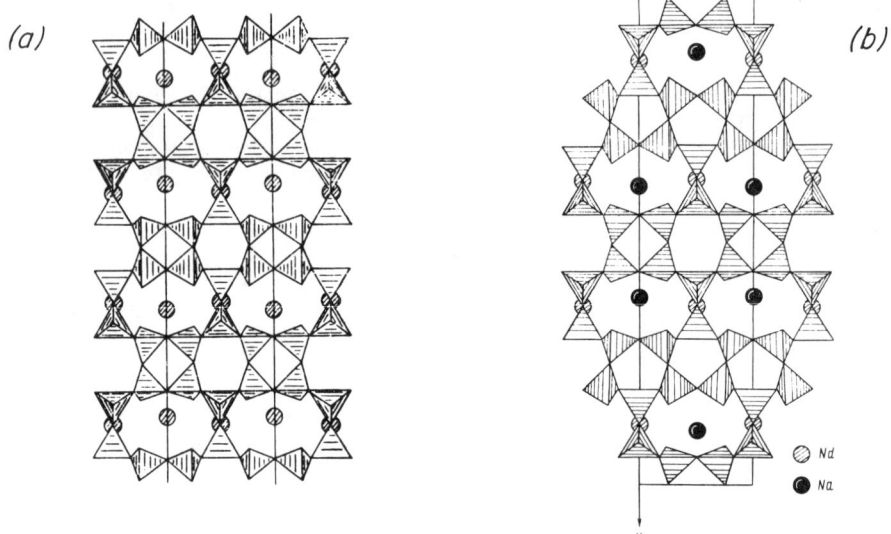

Figure 16. Anionic tetrahedron complexes of two layer structures.

Problem 2 : The compounds $R_2Si_2O_7$ (R indicates a lanthanide element), $In_2Si_2O_7$ and $Sc_2Si_2O_7$ all have the same crystal structure (monoclinic D-form), characterized by $[Si_2O_7]$ pyrogroups. At high pressure (180 kbar) $In_2Si_2O_7$ and $Sc_2Si_2O_7$ transform into the cubic P-modification (isostructural with pyrochlore) having Si,O octahedra, while for $R_2Si_2O_7$ compounds this modification is not known. Explain this fact.

Problem 3 : In silicates with condensed Si,O tetrahedra the distances Si - O_{br} are longer than Si - O_{nbr}. Why ?

8.1. PROBLEM ANSWERS

Problem 1 : Two types of Si,O tetrahedron layers are shown in Figure 16. The layer in Figure 16a contains 8-, 6- and 4-membered rings and the layer in Figure 16b 8-, 6-, 5- and 4-membered rings. Every tetrahedron in both layers has three bridged and one non-bridged oxygen atom. Thus, the formula of both anionic tetrahedron layers is $SiO_{1+3\cdot 0.5} = SiO_{2.5}$ or $[Si_2O_5]^{2-}$.

Problem 2 : The electronegativity values of In (1.7) and Sc (1.3) are higher than the values of the lanthanide elements R (1.1). This fact means that in Si - O - M bonds the electrons from the O atoms are more strongly attracted to the M cations if M = In or Sc. As a consequence the Si - O bonds in $In_2Si_2O_7$ and $Sc_2Si_2O_7$ are weaker and Si,O tetrahedra can be transformed more easily into Si,O octahedra.

Problem 3 : Pauling's electrostatic valence rule states that the sum of the bond strengths of all electrostatic bonds which originate from surrounding cations and reach a particular anion should be equal to the charge of the anion. In the case of a bridging oxygen atom O_{br} in a silicate structure, that sum is given by $2\cdot 4/4$ plus the contribution from the nontetrahedral cations. To avoid the "overload" of bond strengths, the distance O_{br} - Si should be longer in comparison to the distance O_{nbr} - Si.

CRYSTAL CHEMISTRY OF HIGH Tc SUPERCONDUCTING OXIDES

BERNARD RAVEAU, CLAUDE MICHEL AND MARYVONNE HERVIEU
Laboratoire de Cristallographie et Sciences des Matériaux
CRISMAT-ISMRA
URA CNRS 1318
Boulevard du Maréchal Juin
F - 14050 Caen cédex
France

ABSTRACT. The complex crystal chemistry of high Tc superconducting oxides is based on the ability of copper to exhibit the mixed valency Cu(II)-Cu(III) and on its Jahn-Teller effect, leading to various coordinations ranging from 4 to 6. The structural principles of the different oxides are described leading to the present formulation $(ACuO_{3-y})_m(A'O)_n$. The oxygen non-stoichiometry in the orthorhombic superconductor $YBa_2CuO_{7-\delta}$ is studied, as well as the intergrowth extended defects, and interconnections in thallium, bismuth and lead cuprates. The incommensurate structure of bismuth cuprates is discussed. Relationships between redox mechanisms and superconductivity are presented.

1 - Copper chemistry in oxides : oxidation state and coordination

The generation of superconductivity in copper oxides requires two conditions to be satisfied, the existence of semi-metallic or metallic conductivity, and a strong anistropy of the structure allowing magnetic interactions to be decreased.

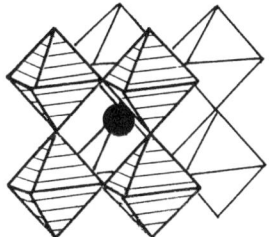

Figure 1 : Perspective view of the perovskite structure : representation of the corner-sharing CuO_6 octahedra.

The first condition, is the classical mixed valency necessary to build oxygen bronzes. It corresponds to an overlapping of the d orbitals of copper with the p ortibals of oxygen, allowing a band to be built ; in the latter, holes resulting from the existence of the mixed valency Cu(II)-Cu(III) are delocalized leading to a p type conductivity. In order to realize a high conductivity, the CuO_n polyhedra forming the copper-oxygen framework must share at least their corners. This is for instance the case of the perovskite-type structure $LaCuO_3$ (Fig. 1), whose $[CuO_3]_\infty$ framework consists of corner-sharing CuO_6 octahedra, the

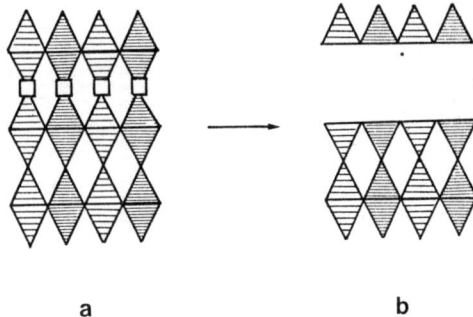

a   **b**

Figure 2 : Elimination of rows of oxygen atoms along [110] in the perovskite framework (a) leading to the formation of a layer structure (b).

conduction band being only partly occupied owing to the trivalent oxidation state of copper. However such an oxide could only be synthesized under high oxygen pressure due to the fact that Cu(III) is not stable at high temperature under normal pressure

a   **b**   **c**

Figure 3 : The structure of different members of the family of the titanates $SrO(SrTiO_3)_m$. (a) m = 1, (b) m = 2, (c) m = 3. They correspond to the intergrowth of single SrO rock-salt-type layers with multiple perovskite layers $(SrTiO_3)_m$.

conditions. This unusual oxidation state can be stabilized by introduction of basic ions in the structure, such as alkaline earth elements. Those divalent elements should theoritically, destabilize the perovskite structure by creation of anonic vacancies. But this effect is compensated by the well-known Jahn Teller effect of copper, which allows coordinations smaller than six to be realized.

The anisotropy of the structure can easily be realized in the perovskite family owing to the great flexibility of its octahedral framework. This second condition can be satisfied in two different ways. The first way consists in the elimination of rows of oxygen atoms along the [110] direction of the cubic perovskite structure forming layers of CuO_5 corner-sharing pyramids (Fig. 2). The second manner to create bidimensionality is based on the possible intergrowths between the rock salt type and perovskite structures. The two dimensional accord between these two structures in the (001) plane of their cubic cells allows perovskite layers with variable thicknesses to be adapted with single rock salt layers (Fig. 3).

All the structures of the superconducting copper oxides isolated up to now are based on these principles.

2 - The different structural families

Most of the superconducting layered cuprates can be described by the general formula $[ACuO_{3-x}]_m [AO]_n$, where m and n represent the number of copper layers and $[AO]_\infty$ layers in the perovskite and rock salt slabs respectively. The different members of this

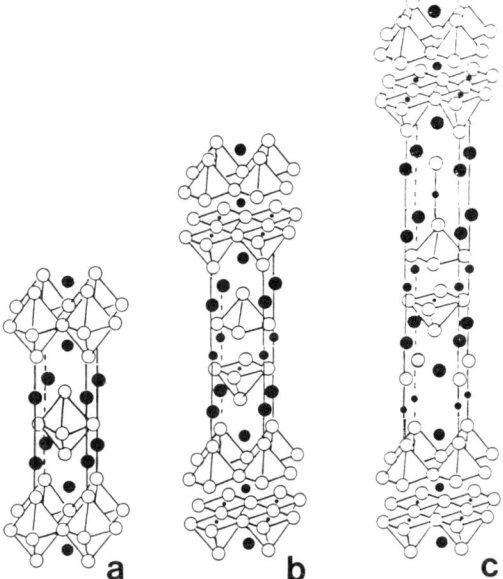

Figure 4 a,b,c : Crystal structure of cuprates characterized by single rock salt layers (n=1) : (a) m=1 member, (b) m=2 member, (c) m=3 member (the structure of this member which corresponds to the oxide $BaPbYSrCu_3O_8$ differs from the general description by the location of oxygen vacancies in one sheet of the triple perovskite layer).

large family are summarized in table 1. The n = 1 members are characterized by single rock salt layers built up themselves of two AO planes (Fig. 4a-b-c) ; the n = 2 cuprates consist of double rock salt layers, in which one $[TlO]_\infty$ monolayer is sandwiched between two $[SrO]_\infty$ or $[BaO]_\infty$ layers (Fig. 4 d-e-f-g) ; the n = 3 members exhibit double $[(BiO)_2]_\infty$ or $[(TlO)_2]_\infty$ layers sandwiched between $[SrO]_\infty$ or $[BaO]_\infty$ layers (Fig. 4 h-i-j-k). The coordination of copper in the different compounds depends on the thickness of the perovskite slab, i.e. on the number m of copper layers forming the latter. For single perovskite layers (m = 1) one observes CuO_6 octahedra sharing their corners (Fig. 4) ; the m = 2 cuprates consist of double pyramidal copper layers deduced from the double octahedral copper layers by elimination of rows of oxygen atoms along [110] (Fig. 4). In order to build disconnected copper layers for m ≥ 3, one has to eliminate rows of oxygen atoms on both sides of the basal plane of the CuO_6 octahedra ; as a result the different oxides with m ≥ 3, exhibit also ordered oxygen different perovskite slabs, in which (m - 2) $[CuO_2]_\infty$ layers corner-sharing CuO_4 square planar groups are sandwiched by two pyramidal copper layers (Fig. 4). Thus the ideal structure of all these layered cuprates can be represented by the symbol [m, n] ; in this representation m and n are integral numbers for single intergrowths, and can take non integral values for multiple intergrowths. It is also worth pointing out that one goes from one "n" column to the "n + 1" one (table 1) by adding one AO layer, whereas one goes from one "m" horizontal line to the "m + 1" one by adding one $ACuO_2$ layer, for integral m and n values.

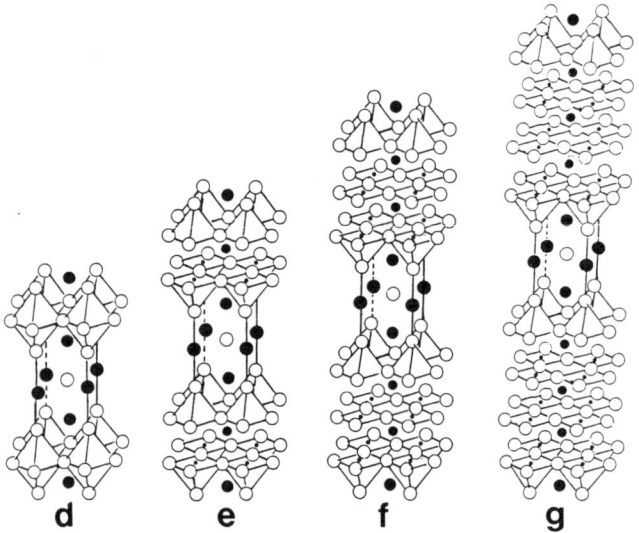

Figure 4 d,e,f,g : Crystal structure of cuprates characterized by double rock salt layers (n=2) : (d) m=1 member, (e) m=2 member, (f) m=3 member, (g) m=4 member.

The classification of these oxides according to the three different sorts of rock salt layers shows the following points :

(i) single rock salt type layers are obtained mainly for alkaline earth cuprates and some lead cuprates, so that they can be formulated $A_2(Ca, Ln)_{m-1}Cu_mO_{2m+2}$. The oxides

TABLE 1 : The [m, n] layered cuprates : P/RS-intergrowths $(ACuO_{3-x})_m(AO)_n$

m/n	1	2	3
1	[1.1] $La_{2-x}A_xCuO_4$ (A=Ca, Sr, Ba) T_c=20-40 K La_2CuO_{40} $Pb_2Sr_{2-x}La_xCu_2O_{6+\delta}$ T_c=30 K	[1.2] $TlBa_2CuO_{5-\delta}$ Nonsuperconducting $TlSr_2CuO_{5-\delta}$ traces $Tl_{1-x}Pr_xSr_{2-y}Pr_yCuO_{5-\delta}$ $T_c \approx 40$ K (x = 0.2, y = 0.4) $Tl_{0.5}Pb_{0.5}Sr_2CuO_{5-\delta}$ Nonsuperconducting to $T_c \approx 40$ K $TlBa_{1+x}La_{1-x}CuO_{5-\delta}$ $T_c \approx 50$ K (x = 0.2)	[1.3] $Tl_2Ba_2CuO_6$ Nonsuperconducting to $T_c \approx 92$ K $Bi_2Sr_2CuO_6$ $T_c \approx 10-22$ K
1.5	[1.5, 1] $Pb_2Sr_2Y_{1-x}Ca_xCu_3O_8$ T_c=50 K (x = 0.5) to 80 K $Pb_{2-x}Bi_xSr_2Y_{1-x}Ca_yCu_3O_8$ T_c=79 K (x=0.6 ; y=0.5, 1)		
2	[2, 1] $La_{2-x}A_{1+x}Cu_2O_6$ A=Ca, Sr Nonsuperconducting to $T_c \approx 60$ K	[2, 2] $TlBa_2Ca_2Cu_2O_7$ T_c=50 to 112 K $TlSr_2CaCu_2O_7$ $T_c \approx 50$ K $Tl_{0.5}Pb_{0.5}Sr_2CaCu_2O_7$ $T_c \approx 85$ K $TlBa_2LnCu_2O_7$ Ln=Pr, Y, Nd Nonsuperconducting $Pb_{0.5}Sr_{2.5}Y_{0.5}Ca_{0.5}Cu_2O_{7-\delta}$ $T_c \approx 59$ K	[2, 3] $Tl_2Ba_2CaCu_2O_8$ $T_c \approx 97$ to 108 K $Tl_{3-4x/3}Ba_{1+x}LnCu_2O_8$ (x = 0.25, Ln = Pr, Nd, Sm) Nonsuperconducting $Bi_2Sr_2CaCu_2O_8$ $T_c \approx 85$ K $Tl_2Sr_2CaCu_2O_8$ $T_c \approx 44$ K $Bi_{2-x}Pb_xSr_2Ca_{1-x}Y_xCu_2O_8$ $T_c \approx 85$ K to nonsuperconducting
3	[3, 1] $PbBaYSrCu_3O_{8-\delta}$ Nonsuperconducting $PbBaSrY_{0.7}Ca_{0.3}Cu_3O_7$ T_c=37 K	[3, 2] $TlBa_2Ca_2Cu_3O_9$ $T_c \approx 120$ K $Tl_{0.5}Pb_{0.5}Sr_2Ca_2Cu_3O_9$ $T_c \approx 120$ K	[3, 3] $Tl_2Ba_2Ca_2Cu_3O_{10}$ $T_c \approx 120$ K $Bi_{2-x}Pb_xSr_2Ca_2Cu_3O_{10}$ $T_c \approx 110$ K
4	[4, 1]	[4, 2] $TlBa_2Ca_3Cu_4O_{11}$ $T_c \approx 110$ K	[4, 3] $Tl_2Ba_2Ca_3Cu_4O_{12}$ $T_c \approx 104-108$ K
5	[5, 1]	[5, 2] $TlBa_2CaCu_6O_{13}$ T_c=105 K	[5, 3] $Tl_2Ba_2Ca_4Cu_5O_{14}$ $T_c \approx 105$ K

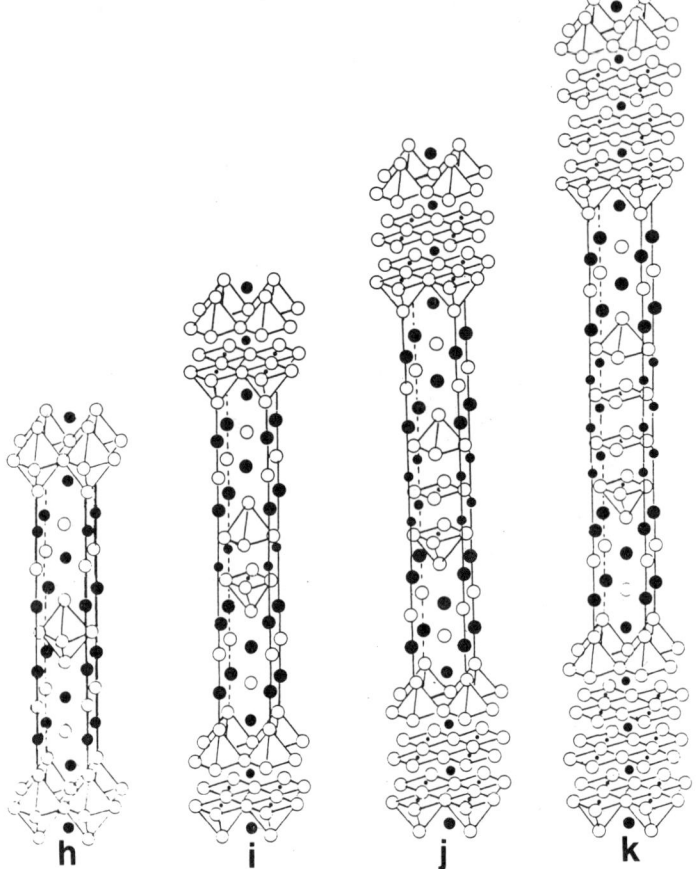

Figure 4 h,i,j,k : Crystal structure of cuprates characterized by triple rock salt layers (n=3) : (h) m=1 member, (i) m=2 member, (j) m=3 member, (k) m=4 member.

La_2CuO_4 or $La_{2-x}A'_xCuO_4$ (A = Sr, Ba, Ca) corresponding to m = 1, $La_{2-x}A_{1+x}Cu_2O_6$ (A = Ca, Sr) corresponding to m = 2 ; $BaPbYSrCu_3O_8$ (m = 3) belong to this series.

(ii) compounds with double rock salt type layers are represented by thallium cuprates containing thallium mono-layers according to the general formula $TlBa_2Ca_{m-1}Cu_mO_{2m+3}$ and $TlSr_2Ca_{m-1}Cu_mO_{2m+3}$ with m ranging from m = 1 to 5. A series of lead cuprates $Pb_{1-x}A_xSr_2Y_{1-y}Ca_yCu_2O_7$, with A = Sr, Ca, Cu corresponding to m = 2 belongs also to this family.

(iii) triple rock type layers are obtained for thallium and bismuth cuprates only. They imply the existence, in the rock salt slabs of thallium or bismuth bilayers leading to the general formula $Tl_2Ba_2Ca_{m-1}Cu_mO_{2m+4}$ ($1 \leq m \leq 5$) and $Bi_2Sr_2Ca_{m-1}Cu_mO_{2m+4}$ ($1 \leq m \leq 3$).

(iv) the ordered oxygen deficient perovskite $YBa_2Cu_3O_7$, whose structure will be described later, corresponds to the [0, ∞] member of this large family of cuprates.

As shown in table 1, most of these oxides exhibit superconducting properties, with critical temperatures ranging from 10 K to 125 K. However the critical temperature may vary dramatically with the stoichiometry of the phase as will be discussed further.

The formula given here for the different compounds correspond to the ideal structures. In fact, most of these oxides are characterized by an oxygen non-stoichiometry, corresponding either to an oxygen deficiency as for instance in $YBa_2Cu_3O_{7-\delta}$ or $TlBa_2CaCu_2O_{7-\delta}$ or to an oxygen excess as for instance in the bismuth cuprates $Bi_2Sr_2Ca_{m-1}Cu_mO_{2m+4+\delta}$. This deviation from oxygen stoichiometry with respect to the ideal composition affects strongly the superconducting properties of these materials. Besides this type of non-stoichiometry, there exists a cationic non-stoichiometry dealing especially with thallium, bismuth, lead and structures in the rock salt type layers; the non-stoichiometry is complicated by order-disorder phenomena appearing between these cations within the rock salt layers.

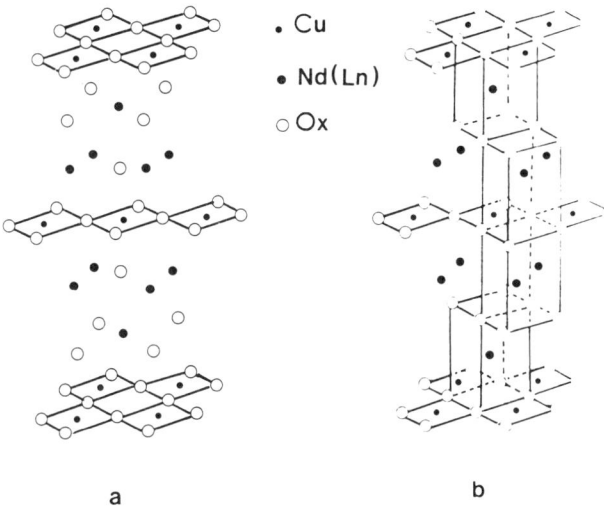

Figure 5 : Structure of Nd_2CuO_4 showing its relationships with (a) La_2CuO_4 and (b) the fluorite structure.

Besides this large family of layered cuprates there exist two other structural types which are closely related to these oxides. The first one, $YBa_2Cu_4O_8$ is a p-type superconductor with a critical temperature of 70 K, whose structure derives from that of $YBa_2Cu_3O_7$ and will be described later.

The second type exhibits the Nd_2CuO_4 structure (Fig. 5). These oxides $Nd_{2-x}Ce_xCuO_4$ and $Nd_{2-x}Th_xCuO_4$ are the only n-type superconducting cuprates isolated up to now ; their critical temperature is smaller than 24 K. The Nd_2CuO_4 type structure is closely related to that of La_2CuO_4, member [1, 1] of the above series. It differs in the positions of some oxygen atoms leading to a square coordination of copper. Both structures have similar positions for Cu, Ln and O atoms forming the basal planes of the CuO_6 octahedra in La_2CuO_4 (Fig. 5a), i.e. the $[CuO_2]_\infty$ planes of CuO_4 groups are identical in the two phases. The remaining oxygen atoms are located in very different positions; the oxygen atoms forming the [LaO]$_\infty$ layers in La$_2$CuO$_4$ (Fig. 5a), have migrated between two Nd layers in Nd$_2$CuO$_4$ (Fig. 5b) in such a way that they form a fluorite type Nd$_2$CuO$_4$ lattice with the oxygen atoms of the [CuO$_2$]$_\infty$ layers. As a result, this structure consists of double [Nd$_2$O$_4$] fluorite type layers sharing the face of their NdO$_8$ cube in the (001) plane.

3 - The complex non-stoichiometry of the "123" structure YBa$_2$Cu$_3$O$_{7-\delta}$

The oxygen non-stoichiometry in the "123" oxide YBa$_2$Cu$_3$O$_{7-\delta}$ is one of the most complex phenomena which has been observed in these cuprates. Its study is of capital importance since it influences drastically the critical temperature of this phase which ranges from Tc = 92 K (δ = 0) to non superconducting (δ = 1).

3-1 - THE PROGRESSIVE PASSAGE FROM THE ORTHORHOMBIC 92 K SUPERCONDUCTOR YBa$_2$Cu$_3$O$_7$ TO THE TETRAGONAL INSULATOR YBa$_2$Cu$_3$O$_6$

The structure of the 92 K-superconductor YBa$_2$Cu$_3$O$_7$ (Fig. 6a) is orthorhombic with all parameters closely to the a$_p$ parameter of the cubic perovskite (a $\approx$ b $\approx$ a$_p$; c $\approx$ 3a$_p$). It is derived from the stoichiometric perovskite by removing rows of oxygens parallel

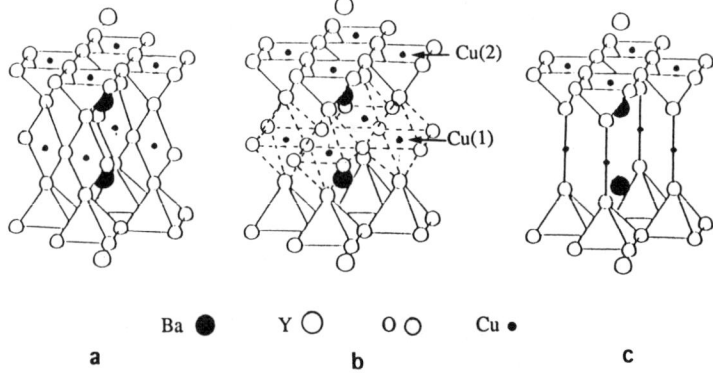

Figure 6 : Crystal structures of the 92 K superconductor YBa$_2$Cu$_3$O$_7$ (a) and of the non superconducting materials YBa$_2$Cu$_3$O$_{6.2}$ (b) (the oxygen sites at the level Cu$_1$ are partially occupied) and YBa$_2$Cu$_3$O$_6$ (c).

to $\vec{b}$ at the levels z = 0 and z = 1/2 of the orthorhombic cell, and by ordering Ba^{2+} and Y^{3+} cations in such a way that one yttrium plane alternates with two barium planes along $\vec{c}$. Thus, the structure can be described as built up from triple copper layers [Cu$_3$O$_7$]$_\infty$ perpendicular to $\vec{c}$, whose cohesion is insured by planes of yttrium cations. Each [Cu$_3$O$_7$]$_\infty$ layers consists of two layers of corner-sharing CuO$_5$ pyramids (Cu$_2$), connected to each other through [CuO$_2$]$_\infty$ rows of corner-sharing CuO$_4$ square planar groups (Cu$_1$). This phase, which exhibits the mixed valency Cu(II)-Cu(III), can only be obtained by annealing or slow cooling at low temperature (400°C-500°C) in an oxygen flow.

By heating the orthorhombic 92 K-superconductor in air at 750°C-900°C and quenching at room temperature, a tetragonal form denoted T1 ($a \approx a_p$; $c \approx 3a_p$) is obtained. This transition corresponds to an oxygen loss leading to the composition range $YBa_2Cu_3O_{6.2}$ to $YBa_2Cu_3O_{6.4}$. The structure of this phase (Fig. 6b) differs from the orthorhombic superconductor only in the position of some oxygen atoms. The pyramidal copper layers corresponding to the (Cu_2) atoms, as well as the position of Y^{3+} and Ba^{2+} and (Cu_1) are similar to those observed for $YBa_2Cu_3O_7$. The main difference deals with the fact that at the level of (Cu_1), the CuO_4 groups are replaced by CuO_6 octahedra, whose basal plane corners are less than half occupied by oxygen. This tetragonal phase does not superconduct, and exhibits poor semi-conducting properties.

By heating the orthorhombic superconductor or the T1 phase up to 600°C in vacuum, a complete desoxygenation can be performed leading to the insulating tetragonal phase $YBa_2Cu_3O_6$. The structure of this phase (Fig. 6c), denoted T2 is deduced from that of the orthorhombic form and of T1 by simply removing the oxygen ions located at the same level as (Cu_1). It results that the pyramidal copper layers are occupied by Cu(II), whereas univalent copper exhibits the two-fold linear coordination, in agreement with the formula $YBa_2Cu_2^{II}Cu^{I}O_6$.

The existence of these three phases has suggested that a possible range of non stoichiometry $YBa_2Cu_3O_{7-\delta}$ with $0 \leq \delta \leq 1$ could exist. This is indeed observed, and the different values of Tc plotted versus δ (fig. 7) show the drastic influence of the oxygen non stoichiometry upon the critical temperature of these phases. Moreover, the

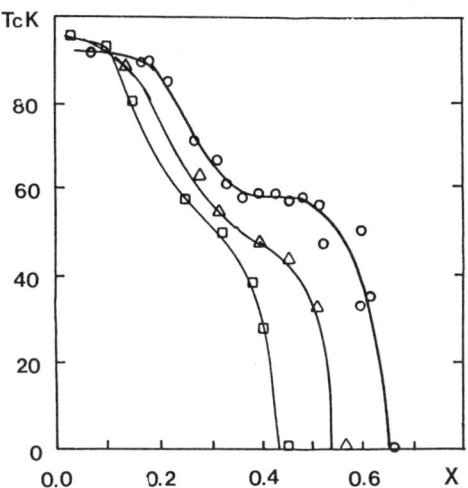

Figure 7 : Variation of Tc as a function of δ for long-annealed $LnBa_2Cu_3O_{7-\delta}$ samples : Ln = Y (o), Gd (△), and Nd (□).

experimental value of Tc varies according to the method of synthesis, owing to order-disorder phenomena on the anionic sites. One example of such phenomena will be described later for the 60 K superconductor.

It must also be noted that besides the tetragonal phases T1 and T2, there exists an oxygen rich tetragonal phase $YBa_2Cu_3O_{6.8}$ which does not superconduct. This phase is metastable and can only be obtained by ceramic techniques using a wet medium for grinding and heating at 830-950°C. Its mean structure is similar to T2 (Fig. 5b) except that c/a is rigourously equal to three. However the examination of this oxide by high resolution electron microscopy shows the existence of microdomains oriented at 90° as illustrated for instance in Fig. 8. The latters are due to the adaptability of the "a" and "c" parameters.

3-2 - TWINS AND ORIENTED DOMAINS

A characteristic feature of the orthorhombic superconductor deals with the fact that all the crystals are systematically twinned. This phenomen results from the phase transition

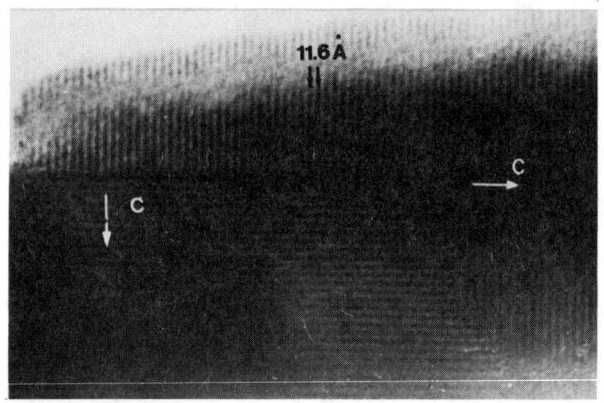

Figure 8 : Orthogonal arrangement of two c-axes : note that the boundaries are straight and sharp.

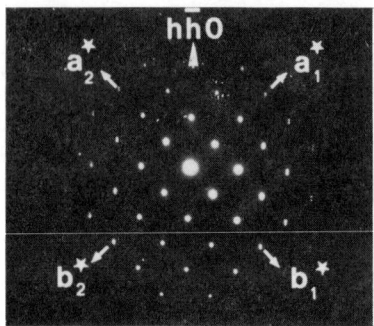
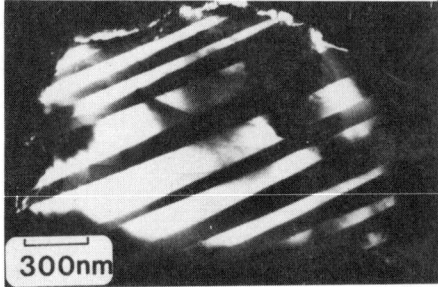

Figure 9 : (a) Typical electron diffraction pattern and (b) dark-field [001] image of an orthorhombic crystal $YBa_2Cu_3O_{7-\delta}$.

betweeen the tetragonal high-temperature phase and the orthorhombic low temperature oxide. It is evidenced from X-ray and electron diffraction patterns. The electron microscopy images allow such microtwinning domains to be observed,; with sizes ranging from 500 to 1000 Å as shown in Fig. 9. This microtwinning implies that the CuO_4 square planar groups in one domain are oriented perpendicular to those in the next one (Fig. 10). Several models of boundaries can be proposed to explain the junction

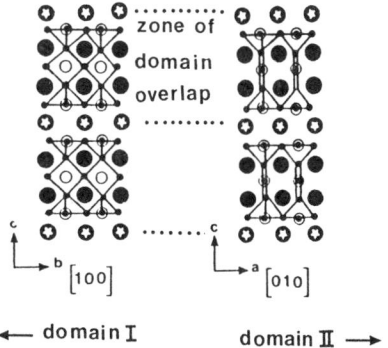

Figure 10 : Projection of twinning domains in a crystal. The change in the orientation of the CuO_4 groups is observed in one perovskite triple layer.

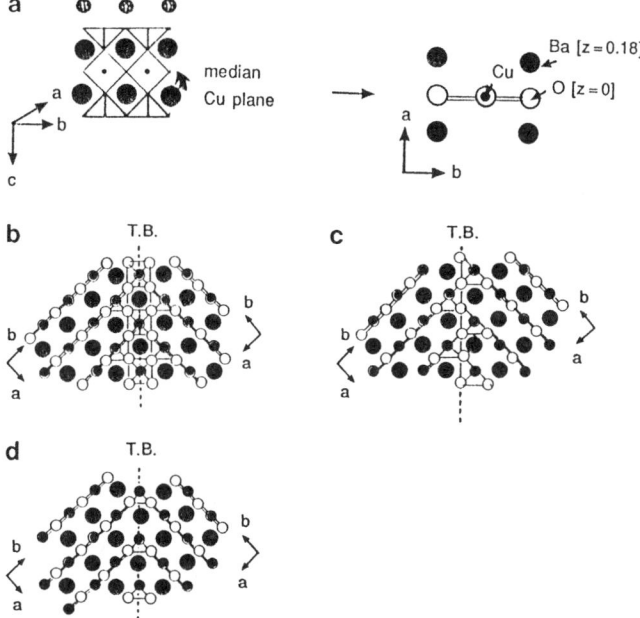

Figure 11 : (a) principle of schematization of the drawing of the twin boundaries (T.B.s) (b-d). Triple idealized models of junctions between the twinning domains in the $[CuO_2]_\infty$ layers : (b) junction in a mirror plane through CuO_6 octahedra and CuO_5 pyramid. (c) Junction through CuO_5 pyramids. (d) Junction in a mirror plane through CuO_4 tetrahedra.

between the CuO_4 square planar groups (Fig. 11). Two of them (Fig. 11b-c) imply the presence of additional oxygen at the boundary through the existence of CuO_6 octahedra or CuO_5 pyramids, but no drastic displacement of the cations. The third model (Fig. 11d), which implies CuO_4 tetrahedra at the junction does not change the oxygen stoichiometry but leads to a significant displacement of the copper atoms. The role of microtwinning in the superconducting properties is at the present time always subject to controversy. There is no doubt that the twin boundaries do not influence the critical temperature, but their role in pinning is still an enigma.

The HREM observations, show a second type of domains, which slice the crystal perpendicularly to the $\vec{c}$ axis (Fig. 12a). From the enlargement of the HREM images (Fig. 12b) it is possible to distinguish a variation of the contrast from one brim of the crystal to

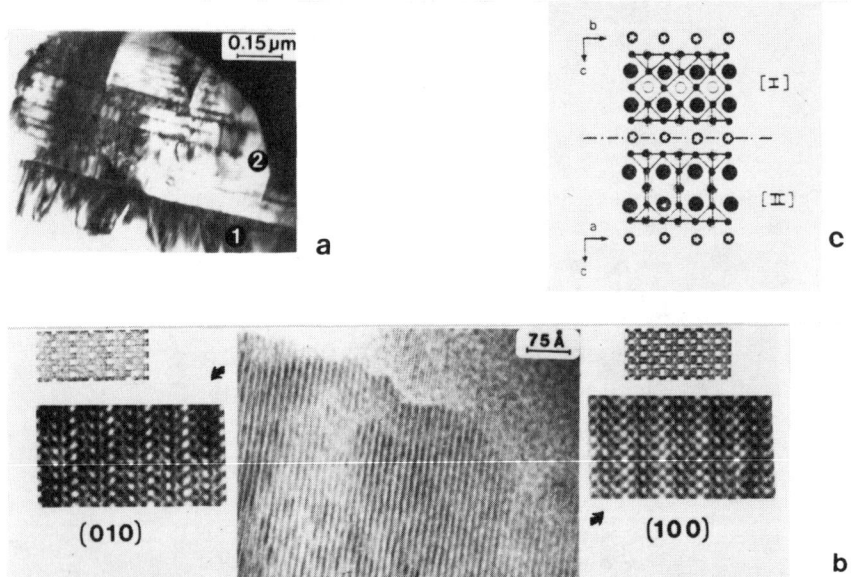

Figure 12 : (a) Low resolution image of oriented domains. (b) High resolution image of "oriented" domains ; areas labelled 1 and 2 show contrast corresponding to [100] and [010] images respectively ; calculated images included in the enlargements were calculated for a defocus of -430Å and a thickness -30Å. (c) Idealized model showing the orientations of the CuO_4 groups.

the other which can be interpretated by the idealized model of Fig. 12c. From one domain to the other along $\vec{c}$, the $[CuO_2]_\infty$ rows of square planar group are turned of 90°, the boundary between two domains being the yttrium plane parallel to [001]. Such a model has been confirmed by the simulation of the images, taking into consideration the view of the structure along [010] and [100] respectively (Fig. 12b). Such 90° domains, imply a coincidence between "a" of one domain and "b" of the adjacent one, leading to microfractures owing to the mismach between these two cell parameters. Indeed strains and moiré patterns due to misorientation are often observed.

3-3 - OXYGEN OVER STOICHIOMETRY

Contrary to oxygen deficiency, no phase $YBa_2Cu_3O_{7+\delta}$ with an excess oxygen has been isolated. However, HREM images have shown a variation of the contrast (Fig. 13) which could be interpreted by the presence of additional oxygen between the rows of CuO_4 groups, forming again the CuO_6 octahedra. Such a model (Fig. 14) which corresponds to the association of CuO_5 pyramids and CuO_6 octahedra is in agreement with the calculated images (Fig. 13). Nevertheless such a phenomenon appears only in the form of small domains (< 200 Å), and no accurate occupancy of oxygen site can be concluded from the HREM simulations.

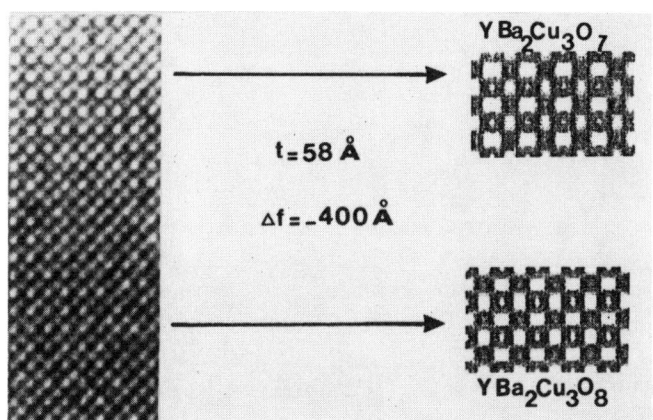

Figure 13 : Enlargement of a defect area. At the top of the micrograph the bright spots are spaced at 3.8Å, in agreement with the calculated image for $YBa_2Cu_3O_7$. At the bottom, all the spots are equivalent. The images were calculated for a focus value close to -400Å and a crystal thickness of ≈ 85Å.

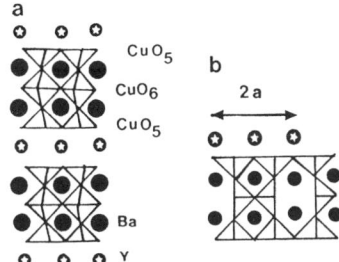

Figure 14 : The presence of additional oxygens involves the formation of (a) octahedra (one oxygen per cell) or (b) pyramids (for 0.5 oxygen per cell). A regular model built up from pyramids involves a doubling of the a parameter.

3-4 - INTERGROWTHS BETWEEN $YBa_2Cu_3O_7$ AND $YBa_2Cu_4O_8$ STRUCTURES

The structure of the 70 K-superconductor $YBa_2Cu_4O_8$ (Fig. 15a) derives from that of the orthorhombic 92 K-superconductor (Fig. 6a) by just replacing the single $[CuO_2]_\infty$ rows of square planar groups, by double $[Cu_2O_3]_\infty$ rows of edge-sharing CuO_4 square planar groups. Thus, the two structures exhibit a bidimensional accord in the [001] plane, i.e. parallely to the copper layers. Consequently intergrowths between the two structures corresponding to the general formula $[YBa_2Cu_3O_7]_n [YBa_2Cu_4O_8]_{n'}$ can be generated.

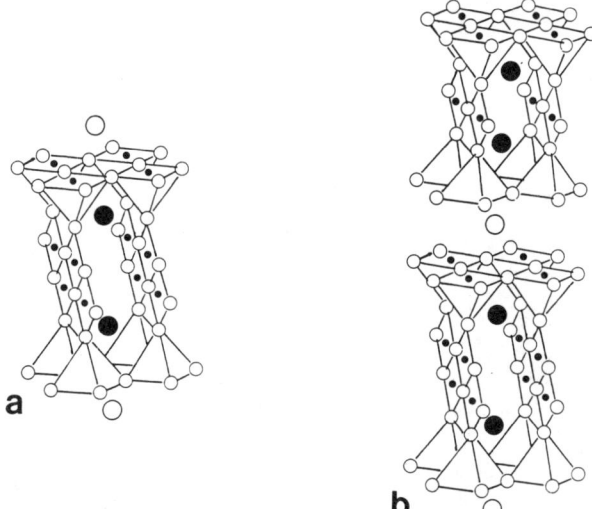

Figure 15 : Idealized structures of (a) $YBa_2Cu_4O_8$ and (b) $Y_2Ba_4Cu_7O_{15}$.

This is the case of the first member of the series $Y_2Ba_4Cu_7O_{15}$ (n = n' = 1) whose structure (Fig. 15b) consists of a stacking of $YBa_2Cu_3O_7$ and $YBa_2Cu_4O_8$ type layers. Consequently $YBa_2Cu_4O_8$, can appear in the form of extended defects in a $YBa_2Cu_3O_7$ matrix as shown from the HREM micrograph of Fig. 16, where the rows of large bright dots in double rows are correlated to barium, and the smaller white dots correspond to copper.

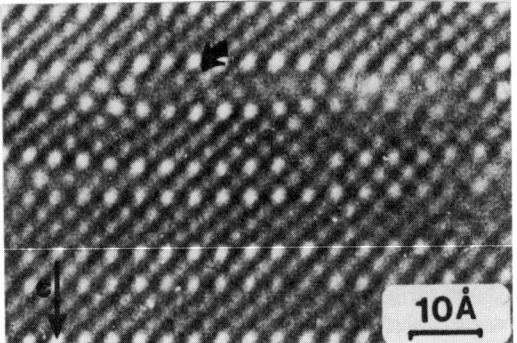 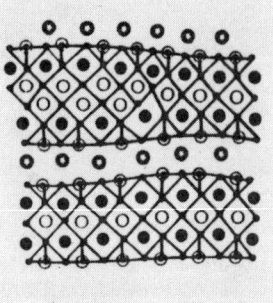

Figure 16 : (a) HREM image of a defect corresponding to the insertion of an extra row of copper. (b) Idealized model corresponding to a local $YBa_2Cu_4O_8$ stoichiometry.

It is worth pointing out that such defects should not affect dramatically the critical temperature of these phases owing to the Tc's of $YBa_2Cu_4O_8$ close to that of $YBa_2Cu_3O_7$.

3-5 - CATIONIC DISORDERING BETWEEN YTTRIUM AND BARIUM

Disordering between Y and Ba is often observed by HREM in the $YBa_2Cu_3O_7$ structure as shown from the micrograph of Fig. 17. One observes a c/3 shifting of the images corresponding to the triple $[Cu_3O_7]_\infty$ layers ; this defect which crosses the image

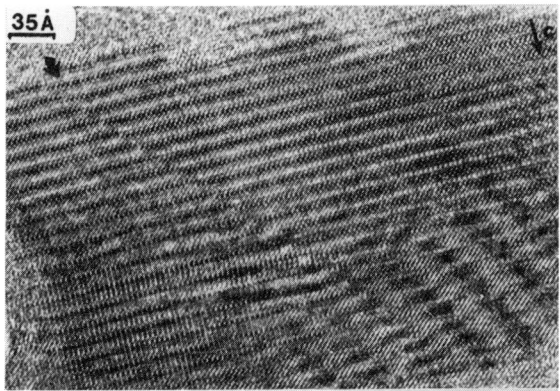

Figure 17 : HREM image of cation disorder corresponding to a reversal of one barium layer and one yttrium layer.

diagonally is explained by a reversal of one barium and one yttrium layer from one part of the defect to the other, whereas the second barium layer remains unchanged (Fig. 18). Such a defect keeps the ratios Y/Ba and Y+Ba/Cu constant, but implies a local variation of the oxygen content at the domain boundary. One can propose either an oxygen deficiency (Fig. 18a), or an excess oxygen by forming CuO_6 octahedra (Fig. 18b).

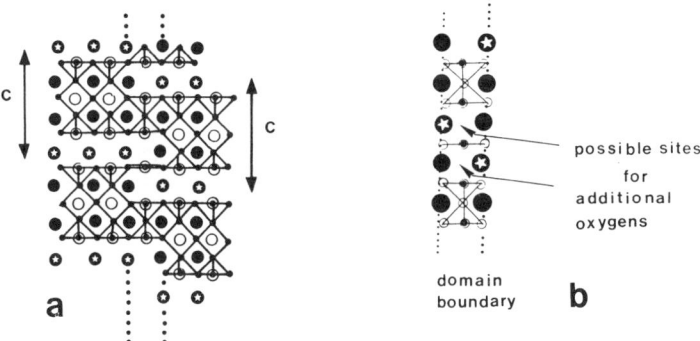

Figure 18 : (a) Idealized model of the translation of the triple perovskite layers along the c axis. (b) Example of possible sites for additional oxygen at the domain boundary.

3-6 - LOCAL ORDERING AND DISORDERING OF OXYGEN IN THE 60 K-SUPERCONDUCTOR $YBa_2Cu_3O_{7-\delta}$ ($0.37 \leq \delta \leq 0.45$)

As soon as δ differs from zero, various ordering and disordering of oxygens and vacancies on the anionic sublaltice appear, which vary according to the experimental conditions. These phenomena have been studied by many authors, as well in the orthorhombic phases as in the tetragonal ones. They are at the origin of the dramatic variation of the superconducting properties with oxygen non stoichiometry and thermal treatment.

It is not possible to present here an exhaustive study of this phenomenon. Let us consider as an example the HREM study of the 60 K-superconductor $YBa_2Cu_3O_{7-\delta}$ ($0.30 \leq \delta \leq 0.50$) obtained by the controlled reoxidation of a reduced precursor $YBa_2Cu_3O_{6.1}$. This oxide, although it is characterized by a sharp transition, exhibits numerous microstructures and modulations, but also highly disordered regions in the microcrystals.

Fig. 19 a shows such strains and modulations of the contrast corresponding to a beginning of ordering of oxygen. The enlargement of the latter micrograph (Fig. 19b) shows variations of the spacing of rows of white spots, their modulations and superdislocation like defects. These observations are typical of many crystals where the local ordering of oxygen is not achieved.

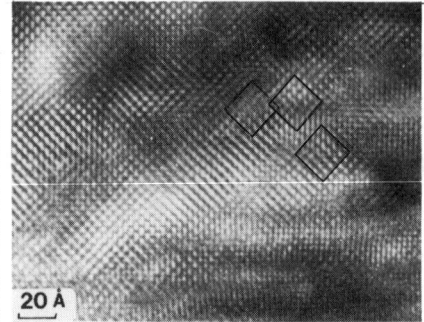

Figure 19 : (a) [001] HREM image of $YBa_2Cu_3O_{6.66}$, characterized by strains and contrast modulations. (b) Enlargement showing variations in the spacing of the rows and their undulation.

Besides the highly disordered regions, there exist microstructures corresponding to oxygen ordering in different directions such as [100]p, [110]p, [210]p and [310]p. An example of such an ordering which involves to a tripling of the "a" parameter is shown in Fig. 20. Another complicated example, is shown in Fig. 21 where it can be seen that various superstructures are set up in different directions. In fact they are not well established over large domains, but they exist in very tiny areas with different periodicities. For instance in the zone labelled 1 in Fig. 20, one observes contrast periodicities of "$\sqrt{2}a_p \times \sqrt{5}a_p$" corresponding to the directions [110]p and [2$\bar{1}$0]p, whereas in the zone labelled 2, the [3$\bar{1}$0]p direction is combined with local $2a_p$ superstructure.

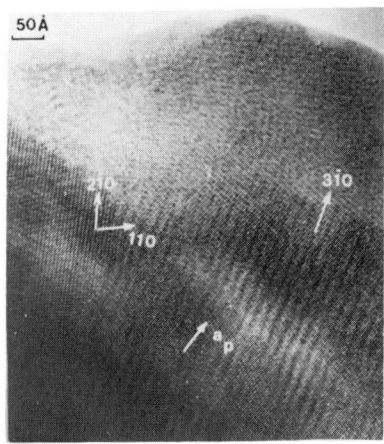

Figure 20 : Complex modulations observed in an orthorhombic crystal $YBa_2Cu_3O_{6.66}$. New periodicities appear along the [210]* in the areas "1", [110]* and [310]* in the areas "2" directions of the perovskite subcell.

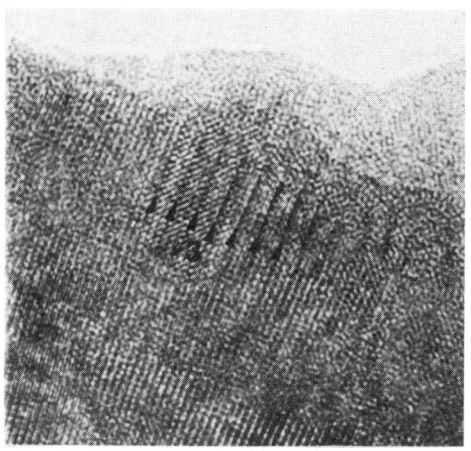

Figure 21 : [001] HREM image of a local superstructure corresponding to a new periodicity a' = 3a.

These oxygen and vacancy orderings are explained on the basis of copper disproportionation, according to the equation : $2Cu(II) \rightleftarrows Cu(III) + Cu(I)$

One can indeed propose structural models, taking into consideration that the three fold coordination for Cu is unlikely, and that Cu(I) exhibits a twofold coordination, whereas

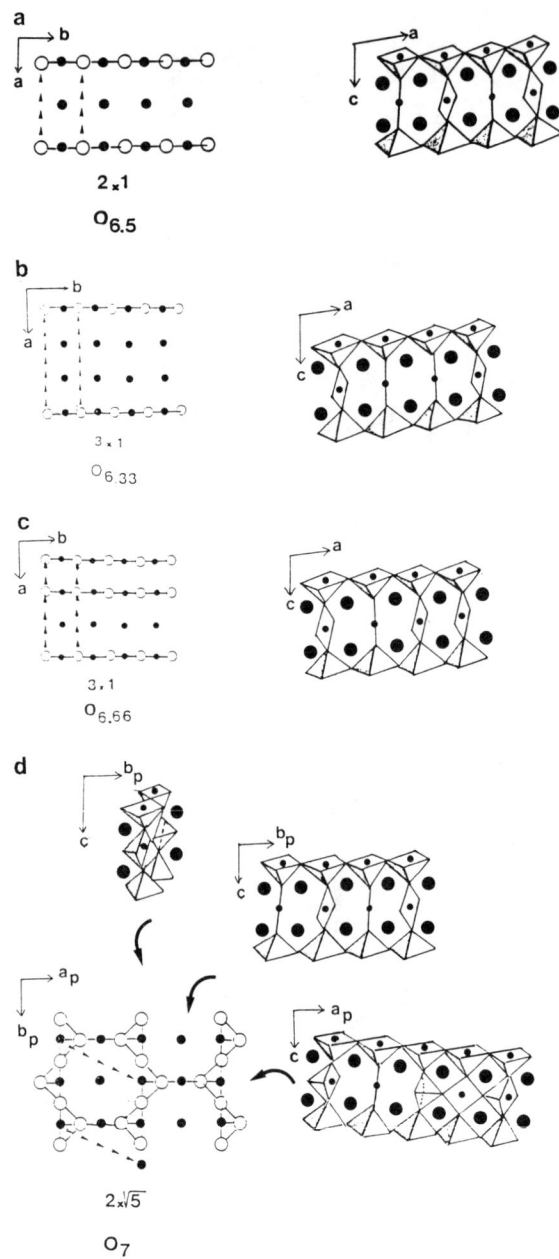

Figure 22 : Some idealized models of oxygen and vacancy ordering in the median copper plane. The new periodicities and oxygen contents are specified. The projected and perspective views along the different chains are drawn.

Cu(II) and Cu(III) are susceptible to present coordination 4 or 5 or 6. Some idealized models are shown in Fig. 22. The doubling of the "a" parameter is interpreted by the composition "$O_{6.5}$" in which one $[CuO_2]_\infty$ row of CuO_4 groups alternates with one $[Cu^IO]_\infty$ of Cu^IO_2 sticks (Fig. 22a). In the same way the tripling of the "a" parameter corresponds either to the compounds $O_{6.33}$ characterized by double $[Cu^IO]_\infty$ rows alternating with single $[CuO_2]_\infty$ rows (Fig. 22b) or to the composition $O_{6.66}$ in which double $[CuO_2]_\infty$ rows alternate with single $[Cu^IO]_\infty$ rows (Fig. 22c). Much more complex orderings involving interconnections between CuO_5 pyramids and CuO_4 groups can be proposed for the various local superstructures, as shown for instance for the "$2a_p \times \sqrt{5} a_p$" superstructure (Fig. 22d).

Besides these complex local superstructures which take place in a more or less random way, there exist modulations of the contrast along the a direction of the orthorhombic cell. One also observes that their direction varies from one twinning domain to the next. Such a phenomenon corresponds to a doubling of the "a" parameter and to the stacking of one $[CuO_2]_\infty$ row with one $[Cu^IO]_\infty$ row alternately in agreement with the model of Fig. 22a. However the modulations are not regular as in the latter idealized model, so that one has to consider either the introduction of additional oxygen forming locally CuO_5 pyramids (Fig. 22b) or the formation of copper vacancies involving a chain interruption (Fig. 22c).

There is no doubt that all these order-disorder phenomena affect dramatically the critical temperature of these phases, even if one admits that the "2a" superstructure is mainly responsible for the 60 K-superconductivity.

4 - Intergrowth mechanisms : thallium and bismuth cuprates

The fact that rock salt layers can adapt to perovskite layers suggests the possibility of formation of extended defects due to a variation of the thickness of these layers in the original matrix. Such extended defects are frequently observed by high resolution electron microscopy in bismuth cuprates and especially in thallium cuprates. Moreover the relationships between the two structural types and the fluorite structure makes that fluorite type defects are also observed.

4-1 - PEROVSKITE INTERGROWTH DEFECTS

The formation of perovskite layers whose thickness is different from that of the matrix is sometimes observed in bismuth cuprates, and is very current in thallium cuprates. For instance the formation of single octahedral perovskite layers instead of double pyramidal copper layers, leading to $Bi_2Sr_2CuO_6$ type defects in a $Bi_2Sr_2CaCu_2O_8$ matrix are effectively observed by HREM (Fig. 23). In the case of thallium cuprates such defects are much more numerous ; they depend upon the thickness of the perovskite slab of the matrix, i.e. of m. Indeed for low m values, the faulty perovskite layer corresponds to m' values close to m, i.e. m' = ± 1 ; on the opposite for large m values, m' can differ from m by a large value. The latter view point is illustrated by the HREM micrograph of the cuprate $Tl_2Ba_2Ca_3Cu_4O_{12}$ (m = 4) which exhibits faulty perovskite layers corresponding to m' = 5 and 7 (Fig. 24). In a general way it is worth pointing out that the number of such defects increases as m increases, depending on the thermal treatments which govern the order-disorder phenomena of the stacking of the layers.

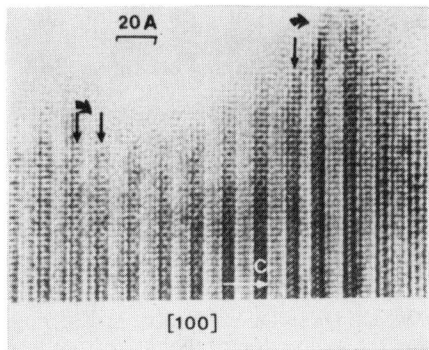 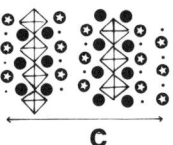

Figure 23 : (a) [100] image of $Bi_2Sr_2CaCu_2O_8$ showing the formation of $Bi_2Sr_2CuO_6$ layers. (b) Idealized drawing of $Bi_2Sr_2CuO_6$ structure (n = 3, m = 1).

4-2 - ROCK SALT INTERGROWTH DEFECTS

These types of defects appear mainly in the thallium cuprate, owing to the great mobility of thallium. This allows a variation of the rock salt layer thickness which can contain either a thallium monolayer (n = 2) or a thallium bilayer (n = 3). A typical example is shown in Fig. 25a, where the quadruple rows of white dots correlated to the rows of cations according to the cationic sequence "BaTlTlBa" are accidentally replaced by triple rows according to the cationic sequence "BaTlBa". This type of contrast can be interpreted (Fig. 25b) as the formation of n' = 2 defects, i.e. $[(TlO)(BaO)]_\infty$ layers in the n = 3 matrix of the oxide $Tl_2Ba_2CaCu_2O_8$. In the same way, starting from a n = 2

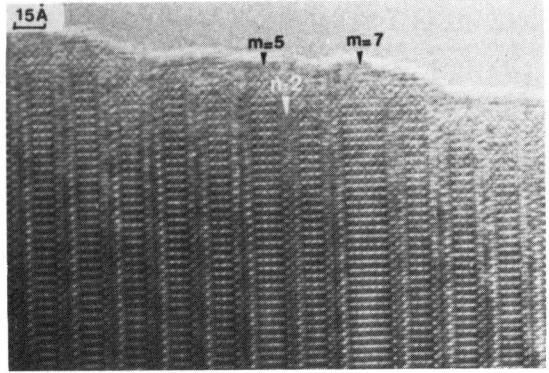

Figure 24 : Perovskite layers with defects in the thallium oxides. The number of defects increases with m and the m' value can then deviate greatly from the nominal composition : example of m' = 7 in a nominally m = 4 matrix.

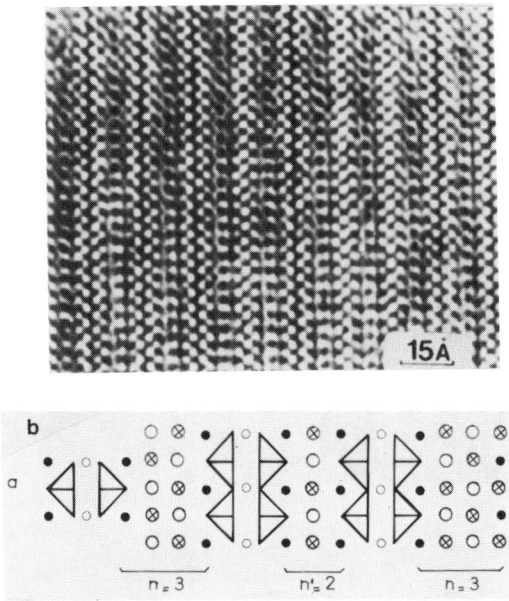

Figure 25 : (a) Defective rock-salt-type layer stacking. n' = 2 in an n = 3 nominal matrix. (b) Idealized drawing of the defect.

matrix such as $TlBa_2CaCu_2O_7$, one sometimes observes defects (Fig. 26a), corresponding to the formation of $[BaO]_\infty$ single rock salt layer (i.e. n' = 1) as interpreted in Fig. 26b.

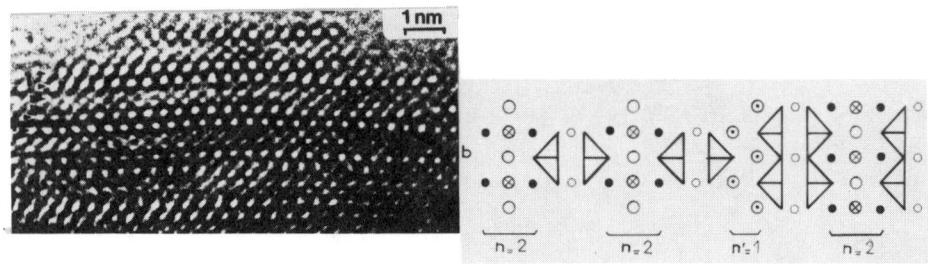

Figure 26 : (a) Single rock-salt-type layers appear as defects in $TlBa_2CaCu_2O_7$. (b) Idealized drawing of the defect.

Besides these defects which are currently observed, one observes, much more rarely, the intercalation of additional $[CaO]_\infty$ layers between the thallium layers. Such an example of double $[CaO]_\infty$ layers, appearing in a matrix of the cuprate $Tl_2Ba_2CaCu_2O_8$ is depicted in Fig. 27. The regular sequence of four rows of bright dots, corresponding to the triple rock salt layers, i.e. to the sequence "Ba-Tl-Tl-Ba" is replaced, at the level of the defect by a sequence of two rows of weaker dots sandwidched by two rows of bright dots (Fig. 27a) leading to the sequence "Ba-Tl-Ca-Ca-Tl-Ba" (Fig. 27b).

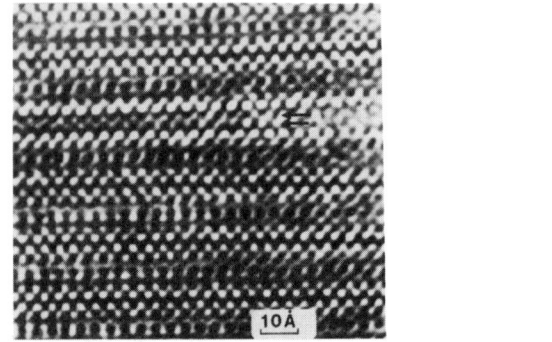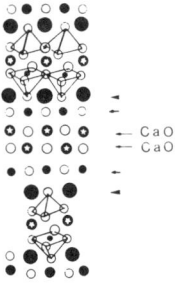

Figure 27 : (a) Additional AO layers between the two TlO layers. (b) Idealized model of the defect.

4-3 - FLUORITE TYPE DEFECTS

The consideration of the coordination of calcium in the oxides characterized by double pyramidal copper layers shows that the CaO_8 cubes share their edges forming a single fluorite type layer (Fig. 28a). This suggests the possibility of replacing these single fluorite type layers by double or triple fluorite type layers (Fig. 28b). Indeed

Figure 28 : Schematic drawing of single fluorite layers (a) and of double fluorite layers (b) (edge sharing AO_8 cubes).

different series of oxides were synthesized which all correspond to the replacement of the single $[CaO_2]_\infty$ fluorite layers by $[Ln_2O_4]_\infty$ double fluorite layers. All these structures will not be described here. Fig. 29 shows one example, obtained from

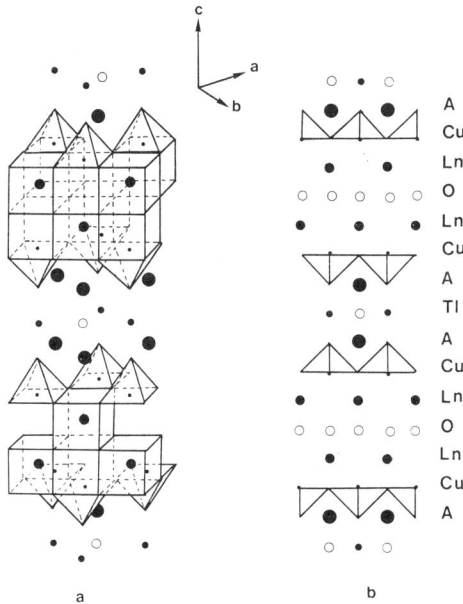

Figure 29 : Crystal structure of the oxides $Tl_{1+x}A_{2-y}Ln_2Cu_2O_9$ (A = Sr, Ba ; Ln = Pr, Nd, Ce).

the oxides $Tl_{1+x}A_{2-y}Ln_2Cu_2O_9$ (A = Sr, Ba ; Ln = Pr, Nd, Ce). It can be seen that such a structure derives from that of $Tl_2Ba_2CaCu_2O_7$ by just replacing the $[CaO_2]_\infty$ fluorite layers by $[Ln_2O_4]_\infty$ double fluorite layers. Besides these regular intergrowths of the three structures -perovskite, rock salt and fluorite- leading to well defined single phases, fluorite type defects may appear in the form of defects in the matrix of layered cuprates containing double pyramidal copper layers. An example is shown in the HREM micrograph of the oxide $TlBa_2NdCu_2O_7$ (Fig. 30). Two rows of double copper layers are observed (arrowed Fig. 30a) ; adjacent to these double rows, two series of three rows of white dots (three small black arrows) are correlated to the sequence "BaO-TlO-BaO" (Fig. 30a). Between these two triple rows of white spots, one observes four rows of less bright spots

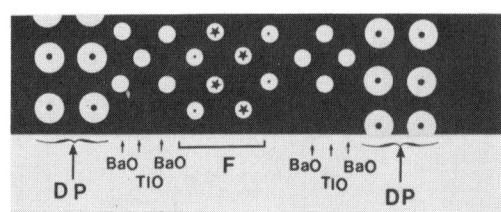

Figure 30 : (a) [010] image of an original defect observed in $TlBa_2NdCu_2O_7$. Symbols are DP for the Double Perovskite layers, three small arrows for the BaO-TlO-BaO layers and F in a square bracket for the structural unit ($Nd_2Cu_2O_6$). (b) Schematic representation of the image contrast. The sizes of the white spots are not strictly proportional to those observed in the experimental image ; they only indicate the nature of the different rows.

(labelled F). The corresponding structural model (Fig. 30c) corresponds to the intercalation of an additional neodymium layer between the pyramidal copper layers of the $TlBa_2CaCu_2O_7$ structure (Fig. 30d). Extended defects involving triple fluorite type layers are also observed.

The comparison of the different [m, n] superconducting cuprates suggests that the formation of rock salt or perovskite type defects will not affect dramatically the superconducting properties of these materials. On the opposite, one can expect a dramatic decrease of the superconductivity due to the appearence of double or triple fluorite type defects, since the corresponding regular intergrowths do not superconduct.

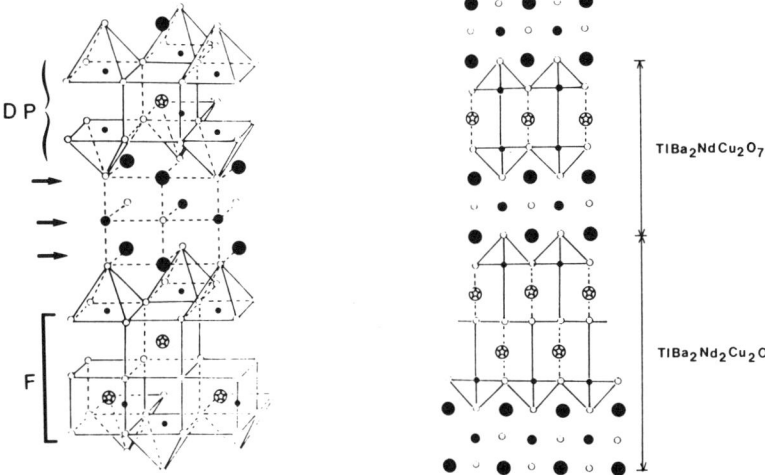

Figure 30 : (c) Perspective drawing of the model of the defect. (d) Schematic projection along [010] of the structural model.

5 - Layer interconnections

Layers of different nature can be connected to each other leading to the formation of boundaries in the crystal. Such phenomena appear currently in the lead cuprates. A first example is shown in Fig. 31a for the cuprate $Pb_{1.4}Bi_{0.6}Sr_2Ca_{0.5}Y_{0.5}Cu_3O_8$, where we can observe an antiphase boundary (AB) inclined at 45° with respect to the rows of white dots corresponding to the $[PbO]_\infty$ and $[SrO]_\infty$ layers ; from one domain to the other there exists a shift of about c/2 across the boundary (marked in the HREM micrograph by small arrows). The structure of this phase (Fig. 31b) which consists of double pyramidal copper layers, alternating with layers of Cu^IO_2 sticks is particularly favourable to the formation of such boundaries. In this shift, one SrO layers (labelled S) is not interrupted in the two domains, whereas $[CuO_2]_\infty$ layers are replaced by PbO layers crossing over the boundary (Fig. 31b) ; in the same way the layers of Cu atoms of the CuO_2 sticks are interrupted, and replaced by $[Ca, Y]_\infty$ planes. It is worth pointing out that the boundary, consists of a pure oxygen plane.

Fig. 32 shows another example of antiphase boundary in the phase $Pb_2Sr_2Ca_{0.5}Y_{0.5}Cu_3O_8$. In that case, the boundary is parallel to $\vec{c}$, i.e. perpendicular to the layers as shown from the HREM micrograph (Fig. 32a) ; the structural model (Fig. 32b) shows that the boundary corresponds to a $[AO]_\infty$ plane (A = Y, Ca, Sr). Contrary to the first type of boundary, all the $[CuO_2]_\infty$ and $[PbO]_\infty$ layers are not interrupted by crossing over the boundary. One $[CuO_2]_\infty$ and one $[PbO]_\infty$ layer out of two remains unchanged, whereas the second $[CuO_2]_\infty$ and $[PbO]_\infty$ layers are correlated to each other through the boundary.

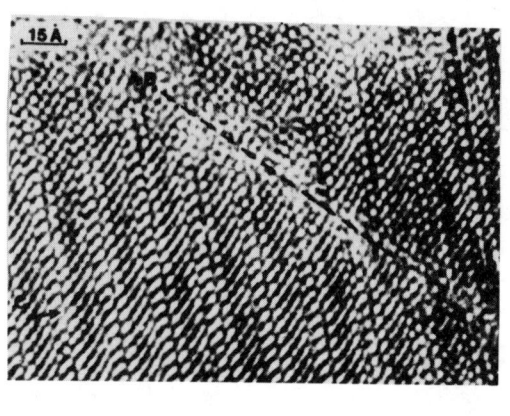

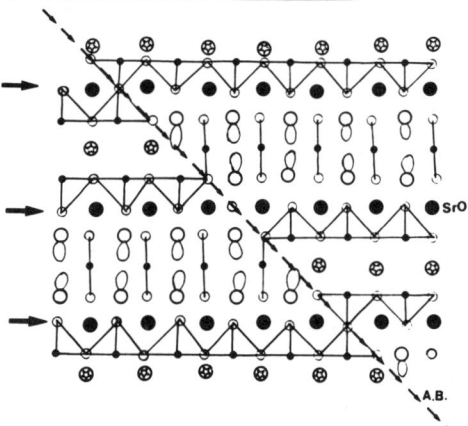

Figure 31 : $Pb_{1.4}Bi_{0.6}Sr_2Ca_{0.5}Y_{0.5}Cu_3O_8$: (a) [110] HREM image of a 45° antiphase boundary (AB). The nature of the cationic strontium and lead layers is indicated as S and P respectively ; the boundary is marked by a row of small black arrows. (b) Idealized drawing of the connection of the different layers through the 45° antiphase boundary. The SrO layers, which remain unchanged across the boundary, are shown with large black arrows. The 45° boundary is parallel to an oxygen plane (plane of small arrows).

Such interconnections are sometimes observed only in one slice of the crystals. An example is shown in the HREM micrograph of Fig. 33a. The contrast at the edge of crystal, characterized by double rows of white dots corresponds to a triple slab "PbO-Cu-PbO" ; at the level of the defect such a double row disappears and is replaced by a triple row of small spots. This defect is explained by the fact that a "45°" antiphase boundary is interrupted at the level of the "SrO" planes ; as a consequence, this limited defects can be interpreted as the replacement of triple "PbO-Cu-PbO" layer by a "CuO_2-Y-CuO_2" layer (Fig. 33b).

All these types of defects interrupt the propagation of the carriers along the $[CuO_2]_\infty$ planes and consequently are susceptible to affect drastically the superconducting properties of these materials.

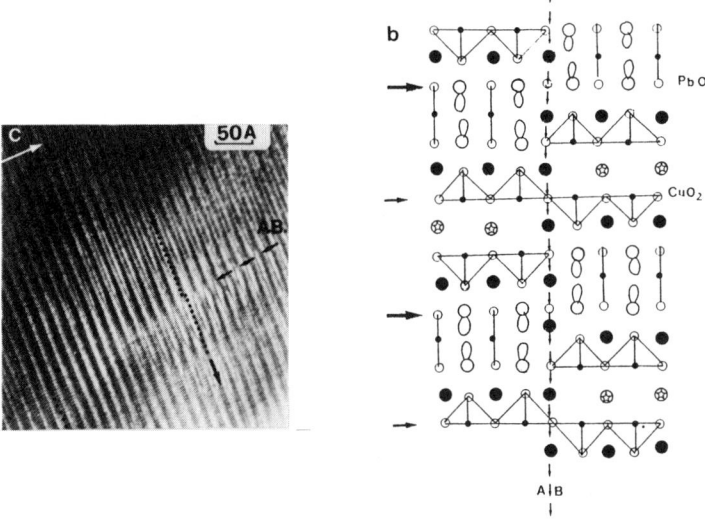

Figure 32 : (a) [110] image of an antiphase boundary parallel to the c axis in the thick part of the crystal. The shift of the layers accross the boundary is close to c/4 (indicated by rows of small dots parallel to the layers). (b) Idealized drawing of the composition of the layers through the antiphase boundary parallel to (110). The unchanged planes ; $[PbO]_\infty$ and $[CuO_2]_\infty$, are indicated by large and medium arrows. The boundary appears in an [AO] plane (row of small arrows).

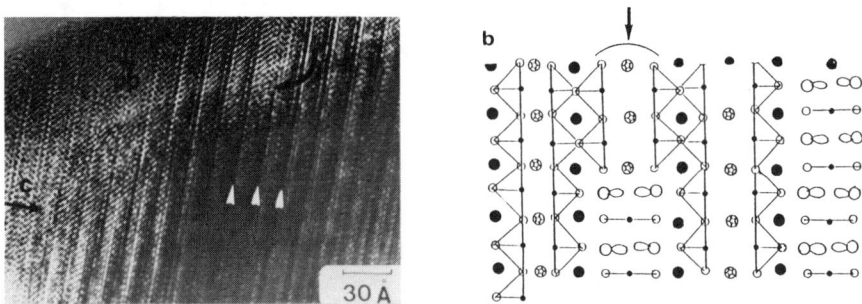

Figure 33 : (a) HREM image of a defect where the two rows of lead atoms, which appear as double white rows in the thick part of the crystal (white arrows), disappeared ; they are replaced by three rows of staggered dots. (b) Drawing of the idealized structural model : the defect would consists of the local replacement of a "PbO-Cu-PbO" triple layer by a "CuO_2-Y-CuO_2" triple layer. The defective layer is marked by a black arrow.

6 - Satellites and structure modulation

A particular feature of the bismuth cuprates deals wih the existence of incommensurate satellites on their electron diffraction patterns as shown for instance for $Bi_2Sr_2CaCu_2O_{8+\delta}$ (Fig. 34). This modulation of the structure is also observed on the single crystal X-ray

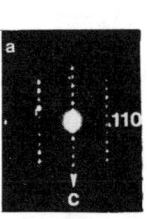 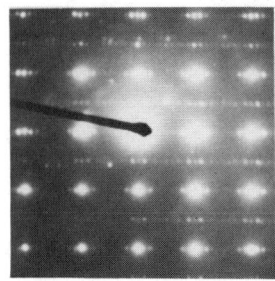 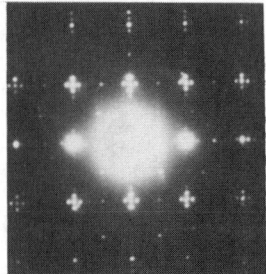

Figure 34 : ED patterns observed in $Bi_2Sr_2CaCu_2O_{8+\delta}$. (a) [010]. (b) [001] with incommensurate satellites along [100]*, $q \approx 4.6$. (c) Superposition of two areas [001] with extra spots in two perpendicular directions.

diffraction patterns. It is also worth pointing out that the intensity of the satellite spots are very intense. The high resolution electron microscopy studies confirm the existence of incommensurate superstructures along one or two directions. This feature is illustrated on Fig. 35, for $Bi_2Sr_2CuO_{6+\delta}$, where it can be seen that modulations appear along the $\vec{a}$ direction of the structure i.e. along the [110] direction of the perovskite subcell, on the right hand side of the micrograph. On the opposite, on the left hand side the modulations are no longer observed. This is explained by the coexistence in the same crystal of two 90° oriented domains, corresponding to the two [100] and [010] orientations with respect to the electron beam. In fact, the resolution of the modulated structure of these oxides from single crystal by X-ray diffraction shows the displacive character of the modulation. The different layers - $[BiO]_\infty$, $[CuO_2]_\infty$ and $[SrO]_\infty$ - are waving in the crystal as shown for instance for the cuprate $Bi_{2.08}Sr_{1.84}CuO_6$ (Fig. 36). The origin of this phenomenon is always subject to controversy. However, it can now be established that it results from a buckling of the double $[(BiO)_2]_\infty$ layers ; moreover, it has been observed that the latter contain an oxygen excess with respect to the ideal "BiO" composition. Two hypothesis can be proposed for the formation of such corrugated layers. In the first one, the extra oxygen is at the origin of the modulation. In the second hypothesis, the stereoactivity of the $6s^2$ lone pair of Bi(III) leads to a corrugation of the "BiO" layer since it tends to take the place of an anion, leading to a mismatch between the lattices of the $[CuO_2]_\infty$ layers and $[BiO]_\infty$ layers ; in this second hypothesis the oxygen uptake is only a consequence of the lone pair effect. The fact that the stoichiometric phase $Bi_2Sr_2CaCu_2O_8$, exhibits also a structure modulation supports strongly the second hypothesis. Besides the modulated displacement of the cations, there exists a modulation in the distribution of the bismuth and strontium cations within the rock salt layers. The recent structure determination of the cuprate $Bi_{2.08}Sr_{1.84}CuO_{6+\delta}$, has indeed shown that this phase exhibits a significant strontium deficiency and that the strontium sites are partly occupied by bismuth with a modulation of the occupancy of these catonic sites. It is also worth pointing out that the modulation vector q varies with the oxygen content. For instance, it has recently been shown in the $Bi_2Sr_2CaCu_2O_{8+\delta}$ structure that it decreases from q = 4.90 for δ = 0 to 4.65 for δ = 0.10.

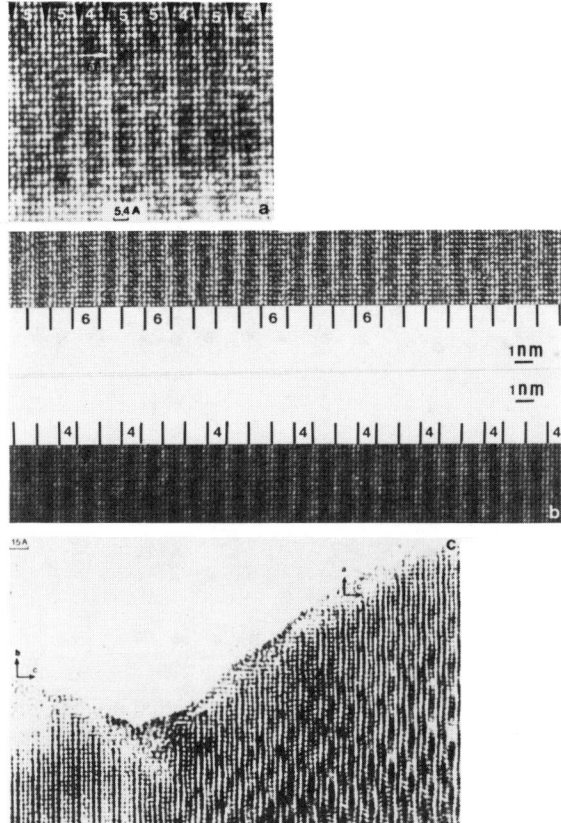

Figure 35 : Modulations in bismuth oxides. (a) $Bi_2Sr_2CuO_6$: pseudoperiodicity of blocks corresponding to $4(\sqrt{2}/2)a_p$ and $5(\sqrt{2}/2)a_p$ in an [001] image. (b) [001] projections of $Bi_2Sr_2CaCuO_8$ crystals. The insertion of blocks of x4 and $x6(\sqrt{2}/2)a_p$ is not perfectly periodic. (c) HREM image of 90° oriented domains in $Bi_2Sr_2CuO_6$. The undulations of the bismuth layers are clearly visible at the right-hand side, in agreement with an [010] orientation of the crystals. The left part of the micrograph exhibits a contrast which agrees with a [100] orientation. Bismuth positions are correlated with the black spots of the image. Note the perfect interface.

Contrary to bismuth cuprates, the thallium cuprates exhibit rarely incommensurate satellites. Moreover, the intensity of the latters is weak. In fact the existence of

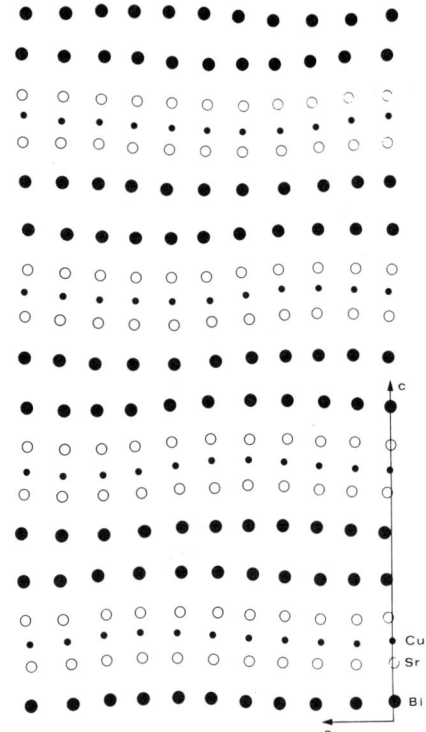

Figure 36 : Waving of the Bi, Sr and Cu sheets in the actual crystal, projected onto the (010) plane. The atomic displacements from the basic positions are not magnified.

Such a feature is illustrated in Fig. 37 which shows two kinds of E.D. patterns observed for the tetragonal and orthorhombic forms of the cuprate $Tl_2Ba_2CuO_6$ respectively. The modulated orthorhombic phase was performed at a lower oxygen pressure. This type of modulation has nothing to do with that observed for bismuth cuprates and it does not affect the structure significantly. Its origin is not explained up to now. It can be explained by a small thallium deficiency leading to a modulated distribution of thallium and vacancies on the cationic sites ; it can also be due to the presence of Tl(I) which may affect the $[TlO]_\infty$ layers due to its size, bigger than Tl(III), but also to its $6s^2$ lone pair.

Numerous extraspots are also observed in the ED patterns of lead cuprates. For instance, the oxide $Pb_{0.5}Sr_{2.5}Ca_{0.5}Y_{0.5}Cu_2O_{7-\delta}$ exhibits satellites whose intensity varies with the thermal treatment. One typical [010] ED pattern is shown in Fig. 38a, in which satellites are observed along [102]* with a modulation factor of 4. Another kind of satellites are observed along [203]* (Fig. 38b), whose intensity is weak with a q vector of 6. The corresponding high resolution images show contrast modulations which appear in

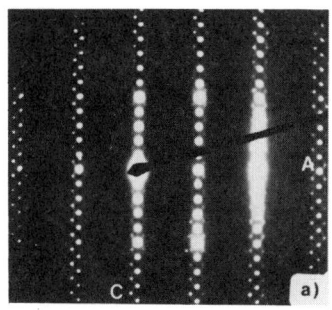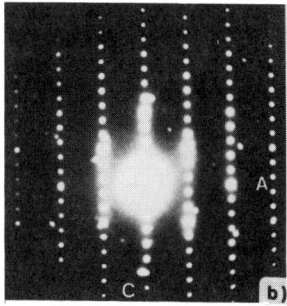

Figure 37 : $Tl_2Ba_2CuO_6$ phase. (a) [100] ED pattern of the tetragonal phase. (b) [110] ED pattern of the orthorhombic form.

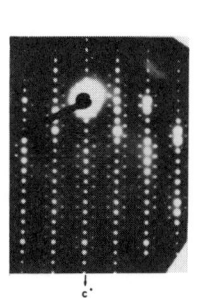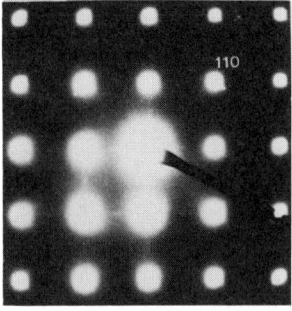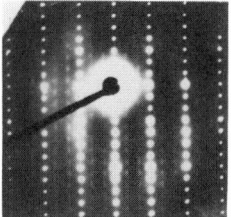

Figure 38 : $Pb_{0.5}Sr_{2.5}Y_{0.5}Ca_{0.5}Cu_2O_{6.75}$. (a, b) [010] ED patterns showing the existence of satellites along [102]* and [203]*, respectively. (c) [001] ED pattern exhibiting weak extra spots and streaks.

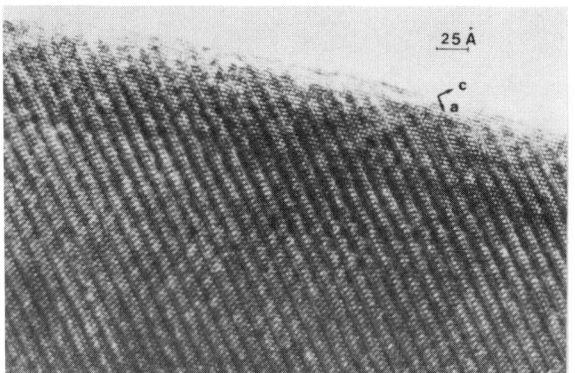

Figure 39 : (a) [010] image showing an extended domain where the contrast modulations appear. (b) Enlargement of a modulated area (the [$Pb_{0.5}Sr_{0.5}O$] layers are indicated by black arrows).

extended domains (Fig. 39a). This image and its enlargement (Fig. 39b) show that the origin of the modulation is different from that observed in bismuth cuprates. Indeed, considering that the row of bright dots is correlated to the [$Pb_{0.5}Sr_{0.5}O$]$_\infty$ layers (black arrows in Fig. 39b), it can be deduced that the modulation appears in the [$Sr_2ACu_2O_{6-x}$]$_\infty$ slabs. Thus, the modulation is probably due to the ordering of the oxygen vacancies, taking into account the oxygen deficient character of this phase ($\delta \approx 0.25$). Up to now, nothing can be said about the relationships between the different sorts of modulations and the superconducting properties of these materials, but what is sure is that the modulation of the structure is not at the origin of superconductivity in these compounds.

7 - Redox mechanisms and superconductivity

Very small deviations from oxygen stoichiometry can change drastically the superconducting properties of the thallium cuprates without destroying the structure. This is the case of the thallium cuprates for which a spectacular effect of oxygen

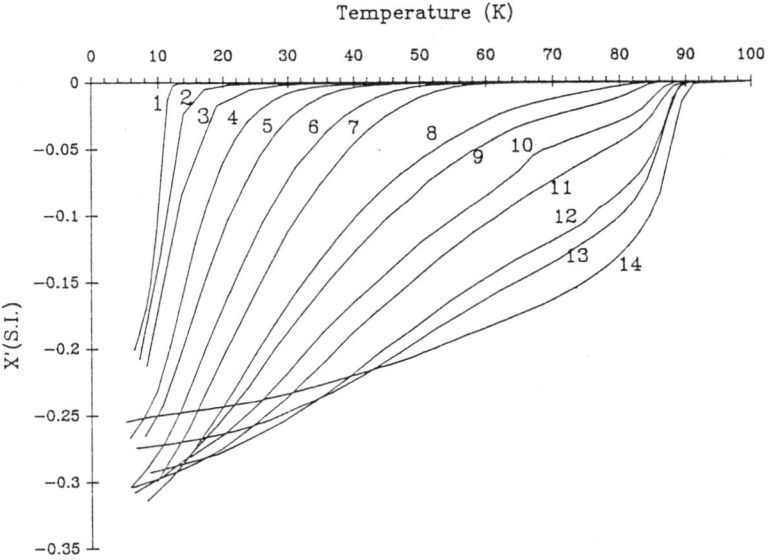

Fig. 40 : Real part $\chi'(T)$ of the magnetic AC susceptible of the "O8-FC" sample, as-synthesized (1) and after the succeeding and variable H2/Ar annealing times (2) : 2 min, (3) : 6 min, (4) : 14 min, (5) : 26 min, (6) : 50 min, (7) : 74 min, (8) : 122 min, (9) : 146 min, (10) : 196 min, (11) : 242 min, (12) : 338 min, (13) : 430 min, (14) : 480 min.

non-stoichiometry upon the critical temperature is observed. For instance, the single layer copper oxide $Tl_2Ba_2CuO_{6\pm\delta}$ was the subject of controversy since it was formed from non superconducting to superconducting with a Tc of 90 K. In fact, this oxide synthesized under an oxygen presure of 10 bars does not superconduct, whereas its annealing under an hydrogen-argon flow (10% H_2) at low temperature (290°C) allows the critical temperature to be increased progressively from 15 K after five minutes annealing up to 92 K (Fig. 40). The HREM investigation of this phase shows that the structure is absolutely not affected by this treatment, i.e. the sequence of the layers remains regular, no extended defects is formed. Moreover the microthermogravimetric analysis, shows that the oxygen loss is very small, i.e. about 3% the total oxygen amount of the phase. This definitely shows that a very small variation of the oxygen stoichiometry leads to a dramatic variation of the Tc's. A similar effect is observed for the double copper layer cuprate $Tl_2Ba_2CaCu_2O_{8+\delta}$ (fig. 41), which, prepared under an oxygen pressure of 10 bars,

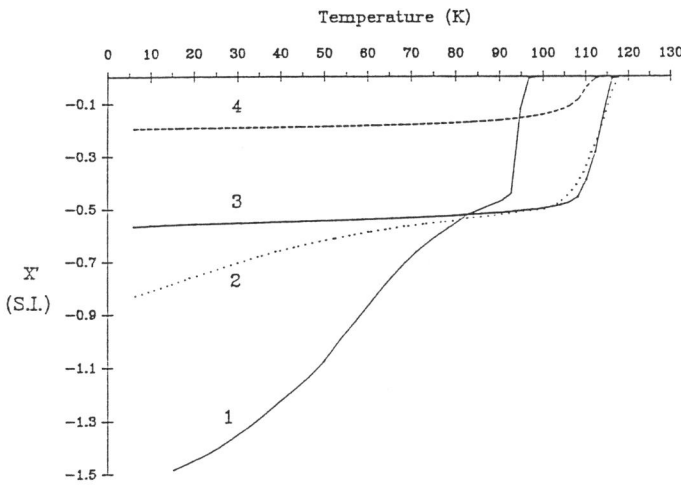

Figure 41 : AC susceptibilities $\chi'(T)$ of the same sample Tl "2212" for different annealing times in a gas mixture (Ar 90% + H_2 10%) : (1) unannealed (Tc = 96K) ; (2) annealed for 15 min (weight loss 0.07%, Tc = 118K) ; (3) annealed for 2h (weight loss 0.4%, Tc = 117K) ; (4) annealed for 12h (weight loss 2.2%, Tc = 115K).

exhibits a Tc of 96 K (curve 1, fig. 41), and whose Tc's can be increased up to 120 after hydrogen-argon annealing at low temperature (curve 2 fig. 41). In the same way, the superconducting volume of this phase does not change as far as the structure is not affected ; when the structure starts to be destroyed, showing amorphous particles by HREM one observes a dramatic decrease of the superconducting volume (curve 4 fig. 41). The neutron diffraction study of the oxides of the series $TlBa_2Ca_{1-x}Nd_xCuO_{7-\delta}$ for x = 0.2, 0.50 and x = 1 shows also the important role of oxygen non stoichiometry in the critical temperature of the phases. Table 2 shows that there exists a very small thallium deficiency in the limit of the error whereas the oxygen deficiency is significant. Moreover all the vacancies appear in the "TlO" layer and it is shown that the coordination of thallium must be considered as tetrahedral rather than octahedral. The mean oxidation state of copper deduced from this study shows that it decreases from 2.32 for the oxide x = 0.20 which exhibits the highest Tc's of 104 K to 2 for x = 1 which does not superconduct, passing through 2.18 for the oxide x = 0.50, which exhibit an intermediate Tc of 40 K.

TABLE 2 : $Tl_{1-\varepsilon}Ba_2Ca_{1-x}Nd_xCu_2O_{7-\delta}$: non stoiciometry deduced from N.D. data and critical temperatures.

x	$T_c(K)$	ε	δ	Cu valency
0.2	104	0.04	0.14	2.32
0.5	40	0.04	0.14	2.18
1	not-supra.	0.02	0.04	2

All these results show that the oxygen non stoichiometry is the key to the optimization of the superconducting properties and especially determines the critical temperature of the thallium cuprates, owing to its great influence upon the hole carrier density of these compounds.

There is also a great probability that in these oxides, the "TlO" layers can play the role of hole reservoirs, extracting electrons from the $[CuO_2]_\infty$ layers according to the equilibrum :

$Tl(III) + Cu(II) \rightleftarrows Tl(III-\varepsilon) + Cu(II+\varepsilon)$.

A similar behaviour can be proposed for bismuth cuprates, for which an X-ray absorption study has shown that the curve at Bi L_{III}-edge was displaced toward low energy as if the oxidation state of the bismuth was less than three. This can be interpreted as the formation in the "BiO" layers, of a narrow band involving Bi-6s and 6p orbitals and O-p orbitals, which allows electrons to be trapped from the $[CuO_2]_\infty$ layers into the $[BiO]_\infty$ layers according to the equilibrium :

$Bi(III) + Cu(II) \rightleftarrows Bi(III-\varepsilon) + Cu(II+\varepsilon)$

These observations also suggest that the $[TlO]_\infty$ and $[BiO]_\infty$ layers are not insulating layers but exhibit a metallic or a semi-metallic character and can become superconductive at low temperature by proximity effect.

Although, as its very beginning, the study of the superconducting lead cuprates shows that their superconducting properties are also greatly influenced by the oxygen non-stoichiometry. This is the case for instance of the oxide $Pb_{0.7}Cu_{0.3}Sr_2Ca_{0.5}Y_{0.5}Cu_2O_{7-\delta}$ whose Tc's goes through a maximum of 40 K as the oxygen pressure during the annealing at low temperature increases from 1 bar up to 80 bars (fig. 42).

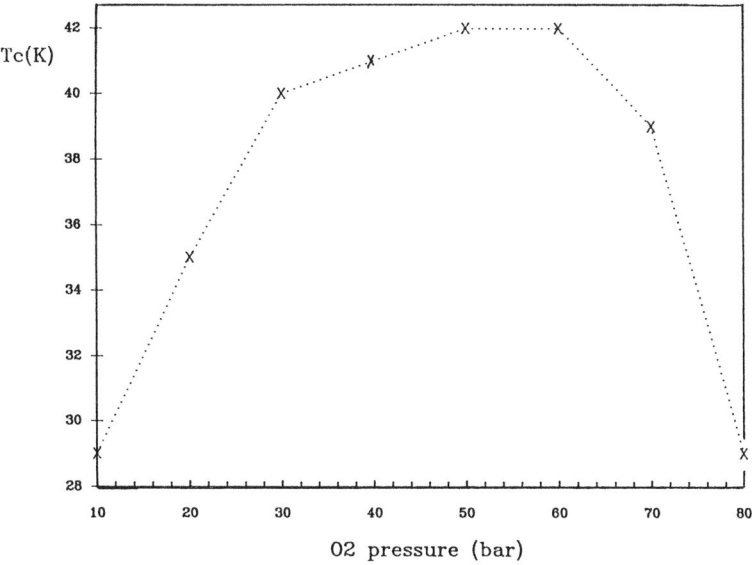

Figure 42 : Oxygen pressure dependence of Tc for $(Pb_{0.7}Cu_{0.3})Sr_2(Ca_{0.5}Y_{0.5})Cu_2O_7$ rangles annealed at 400°C for 10 h.

7 - References

652 References can be found in the book "Crystal Chemistry of high Tc Superconducting copper oxides" by B. Raveau, C. Michel, M. Hervieu, D. Groult. Series in Materials Science 15. Springer Verlag 1991.

8 - Problems

Problem 1 : Propose a structural model for the ordered oxygen deficient perovskite $YBa_2Cu_3O_{6.83}$ which exhibits a symmetry with $a \approx 2a_p$, $b \approx a_p\sqrt{10}$, $c \approx 3a_p$.

Problem 2 : Propose a structural model for the ordered oxygen deficient perovskite $YBa_2Cu_3O_{6.75}$ which exhibits a tetragonal symmetry with $a \approx 2a_p\sqrt{2}$, $c \approx 3a_p$.

Problem 3 : Propose a structural model for the oxides $Bi_2Sr_2Ln_{2-x}Ce_xCu_2O_{10}$ and $Pb_2Sr_2LnCeCu_3O_{10}$ which differ from the oxides $(ACuO_{3-x})_m(A'O)_n$ by the replacement of single fluorite layers by double fluorite layers.

264

8 - 1 - SOLUTIONS OF THE PROBLEMS

Problem 1 :

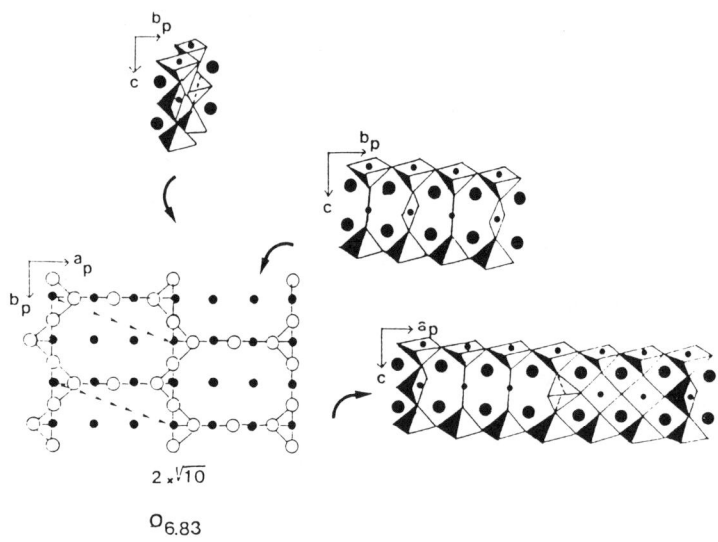

Problem 2 :

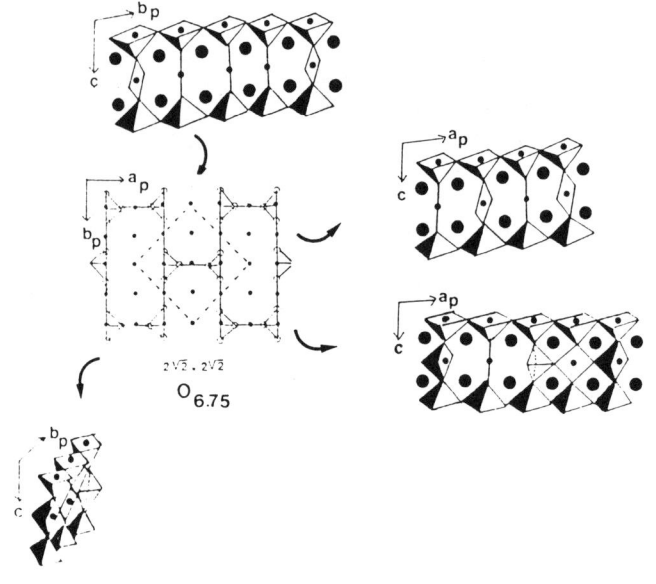

Problem 3 :

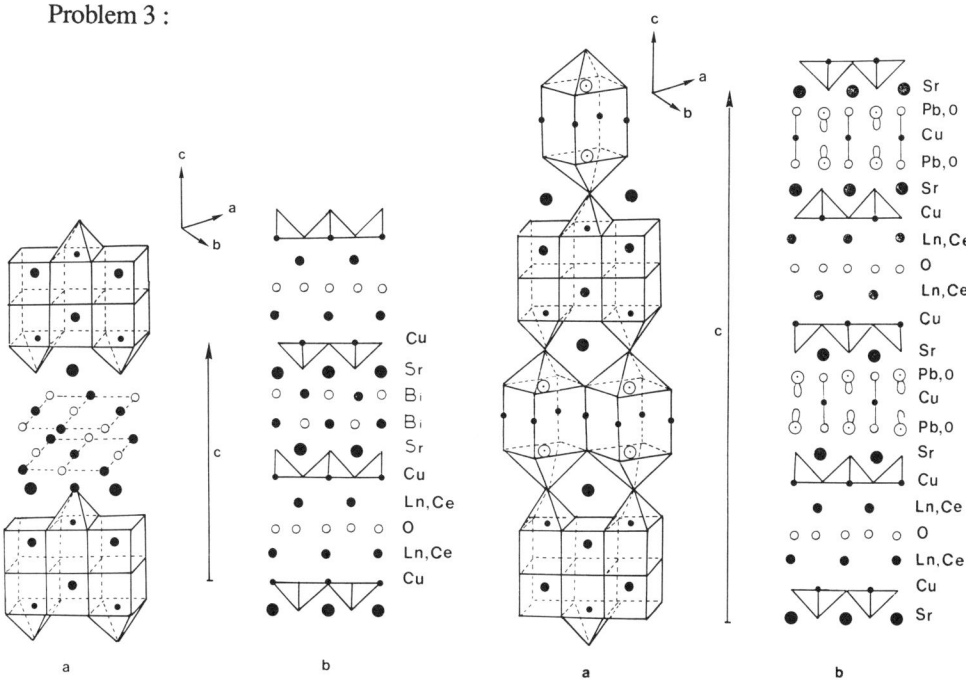

Bi$_2$Sr$_2$Ln$_{2-x}$Ce$_x$Cu$_2$O$_{10}$ Pb$_2$Sr$_2$LnCeCu$_3$O$_{10}$

COMPETITION BETWEEN TRIGONAL PRISMS AND OTHER COORDINATION POLYHEDRA IN BORIDES, CARBIDES, SILICIDES AND PHOSPHIDES

PETER ROGL
Institut für Physikalische Chemie
Universität Wien
Währingerstrasse 42
A-1090 Wien
Austria

ABSTRACT. The crystal structures of metal borides can be classified with respect to the boron aggregation as a function of the metal to boron ratio. The boron centered triangular metal prism is the dominating structural building element. Crystal structures composed of triangular prisms are usually classified by means of the linkage coefficient. In the ternary rare earth - transitionmetal - metametal compounds, the Waist Contact Rule may be employed to reveal transitionmetal - metametal ordering. Competition between trigonal prismatic and octahedral coordination is outlined for the structural series based on units B_3C, AB_6C_3, etc. comprising borides, carbides, silicides and especially pnictides.

1. Introduction

From a geometrical point of view (Pearson, 1972), triangular prismatic coordination A_6B may simply be generated by direct stacking of close packed triangular layers 3^6 of atoms A on top of each other and inserting a second set of layers 3^6 of atoms B between each two layers of A. In this case only half of the triangular prisms are centered and the trigonal prism A_6B is the only coordination figure. Hexagonal or cubic modes of stacking of pairs of layers 3^6 simply result in mixed coordination types of A_6B triangular prisms and A_6B octahedra.

A second form of triangular prismatic coordination is obtained from alternate stacking of hexagonal layer types 3^6 and 6^3 which leads to the crystal structure of AlB_2, where all triangular prisms are centred.

Among the remarkably diversified variety of binary and ternary metal boride structures, the triangular prismatic coordination M_6B has the dominating role, whereas among metal phosphides, silicides and carbides this is merely true for a limited series of structure types.

2. Classification of Metal Borides

Any successful classification of metal borides (Lundström,1977; Rogl,1985; Rogl,1991) has hitherto been based on the pronounced tendency of the electron deficient boron atoms to form strong and essentially covalent boron-boron bonds. At low boron concentrations with a boron to metal ratio B/M < 1/8 isolated boron atoms are often found in octahedral voids of typical intermetallic host lattice structures, where boron atoms may act as stabilizers of ternary solid solutions or of true ternary compounds, such as eta (Ti_2Ni)-type borides, perovskite borides, ß-Mn-type borides or Mn_5Si_3-type borides etc. On the other extreme, with boron atoms being more abundant than B/M > 4, the boron sublattices grow into three-dimensional boron polyhedra (octahedra, cubo-octahedra and icosahedra) with the stabilizing metal donor atoms occupying suitable voids within the rigid boron framework.

The triangular prismatic coordination M_6B is observed to be the prevailing structural building unit for metal borides within the compositional range $1/8 <$ B/M < 4. In far less frequent cases metal coordination around the boron atoms is octahedral $M_6B^{[6o]}$ or an Archimedian antiprism $B^{[8a]}$ (see Figure 1).

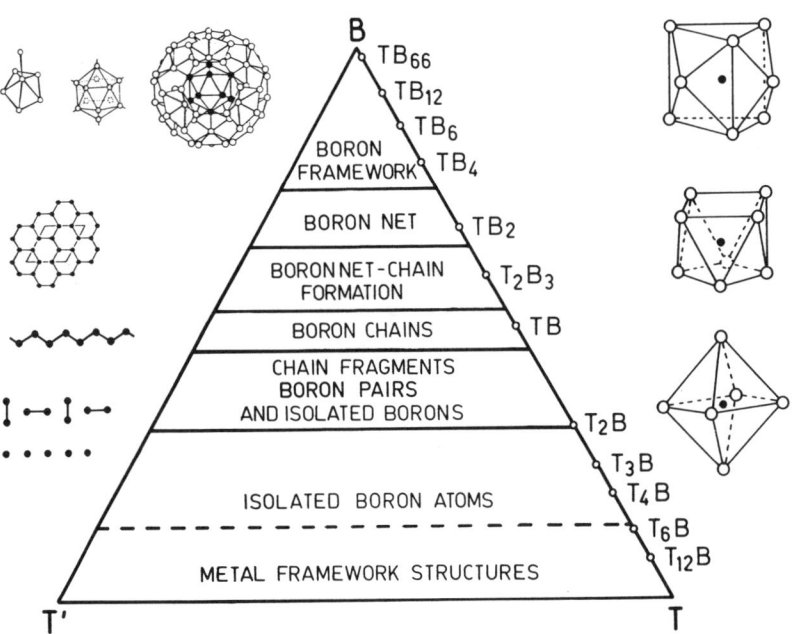

Figure 1: Classification of metal borides with respect to the boron aggregation as a function of the boron to metal ratio in ternary metal borides - the prevailing metal coordination of the B - atom is outlined.

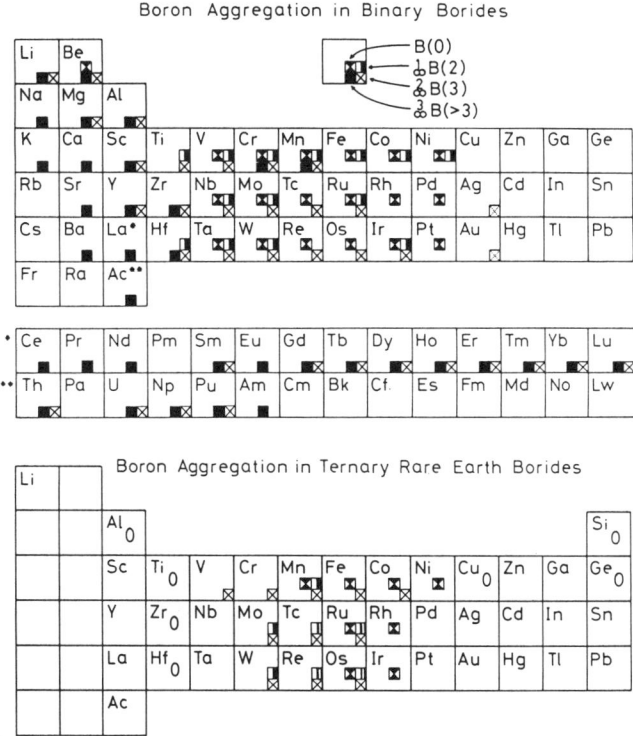

Figure 2 : Boron aggregation in binary and ternary metal borides as a function of the chemical nature of the metal atoms.

At present binary and ternary metal borides are found to crystallize in approximately eighty structure types displaying a remarkably wide variety of modes of linking the sometimes rather distorted triangular prismatic coordination figures around the boron atoms. With the B/M-ratio increasing, linkage between these simple boron subunits grow more and more sophisticated, revealing typical features at defined values of B/M i.e. boron pairs at B/M ~1/2, infinite boron zig-zag chains $\frac{1}{\infty}$B at B/M~1 or infinite two-dimensional plane or puckered boron nets at B/M~2; see also Figure 2. A general classification of metal borides therefore relies on the correlation of both the form and size of the boron aggregation with the boron to metal ratio. Selecting the proper type of boron aggregation for a given B/M ratio may be of valuable help in the design of a structural model for still unknown structure types.

3. Structural Series Based on AlB$_2$- and W-type Units.

Gradual transitions from one to another type of boron aggregation are monitored at best in binary compounds such as e.g. the borides of the transition metals (T) in the concentration range between monoboride and diboride, forming a structural series T_sB_{s+1}, which can formally be derived as the topochemical sum $T_sB_{s+1} = (s-1)TB^{[6p]} + TB_2^{[6p]}$. First observed by (Spear & Gilles, 1969) among vanadium borides for s = 1, 2, 3 and 5, most of the homologous compounds with Nb,Ta have been hitherto discovered (Bolmgren & Lundström, 1990); examples for s = 4 and s > 5, however, are still unknown. The above mentioned vanadium boride structures are exclusively constructed from boron centered triangular metal prisms with prism axes all parallel (i.e. structures of class I; class II in case of prisms with axes in two perpendicular directions etc.). The triangular prisms either form a slab of the AlB_2-type structure with prisms sharing common rectangular faces or slabs appear to be shifted by 1/2z. The interface of prism-blocks in a shifted position essentially corresponds to a body centered subunit i.e. a slab of the W- type structure. This metal atom arrangement provides all boron atoms with an overall tetrakaidekahedral coordination either $B^{[6p,3]}$ or $B^{[6p,2+1]}$ for those at the interface of a shift (see Figure 3).

Crystal structures entirely composed of triangular prisms are usually classified with respect to the mode of linkage of the centered prisms. The linkage coefficient LC defines the average number of prisms in which T-atoms participate. The composition of the compound is then directly related to the LC-value: $T_6B_{LC}^{[6p]}$. In case of a combination of n trigonal prisms and m segments of W-type per unit cell of the crystal structure considered, the linkage coefficient (Grin, Yarmolyuk & Gladyshevskii, 1979) of the vanadium borides may be expressed as $T_6B_{6n/(m+n/2)}$ (Figure 3).

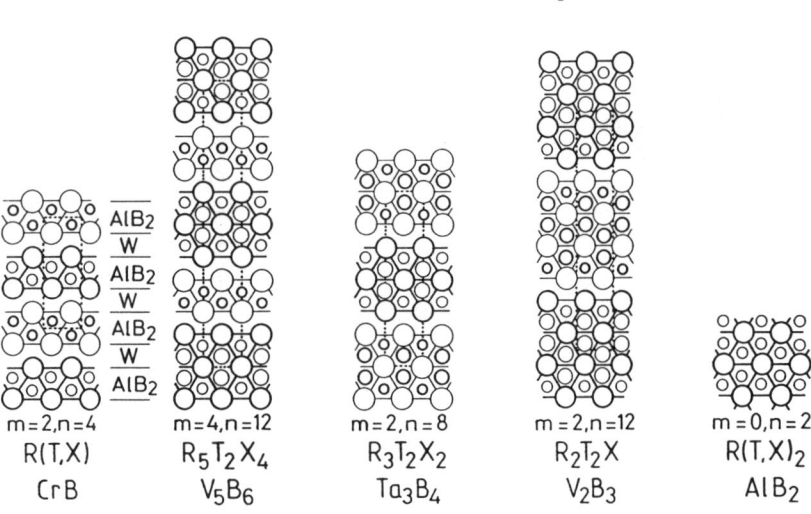

Figure 3: The series of structures T_sB_{s+1} or $T_{m+n/2}B_n$ intermediate between TB- and TB_2-type, seen as intergrowth of W-type slabs and AlB_2-type slabs.

3.1. THE WAIST CONTACT RULE.

Except for T, T-metal ordering there is little variation among the ternary metal borides of these structural series T_6B_{LC}; there are no isotypic representatives known neither for carbides nor for phosphides. Due to the higher stability of triangular prisms R_6X, where X is a metametal M = Si, Ge or one of the iron group metals, it is, however, not surprising to find representatives among ternary rare earth (R)-transitionmetal silicides with a tendency for Si and the iron metals to randomly substitute for each other. In many of these alloys, however, site preferences i.e. T-M metal ordering were found to correspond to the "Waist Contact Rule" (Parthé & Hovestreydt, 1985) i.e. the metametal in contrast to the transition metal rather prefers those sites at the W - type interface, where bonds are formed to one additional rare earth atom around the "waist" of the triangular prism R_6M. The structures in Figure 3 are presented under the assumption of the validity of the waist contact rule in ternary combinations R-T-X, see problem 1.

3.2. STRUCTURAL SERIES BASED ON AIB$_2$- and CsCl-TYPE UNITS.

An interesting case is provided for those of the ternary metal borides which are formed from a combination of transition metals and metametals or main group metals (M), which are to the left of the periodic chart. Only a limited number of structure types has been solved up to now, however, the general understanding of this group of borides is based on the inability of boron to form binary metametal borides. Therefore, wherever metametal - transitionmetal bonds are stronger than transitionmetal mediated metametal boron interactions, we observe a type of crystal structure where boron atoms usually are at the centres of triangular prisms formed by the transition elements avoiding metametal atoms within the first coordination shell. The crystal structure then is virtually separated into boride blocks $T_6B^{[6p]}$ of various linkage and TM_x - units. In all these cases TM - blocks are CsCl-type units, which directly share rectangular faces of the T_6B prisms with boron chains eventually running in two perpendicular directions much the way they are in Ru_2ZnB_2. Examples for these types of structure are found within the homologous structure series $T_{n+m}M_n^{[8c]}B_{2m}^{[6p]} = m(TB_2^{[6p]}) + n(TM^{[8c]})$ for which representatives are known for m=1, n=1 ({Cr,Mn,Fe}$_2$AlB$_2$); m=2, n=1 (Cr$_3$AlB$_4$) and m=1, n=2 (Ru$_3$Al$_2$B$_2$). More complicated arrangements of CsCl- and AlB$_2$ - building blocks are encountered for m=4, n=5 (Ru$_9$Al$_5$B$_6$ $\square_2$ and Ru$_8$ $\square$ Zn$_5$B$_8$) exhibiting defects on the metal or boron sites, depending on steric conditions (see i.e. (Jung & Petry, 1988)).

4. Structures made up of Triangular Prisms

Compact arrangements of edge- or face connected triangular prisms are the base units for structural series such as: $A_{n(n+1)}B_{(n+1)(n+2)}C_{n(n-1)+1}$ or $R_{n^2+3n+2}T_2^{[6,o]}X_{2n^2}^{[6,p]}$

These structures are essentially observed among silicides and pnictides; they are less frequently adopted by borides but with interesting modifications. All these structures are hexagonal with all triangular prisms oriented parallel to the c-axis. The dimensions of the c-axes correspond to the height of the triangular prisms which either form plane networks or appear displaced by c/a in the direction of the prism axis relative to neighbouring prism assemblies. As a consequence of crystal symmetry, columns of octahedra arise at the cell corners parallel to the c-axis. This atom arrangement around the c-axes may also be seen as a column of triangular prisms with additional B-atoms centering the rectangular prism faces providing a tetrakaidekahedral coordination around the prism centres, which then appear to be shifted by 1/4c with respect to the octahedral centres. Naturally there are twice as many octahedra than there are trigonal prisms along the c-axes. Depending on the chemical nature of the filler atom with respect to the host metal, octahedral and tetrakaidecahedral coordination compete. Accordingly we observe pnictide atoms in a prismatic-tetrakaidekahedral coordination rather than at the centerpoints of the octahedra, which in turn are preferentially occupied by small T-atoms such as the iron metals, boron, carbon or silicon. In ternary and higher order compounds the prisms are formed by rare earth (actinoid) or large transition metals usually in an ordered fashion and are centred by the metametals or smaller transition metal atoms. With rare earth (R) atoms at the B-sites and not differentiating between A and C-sites (atoms X) we arrive at the structure series $R_{n^2+3n+2}T_2X_{2n^2}$ where small transition metals T occupy all centre points of the octahedral columns (n=1, $Er_3Ru_2 = Er_6Ru_2Ru_2$; n=2, $Ce_6Ni(Ni,Si)_4$; n=3, $Ce_{10}Ni(Ni,Si)_9$; n=4, $Ce_{15}Ni(Ni,Si)_{16}$; see Figure 4). In case of ternaries, rare earth-transition metal-metametal atomic order may obey the waist contact rule resulting in the modified general formula $R_{n^2+3n+2}X_{n(n+1)}T_{n(n-1)+2}$ (see Figure 4).

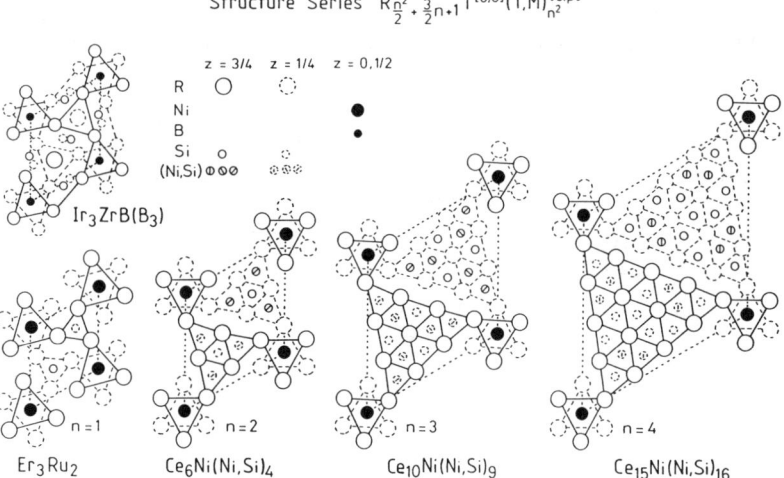

Figure 4: The series of structures $R_{n^2+3n+2}T_2X_{2n^2}$ and the filled up version of the boride member $Ir_3ZrB(B_3)$.

No members have been observed yet for n>4. The ultimate limit of the series is an ∞ layer of centered triangular prisms of the AlB_2-type. $(Sc,Zr,Hf)Ir_3B_4$ are the only borides known to resemble these structures with a metal framework Ir_6Zr_2 corresponding to Er_6Ru_2: two boron atoms center the Ir-prisms resulting in a structural analogy $Ir_6Zr_2B_2 = Er_6Ru_2B_2$. Whereas the structure types $R_{n^2+3n+2}T_2X_{2n^2}$ may be regarded to be composed of AlB_2- and W-type fragments (similar to the vanadiumborides $T_{m+n/2}B_n$) distorted triangular Ir-prisms are in competition with the W-type fragments. These prisms are centered by boron atoms to form an infinite boron chain parallel to the c-axis and thus follow the proper classification of the $ZrIr_3B_4$-type as a monoboride. From the general outlay of the coordination figures, the $ZrIr_3$ $_4$-type may conventionally also be seen as a filled up version of the ZrNiAl-type i.e. to be based on the Fe_2P-type, which finds itself as the nucleus of a large series of structures built by triangular prisms. Slight reorientation of the triangular prisms in the RXT_2-type of structure immediately reveals the structure type of Re_3B as a typical boride or Pu_3Co ($MgCuAl_2$) as a typical intermetallic. Filling of the octahedral voids with small nonmetal atoms (C, N, O) leads to the formation of mixed boron carbides or carbonitrides $Cr_3(B,C)C_{1-x}$ or $Cr_3C(C,N)$, $Cr_3As\{C,N\}$ etc. In all these cases the more electronegative nonmetal atom occupies the octahedral sites (Rogl, Kunsch, Ettmayer, Nowotny & Steurer, 1982).

Restrictions in size may enforce the inoccupation of every second X-site of the prism centres and of every second T-site of the octahedral centres which is summarized in the following formula : $R_{n^2+3n+2}X_{2n^2-n(n-1)}\square_{n(n-1)}\{T_{1/2}\square_{1/2}\}$. Due to this ordered type of defects no direct contacts for the X-atoms occur. The limiting member of this series for n = ∞ is the WC-type. From the above series we already recognize the base units of triangular prisms (see Figure 5), which essentially serve to classify this large group of structures (Parthe & Chabot, 1984; Pivan, Guerin & Sergent, 1987; Chikhrii, Orishchin, Kuzma & Glovyak, 1990). The linkage of up to two of these units (of the same kind) at a common vertex results in simple arrangements in the form of infinite linear or zig-zag chains subdued to various crystal symmetries. Increasing the number of units of the same kind per common vertex for steric reasons is limited to three. Stacking of B_3C-units has been dealt with for the structure series $A_{n(n-1)}B_{(n+1)(n+2)}C_{n(n+1)+1}$ as an example and many modifications exist (see Table 1, Figure 6).

The corresponding example of a structure series based upon AB_6C_3-units is represented by the general formula : $A_{n(n+1)}B_{3(n^2+1)}C_{(2n^2+1)}\square_{(n-1)n/2}$. The combination of the AB_6C_3-units of triangular prisms along common prism edges creates triangular shaped voids occupied by columns of the smaller B_3C-units in form of triangular prisms with the center atoms either in tetrakaidekahedral coordination or in shifted position (by 1/4c) at the center points of octahedra. Some n(n-1) of the triangular prisms stay empty in accordance with the waist contact rule for metametals. For further details see problem 2.

Table 1

Members and representatives of the structural series $A_{n(n-1)}B_{(n+1)(n+2)}C_{n(n+1)+1}$, $R_{(n+1)(n+2)}T_2X_{(2n)^2}$ and the defect series $\square_{n(n-1)}B_{(n+1)(n+2)}C_{n(n+1)+1}$ and $R_{(n+1)(n+2)}T\square X_{2n^2-n(n+1)}\square_{n(n+1)}$.

n	$A_{n(n-1)}B_{(n+1)(n+2)}C_{n(n+1)+1}$		$\square_{n(n-1)}B_{(n+1)(n+2)}C_{n(n+1)+1}$		$R_{(n+1)(n+2)}X_{(2n)^2}T_2$		$R_{(n+1)(n+2)}X^2_{2n-n(n-1)}\square_{n(n-1)}T\square$	
1	$A_0B_6C_3$	Fe_2P		-	$R_6X_2T_2$	Er_3Ru_2	$R_6X_2(T\square)$	-
2	$A_2B_{12}C_7$	$Zr_2Fe_{12}P_7$		$V_{12}P_7$	$R_{12}X_8T_2$	$Ce(Ni,Si)_4Ni$	$R_{12}(X_6\square_2)(T\square)$	-
3	$A_6B_{20}C_{13}$	$Zr_6Ni_{20}P_{13}$		-	$R_{20}X_{18}T_2$	$Ce_{10}(Ni,Si)_9Ni$	$R_{20}(X_{12}\square_6)(T\square)$	$Rh_{20}Si_{13}$
4	$A_{12}B_{30}C_{21}$	$La_{12}Rh_{30}P_{21}$		-	$R_{30}X_{32}T_2$	$Ce_{15}(Ni,Si)_{16}$	$R_{30}(X_{20}\square_{12})(T\square)$	-
5	$A_{20}B_{42}C_{31}$	-		-	$R_{42}X_{50}T_2$	-	$R_{42}(X_{30}\square_{20})(T\square)$	-
6	$A_{30}B_{56}C_{43}$	"$Ho_{20}Ni_{66}P_{43}$"		-	$R_{56}X_{72}T_2$	-	$R_{56}(X_{42}\square_{30})(T\square)$	-
7	$A_{42}B_{72}C_{57}$	-		-	$R_{72}X_{98}T_2$	-	$R_{72}(X_{56}\square_{42})(T\square)$	-
∞	ABC				RX_2	AlB_2	$RX\square$	WC

The series $A_{n(n-1)}B_{(n+1)(n+2)}C_{n(n+1)+1}$ and $R_{(n+1)(n+2)}X_{2n^2}T_2$ are identical insofar as A and C sites are occupied by X-atoms; whereas, however, a C-atom occupies a trigonal prism with tetrakaidekahedral coordination $(B,C)_{6+3}$, T-atoms prefer a position shifted by c/4 at the centerpoints of R_6 octahedra.

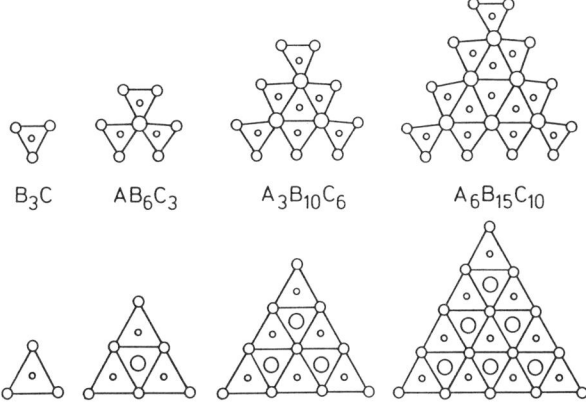

Figure 5: Building blocks of centered triangular prisms as base units for classification.

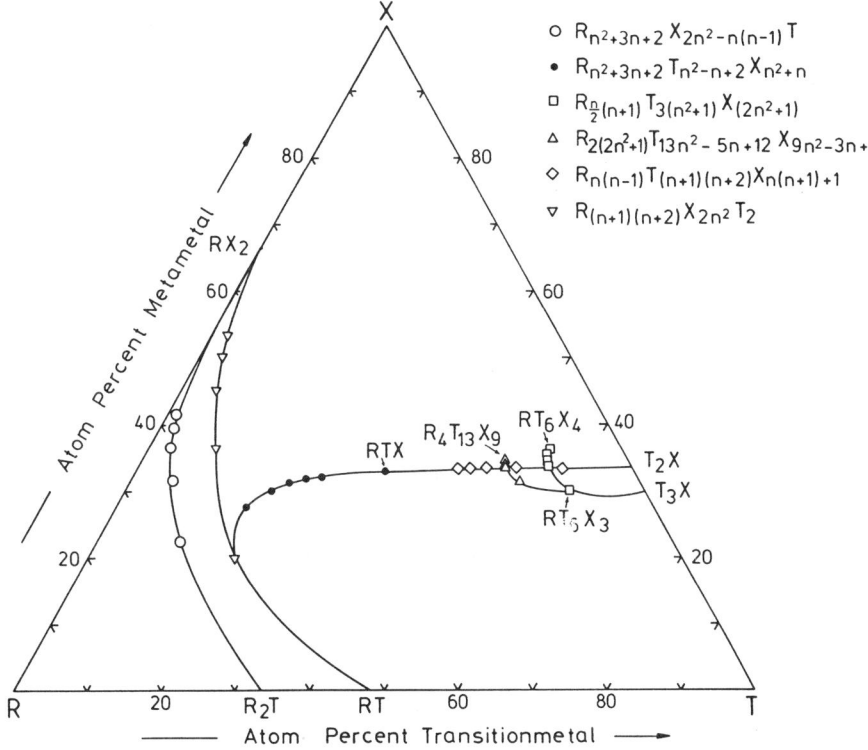

Figure 6: The structure series based on $R_{n(n-1)}T_{(n+1)(n+2)}X_{n(n+1)}$ - subunits and compositions within the R-T-X ternary system.

5. References

Bolmgren, H. & Lundström, T. (1990). 'A New Binary Boride, Nb_5B_6', J. Less-Common Met. **159**, L25-l27

Chikhrii, S.I., Orishchin, S.V., Kuzma, Yu.B. & Glovyak,T. (1990). 'New Ternary Phosphides $R_3Ni_7P_5$ and Their Crystal Structure', Sov.Phys.Crystallogr. **34**, 681-684.

Grin, Yu.N., Yarmolyuk, Ya.P. & Gladyshevskii, E.I. (1979). Sov.Phys.Crystallogr. **24**, 263-266.

Jung, W. & Petry, K. (1988). 'Ternäre Boride des Rutheniums mit Aluminium und Zink, Z. Kristallogr. **182**, 153-154.

Lundström, T. (1977). 'Transition Metal Borides' in "Boron and Refractory Borides", edited by V. I. Matkovich, pp 351-376, Berlin: Springer.

Parthé, E. & Chabot, B. (1984). 'Crystal Structures and Crystal Chemistry of Ternary Rare Earth - Transition Metal Borides, Silicides and Homologues' in "Handbook on the Physics and Chemistry of The Rare Earths", edited by K.Gschneidner, Jr. and L. R. Eyring, Vol. **6**, pp113-334, Amsterdam: North Holland.

Parthé, E. & Hovestreydt, E. (1985). ' The Waist-Contact-Restriction Rule Used to Check Ternary Structures Built From Centered Trigonal Prisms'. J. Less-Common Met. **110**, 307-313.

Pearson, W.B. (1972). "The Crystal Chemistry and Physics of Metals and Alloys". New.York: Wiley-Interscience.

Pivan, J.Y., Guerin, R. & Sergent, M. (1987). 'A New Classification Scheme to Describe and Predict Structure Types in Pnictide and Silicide Chemistry', J.Solid State Chem. **68**, 11-21.

Rogl, P., Kunsch, B., Ettmayer, P., Nowotny & H., Steurer, W. (1982). 'A Neutron Diffraction Study of $Cr_3(B,C)C_{0.85}$ and $Cr_3C(C,N)$', Z. Kristallogr. **160**, 275-284.

Rogl, P. (1985). 'Structural Chemistry and Phase Equilibria of Ternary Rare Earth-Platinummetal Borides'. J.Less-Common Met. **110**, 283-294.

Rogl, P. (1991). 'Existence and Crystal Chemistry of Borides', in "Inorganic Reactions and Methods", edited by A. Hagen, Vol.**13**, pp85-167. New.York: VCH-Publishers.

Spear, K.E. & Gilles, T.W. (1969). 'Phase and Structure Relationships in the Vanadium-Boron System', High.Temp.Sci. **1**, 86-97.

6. Problems

Problem 1: Based on the construction principle of the boride series T_sB_{s+1} or $T_{n/2+m}B_n$, define the hypothetical structures T_4B_5, T_6B_7 and T_7B_8. Assuming validity of the waist contact rule for ternary isotypic rare earth(R)-transitionmetal(T)-metametal(X) compounds, define the possible ordered variants $R_4(T,X)_5$, $R_6(T,X)_7$ and $R_7(T,X)_8$.

Problem 2: Derive the building law for the hexagonal compounds based on $R_3T_{10}X_6$-units with blocks of RT_6X_3 inserted in the trigonal voids of the construction.

6.1. SOLUTION OF PROBLEMS.

Problem 1: (see Figure 7)

Solution: $T_4B_5 = T_{10+6}B_{20}$: m = 6 W-subunits; n = 20 triangular units; $R_8T_3X_7$

$T_6B_7 = T_{14+10}B_{28}$: m = 10 W-subunits; n = 28 triangular units; $R_6T_2X_5$

$T_7B_8 = T_{8+6}B_{16}$: m = 6 W-subunits; n = 16 triangular units; $R_7T_2X_6$

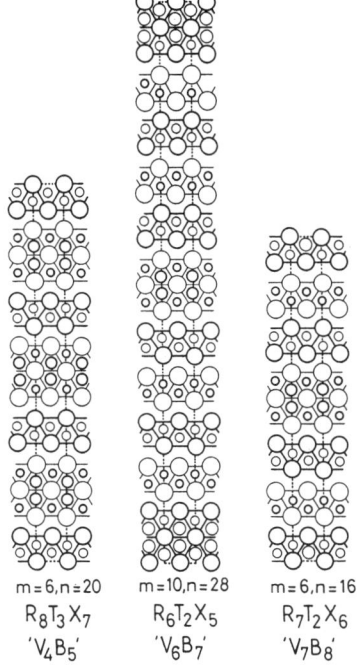

m=6,n=20 m=10,n=28 m=6,n=16
$R_8T_3X_7$ $R_6T_2X_5$ $R_7T_2X_6$
'V_4B_5' 'V_6B_7' 'V_7B_8'

Figure 7: Crystal structures of $R_6T_2X_5$, $R_7T_2X_6$ and $R_8T_3X_7$

Problem 2: *Solution:* $R_{2(2n^2+1)}B_{13n^2-5n+12}C_{9n^2-3n+6}$

$n=0$ $\quad RT_6M_3$

$n=1$ $\quad R_3T_{10}M_6$

$n=2$ $\quad RT_3M_2$

$n=3$ $\quad R_{19}T_{57}M_{39}$

$n=4$ $\quad R_{33}T_{100}M_{69}$

$n=\infty$ $\quad R_4T_{13}M_9$

INDEX

AA	178
Accretional series	135
Adamantane structure compounds	179
Alkaline earth phases	60
Anionic tetrahedron complex	184,186
Antiphase boundary	253
Antiphase domain	7
Archetype	133
Aristotype	4,119
Atomic valence	168
Atomic volume	57
Band structure	99
Band width	100
BaPbYSrCu$_3$O$_8$	234
Base tetrahedra	193
Base tetrahedron	188,192
Bi$_2$Sr$_2$CaCu$_2$O$_8$	247
Bi$_2$Sr$_2$CuO$_6$ type defects	247
Bismuth cuprates	247
Bloch functions	99
Bond length	163
Bond number	171
Bond valence	163
Bond valence parameters	164
Bonding in clusters	35
Bonds	
directional character	119
strength distribution	119
Brillouin zone	99
C'AC'	191
Carbonates	206,208
Cation influence on tetrahedron complexes	211,213
Cationic disordering in superconductors	243
CC	178
C'C'	187
Chain of hydrogen atoms	98
Charge transfer effects	64
"Chimney-ladder" structures	87
Classification codes for base tetrahedra	194
Close packing	119,125
generalization of symbology	125,126
Closest packing	125
cubic c.p.,	125
hexagonal c.p.,	125
Cluster	18
Cluster condensation	47
Cluster connectivity	29
Cluster distortions	42
Cluster electronic configurations	46
Combinatorial series	136
Complex sulfides	131
Condensation of structural units	125,126
Condensed cluster chains	48
Condensed cluster halides	47
Connected units	128
Connectedness	124
Connectivity matrix	165
Constructive layers	125
COOP	109
Coordination number	58,121
Coordination polyhedra in borides, carbides, silicides & phosphides	267
Coordination polyhedron	59,121
Symbology	122
Copper disproportionation	245
Core radius	67
Crystal chemical formula	119,121,179
Crystal orbital overlap population	109
Crystal structure	
Atomic (or close-packed)	120
Chain	120
Crystal chemically isotypic	117
Degrees of similarity	117
Distortion derivatives	118
Framework	120
Group	120
Homeotypic	117
Isoconfigurational	117
Isopointal	117
Polytypic	118
Recombination	118

Relationships	117	Jahn - Teller effect	231
Sheet	120		
Substitution derivatives	118	"Klassengleiche" group	3
Cylindrite family	145	$K_2Pt(CN)_4Cl_{0.3}$	102
Decomposition of the DOS	108	La_2CuO_4	234
Defect adamantane structure	180	Lattice complex notation	119
Defect tetrahedral structure	178	Layer misfit sulfides	132
Density of states	105	Lillianite homologous series	137
Dimensionality regularities	38	Linear structure series	78
DOS	105	Linear symmetry groups	92
Double-metal chains	50	Linkage coefficient	270
		Linkedness	124
e^- - deficient cluster compounds	23	Lone electron pair micelles	137
e^- - rich cluster compounds	23		
Effective atomic number	23	Matrix effects in clusters	42
Elastic effects	64	Maximal subgroup	3
Electron density	67	Maximum gap method	58
Electronegativity	60	Meneghinite homologous series	141
Enthalpy of formation	64	Merlo's approach	68
Equal valence rule	165,170	Metal boride classification	267
Exo bonding	29	Metal cluster halides	27
Extended defects in superconductors	247	Metal - metal bond order	45
Extensive accretional series	136	Metal - metal bonding	18
		Microtwinning in superconductors	239
Family tree	4	Miedema's approach	64
Fermi level	105	Minimal supergroup	3
First order transition	7	Misfit layer structures	141
Fluorite type defects	251	Mixed octahedron and tetrahedron	
Frank-Kasper method	58	complexes	206
		Modulation of the structure	256
Generalized 8 - N rule	178,187	Modulation vector	256
Germanates	206,210,211,224	Molecular tetrahedral structures	183
		Mooser-Pearson diagram	197
H layers	145	Mulliken overlap population	109
Hafner's approach	66	Mulliken population analysis	108
Hettotype	4		
Hole reservoirs	262	$N_{A/M}$	183
Homogeneous intergrowth structures	78	$N'_{A/M}$	185
Homologeous series	135	Nd_2CuO_4	235
		N_{NBO}	178
Incommensurate satellites	256	N'_{NBO}	185
Incommensurate structures	136	Noble metals gallides	86
Inhomogeneous intergrowth structures	78	Non-bonded interactions	169
Interchain bridging	49	Non-commensurability	135
Intergrowth structure	79	Normal adamantane structure	180
Intermetallic compounds	57	Normal tetrahedral structure	178
Interstitials in clusters	30	Normal valence compound	179,187
Isomorphic subgroup	3	Nuclearity	20
Isopointal crystal structures	117	Numeric code notation for	
		intergrowth structures	80

Oligomeric cluster	35
Olivine	223
One-dimensional structure series	78
Orbital topology	104
Oxygen non - stoichiometry in superconductors	236
Oxygen ordering in superconductors	244
Oxygen over stoichiometry	241
Pairs of homologues	136
Parent structure	78
Partial valence electron concentration	178
Pauling bond order	45
Pavonite homologeous series	141
$Pb_{0.7}Cu_{0.3}Sr_2Ca_{0.5}Y_{0.5}Cu_2O_{7-\delta}$	262
$Pb_{1.4}Bi_{0.6}Sr_2Ca_{0.5}Y_{0.5}Cu_3O_8$	253
$Pb_2Sr_2Ca_{0.5}Y_{0.5}Cu_3O_8$	253
Peierls distortion	98
Perovskite - type structure	229
Perovskite intergrowth defects	247
Perowskite family	7
Phase transitions	6
Phosphates	206,207,210,211
Platinocyanides	102
α Po family	6
Polyanionic valence compound	179,187
Polycationic valence compound	179,190
Polymerization of structural units	125,126
Polysomatic series	136
Prism waist contact rule	271
Projections of the DOS	108
Pseudopotential	66
Psi tetrahedron	186,190
$[PtH_4^{2-}]_\infty$ polymer	105
Pt(II) square planar complexes	102
Q layers	141
Quartz	224
Recombination structures	131
Rock salt intergrowth defects	248
Rutile family	4
Sartorite homologous series	141
Second order transition	7
Semi-commensurate structures	136
Sheet structures	52
Si - O bond	203,204
Si - octahedra	223
Silicate classification	205
Square-antiprismatic clusters	51
Structural maps	59
Structural series based on AlB_2- and CsCl-type units	271
Structural series based on AlB_2- and W-type units	269
Structural units	120
Condensation	125,126,127
Constitution	123
Dimensionality	123
Five main categories	120
Multiplicity	124
Packing of subunits	125
Periodicity	124
Polymerization	125,126,127
Structure segment	78
Structure series	78
Structure type	117
Symbology	119
Subgroup of space group	2
Sulfosalts	131
Sulphates	206,208
Superconducting layered cuprates	231
Superstructures	244
Symmetry adaptation	99
Symmetry in linear series	81
Ternary R - T - Ga compounds	89
Ternary systems R - T - M	88
Tetrahedral anion partial structures	184
Tetrahedral structure compounds	177
Tetrahedral structure equation	178
Tetrahedron complex	186,209
Band	215,216,217
Branched	206,213,215
Chain	210,211,212,213,214
Framework	219
Insular	209
Layer	217,218
Linear	209
Ring	209,210
Thallium cuprates	247
Three-dimensional structure series	78
$Tl_2Ba_2CaCu_2O_8$	248
$Tl_2Ba_2Ca_3Cu_4O_{12}$	247
$TlBa_2CaCu_2O_7$	250
$TlBa_2Ca_{1-x}Nd_xCuO_{7-\delta}$	261
Total valence electron concentration	178
Transition - metal compounds	57
"Translationsgleiche" group	3
Trigonal prisms	267

Twin domain	6
Two-dimensional structure series	78
Unit-cell twinning	135
Valence electron concentration	19,178
Vanadium borides	270
Variable - fit series	136
VEC	178
VEC'	184
VEC_A	178
$VEC_{cluster}$	19
VEC_{metal}	20
Vegard's law	62
Volume contraction	63
Voronoi - polyhedron	58
Wave vector	100
Waving of layers	256
Wigner-Seitz cell	58
$YBa_2Cu_3O_6$	237
$YBa_2Cu_3O_{6.2}$	237
$YBa_2Cu_3O_{6.8}$	238
$YBa_2Cu_3O_7$	234,236
$YBa_2Cu_4O_8$	235,242
$Y_2Ba_4Cu_7O_{15}$	242
Zen's law	62
Zeolites	219
Compressibility	221
Ion exchange	219
Molecular sieves	219
Pore parameters	220
Selective shape catalysis	220,221